Treatise on Materials Science
and Technology

VOLUME 14

Metallurgy
of Superconducting
Materials

TREATISE ON MATERIALS SCIENCE AND TECHNOLOGY

VOLUME 14

METALLURGY OF SUPERCONDUCTING MATERIALS

EDITED BY

THOMAS LUHMAN
and
DAVID DEW-HUGHES

Department of Energy and Environment
Brookhaven National Laboratory
Upton, New York

 1979

ACADEMIC PRESS   New York   San Francisco   London

A Subsidiary of Harcourt Brace Jovanovich, Publishers

ACADEMIC PRESS, INC.
111 Fifth Avenue, New York, New York 10003

United Kingdom Edition published by
ACADEMIC PRESS, INC. (LONDON) LTD.
24/28 Oval Road, London NW1 7DX

Library of Congress Cataloging in Publication Data

Main entry under title:

Metallurgy of superconducting materials.

(Treatise on materials science and technology ;
v. 14)
Includes bibliographical references.
1. Superconductors--Metallurgy. I. Luhman,
Thomas. II. Dew–Hughes, David. III. Treatise
TA403.T74 vol. 14 [TN693.S9] 669 79–405
ISBN 0–12–341814–3

PRINTED IN THE UNITED STATES OF AMERICA

79 80 81 82   9 8 7 6 5 4 3 2 1

Contents

LIST OF CONTRIBUTORS   ix
FOREWORD   xi
PREFACE   xv
CONTENTS OF PREVIOUS VOLUMES   xvii

Introduction to Superconducting Materials

David Dew-Hughes

I. Phenomena and Applications of Superconductivity   1
II. Elementary Theories of Superconductivity   10
III. Irreversible Properties of Type II Superconductors   27
References   44

Magnets, Motors, and Generators

Per F. Dahl

I. Introduction   47
II. Magnetic Design Considerations   49
III. Applications   65
References   96

Metallurgy of Niobium–Titanium Conductors

A. D. McInturff

I. Introduction   99
II. The Niobium–Titanium System   100
III. Microstructure and Critical Current Density   102
IV. Theory of Flux Pinning in Niobium–Titanium Alloys   117

v

V. The Nb–Ti Conductors   120
References   135

Physical Metallurgy of A15 Compounds

David Dew-Hughes

I. Introduction   137
II. Superconducting Critical Temperature T_c 142
III. Upper Critical Field H_{c2} 158
IV. Critical Current Density J_c 162
References   167

Superconductivity and Electron Microscopy

C. S. Pande

I. Introduction   171
II. Transmission Electron Microscopy   173
III. Electron Microscopy of Niobium and Its Alloys   188
IV. Transmission Electron Microscopy
of A15 Superconductors   191
V. Observation of Magnetic Flux Lines by Electron Microscopy 203
VI. Superconducting Lenses   207
VII. Conclusion   215
References   216

Metallurgy of A15 Conductors

Thomas Luhman

I. Introduction   221
II. A15 Compound Stability   224
III. Conductor Processing Methods   231
IV. Relationships between Superconducting Properties
and Microstructure in Bronze-Processed Conductors   240
V. Superconducting Critical Currents, Temperatures,
and Magnetic Fields of Nb_3Sn Wire Conductors
under Tensile Strain   254
References   263

Superconductors for Power Transmission

J. F. Bussière

I. Introduction	267
II. Cable Designs and Superconductor Requirements	268
III. Bulk Critical Current of Type II Superconductors	278
IV. Surface Currents in Type II Superconductors	284
V. Theory of AC Losses in Type II Superconductors	296
VI. AC Losses of Pure Niobium	303
VII. AC Losses of Nb_3Sn and Nb_3Ge	309
References	322

Metallurgy of Niobium Surfaces

M. Strongin, C. Varmazis, and A. Joshi

I. Introduction	327
II. Impurities at Nb Surfaces	330
III. Superconducting Properties	339
References	346

Irradiation Effects in Superconducting Materials

A. R. Sweedler, C. L. Snead, Jr., and D. E. Cox

I. Introduction	349
II. Elements	350
III. Body-Centered-Cubic Alloys	359
IV. Non-A15 Compounds	370
V. A15 Compounds	372
References	422

Future Materials Development

David Dew-Hughes and Thomas Luhman

I. Introduction	427
II. Development of Known Materials	428
III. The Possibility of New Superconductors with Higher T_c's	437
IV. Conclusions	442
References	445

INDEX   447

List of Contributors

Numbers in parentheses indicate the pages on which the authors' contributions begin.

J. F. BUSSIÈRE (267), Accelerator Department, Brookhaven National Laboratory, Upton, New York 11973

D. E. COX (349), Physics Department, Brookhaven National Laboratory, Upton, New York 11973

PER F. DAHL (47), Isabelle Project Accelerator Department, Brookhaven National Laboratory, Upton, New York 11973

DAVID DEW-HUGHES (1, 137, 427), Department of Energy and Environment, Brookhaven National Laboratory, Upton, New York 11973

A. JOSHI* (327), Physics Department, Brookhaven National Laboratory, Upton, New York 11973

THOMAS LUHMAN (221, 427), Department of Energy and Environment, Brookhaven National Laboratory, Upton, New York 11973

A. D. MCINTURFF (99), Isabelle Project Accelerator Department, Brookhaven National Laboratory, Upton, New York 11973

C. S. PANDE (171), Department of Energy and Environment, Brookhaven National Laboratory, Upton, New York 11973

C. L. SNEAD, JR. (349), Department of Energy and Environment, Brookhaven National Laboratory, Upton, New York 11973

M. STRONGIN (327), Physics Department, Brookhaven National Laboratory, Upton, New York 11973

A. R. SWEEDLER† (349), Department of Energy and Environment, Brookhaven National Laboratory, Upton, New York 11973

C. VARMAZIS‡ (327), Physics Department, Brookhaven National Laboratory, Upton, New York 11973

* *Present address:* Physical Electronics Industries, Inc., Edina, Minnesota 55424.
† *Present address:* Department of Physics, California State University, Fullerton, California 92634.
‡ *Present address:* Department of Physics, University of Crete, Iraklion, Crete, Greece.

In February 1975, H. H. Kolm wrote in an article in *Cryogenics,* "The most remarkable fact about superconductivity is that sixty years after its discovery, it has still not come of age; superconductivity has yet to be applied to one single purpose which affects every day life!" The commercial availability of multifilamentary NbTi and Nb_3Sn conductors is about to change this situation dramatically.

The publication of this book on the metallurgy of superconducting materials coincides with the general acceptance of superconductors in research and industrial applications. This viewpoint is substantiated by the decision of the Department of Energy (DOE) to build a number of large-scale superconducting devices. These include the Intersecting Storage Accelerator (ISABELLE) and the 100-m ac Superconducting Power Transmission Facility at the Brookhaven National Laboratory (BNL), the Mirror Fusion Test Facility at Lawrence Livermore Laboratory (LLL), and the Tokamak Large Coil Test Facility at the Oak Ridge National Laboratory (ORNL). In addition, a DOE–National Science Foundation (NSF) Project is funding the Michigan State University Heavy Ion Cyclotron, and the Navy, Air Force, and the Electric Power Research Institute are separately sponsoring superconducting generator development.

The scope of the high-energy physics machines points to the confidence that the High Energy Physics Advisory Panel has demonstrated in superconducting materials and technology. The ISABELLE project will require 1100 superconducting magnets. Each magnet is 4 m in length and contains ~100 kg of braided NbTi conductor. Moreover, the refrigerator for ISABELLE will be the largest helium plant contemplated to date, using 2.5×10^6 standard cubic feet of helium. The total cost of the ISABELLE project, which will take seven years to complete, is expected to be $\sim 270 \times 10^6$ dollars.

The single largest projected use of superconducting materials is for plasma confinement in fusion reactors. The mirror fusion test facility at

LLL will use 54,000 kg of NbTi superconductor. This facility will produce a stored energy of 409 MJ. Six prototype plasma confinement coils are being built concurrently for the Tokamak Large Coil Test Facility at ORNL. This is an international effort, three coils coming from manufacturers in the United States, two from Europe, and one from Japan. Of the six D-shaped coils, five coils will be wound from NbTi conductor, the sixth from Nb_3Sn. These coils will be delivered to ORNL by 1981 and are expected to operate at full reactor current densities.

Several applications of superconducting devices have already been developed. Industry has played a key role in the manufacture of superconducting generators and motors. Now being built is a 300-MW 3600-rpm electrical generator with rotating superconducting field windings. A levitated train project in Japan has recently completed a prototype superconducting train running at 350-km/h on a 4-km track.

Superconducting materials and technology are playing a decisive role in the development of the ac superconducting power transmission line. Low ac loss Nb_3Sn tape, now fabricated by industrial firms, will be used in the 100-m demonstration facility at BNL. Operating at 60 Hz, 138 kV, and carrying a current of 4000 A, this cable represents a definitive example of how superconducting devices may influence our everyday life.

In the next decade, superconductors will be used in the magnets of magnetohydrodynamic (MHD) devices. There are special applications of superconducting motors and generators to shipboard use; appropriate motors and generators are being built for the U.S. Navy. These will surely find their way into civilian applications after their unique features have been demonstrated. The need for conserving energy will lead to the use of high-field superconducting magnets for removing weakly magnetic materials from metal scrap prior to remelting, for example, aluminum.

The engineers and scientists who have made these developments possible receive proper recognition in the various chapters of this book. It is appropriate here to give recognition to the funding agencies (in the U.S.), in particular, the Department of Energy (and its predecessors ERDA and the AEC), and to their program managers, W. A. Wallenmeyer, D. K. Stevens, L. C. Ianniello, and M. C. Wittels of the Research Division, B. Belanger, F. F. Parry, and E. W. Flugum of the Electric Energy Systems Division (DOE), and P. Donovan of the NSF. Their realization of the importance of superconductivity to these projects and their early and continued commitment gave the programs the required stability that allowed them to mature to the present state of the art.

The editors are to be congratulated on their recognition of the need for this book. Their energy and enthusiasm in seeing it to a successful completion have provided the reader with a comprehensive view of an emerging technology.

DAVID H. GURINSKY

Preface

This book is intended for the metallurgist who has an interest in the production of superconducting materials, for the physicist who wishes to understand how the physical concepts of superconductivity may be translated into real conductors, and for the engineer who wishes to build superconducting machinery and needs to know the possibilities and limitations of practical superconducting materials. While containing some theory, the book is written primarily with a view toward practical use by metallurgists and engineers; it would also make a useful textbook for a graduate course on superconductors.

An introduction to the phenomenon of superconductivity and an elementary treatment of the theory of superconductors are followed by a discussion of the applications of superconductivity and the demands these applications make on materials' properties and requirements. How these requirements are met in practice is shown in the next chapter, which describes the metallurgy of niobium–titanium alloy conductors from the extraction of the pure metals to the production of multifilamentary wire and cable. The physical metallurgy of A15 compounds precedes a chapter on both the electron microscopy of superconducting materials and the contribution superconducting magnets can make to the improvement of electron microscopes. The metallurgy of conductors made from A15 material is next covered, with emphasis on the bronze process for tapes, multifilamentary conductors, and cables, although not to the exclusion of other technologies. This is followed by a chapter describing the properties required and the development of superconductors for ac power transmission. The next chapter, on the metallurgy of niobium surfaces, is relevant to the use of superconductors in radio-frequency devices. Attention is then given to the effects of radiation on superconductors, a topic of particular importance in view of the fact that the largest foreseeable application for superconductivity is in the generation of power by atomic fusion. In the final chapter, the editors speculate about possible future applications of superconducting materials.

All of the authors are, or were at the time writing, associated with Brookhaven National Laboratory. It is a matter of no little pride to the editors that we were able to commission authoritative contributions on almost every aspect of the metallurgy of superconducting materials entirely from within our own institution.

We believe this book to be an up-to-date presentation of the state of the art in the science and technology of superconducting materials. While much of the material has been published elsewhere, it has never before been collected together in this form, and, as such a collection, this volume represents an invaluable guide to the fabrication and use of superconductors.

We thank Professor Herman for inviting us to edit this volume, our colleagues for their contributions, our secretaries, Lois Arns and Sharon Creveling, for their patience in typing and retyping the manuscript, and the editorial and production staff of Academic Press, whose assistance and advice have made the editing of this volume such a pleasure.

Contents of Previous Volumes

VOLUME 1

On the Energetics, Kinetics, and Topography of Interfaces
    W. A. Tiller
Fracture of Composites
    A. S. Argon
Theory of Elastic Wave Propagation in Composite Materials
    V. K. Tewary and R. Bullough
Substitutional-Interstitial Interactions in bcc Alloys
    D. F. Hasson and R. J. Arsenault
The Dynamics of Microstructural Change
    R. T. DeHoff
Studies in Chemical Vapor Deposition
    R. W. Haskell and J. G. Byrne
AUTHOR INDEX-SUBJECT INDEX

VOLUME 2

Epitaxial Interfaces
    J. H. van der Merwe
X-Ray and Neutron Scattering Studies on Disordered Crystals
    W. Schmatz
Structures and Properties of Superconducting Materials
    F. Y. Fradin and P. Neumann
Physical and Chemical Properties of Garnets
    Franklin F. Y. Wang
AUTHOR INDEX-SUBJECT INDEX

VOLUME 3: ULTRASONIC INVESTIGATION OF MECHANICAL PROPERTIES
    Robert E. Green, Jr.
AUTHOR INDEX-SUBJECT INDEX

VOLUME 4

Microstructural Characterization of Thin Films
    Richard W. Vook

Lattice Diffusion of Substitutional Solutes and Correlation Effects
  J. P. Stark
Solid Solution Strengthening of Face-Centered Cubic Alloys
  K. R. Evans
Thermodynamics and Lattice Disorder in Binary Ordered Intermetallic Phases
  Y. Austin Chang
Metal Powder Processing
  Michael J. Koczak and Howard A. Kuhn
SUBJECT INDEX

VOLUME 5

Solution Thermodynamics
  Rex B. McLellan
Radiation Studies of Materials Using Color Centers
  W. A. Sibley and Derek Pooley
Four Basic Types of Metal Fatigue
  W. A. Wood
The Relationship between Atomic Order and the Mechanical Properties of Alloys
  M. J. Marcinkowski
SUBJECT INDEX

VOLUME 6: PLASTIC DEFORMATION OF MATERIALS

Low Temperature of Deformation of bcc Metals and Their Solid-Solution Alloys
  R. J. Arsenault
Cyclic Deformation of Metals and Alloys
  Campbell Laird
High-Temperature Creep
  Amiya K. Mukherjee
Review Topics in Superplasticity
  Thomas H. Alden
Fatigue Deformation of Polymers
  P. Beardmore and S. Rabinowitz
Low Temperature Deformation of Crystalline Nonmetals
  R. G. Wolfson
Recovery and Recrystallization during High Temperature Deformation
  H. J. McQueen and J. J. Jonas
SUBJECT INDEX

VOLUME 7: MICROSTRUCTURES OF IRRADIATED MATERIALS
  H. S. Rosenbaum
SUBJECT INDEX

VOLUME 8

Equations of Motion of a Dislocation and Interactions with Phonons
  Toshiyuki Ninomiya
Growth, Structure, and Mechanical Behavior of Bicrystals
  C. S. Pande and Y. T. Chou

The Stability of Eutectic Microstructures at Elevated Temperatures
    G. C. Weatherly
Freezing Segregation in Alloys
    Chou H. Li
Intermediately Ordered Systems
    B. Eckstein
SUBJECT INDEX

VOLUME 9: CERAMIC FABRICATION PROCESSES

Powder Preparation Processes
    J. L. Pentecost
Milling
    C. Greskovich
Characterization of Ceramic Powders
    R. Nathan Katz
Effects of Powder Characteristics
    Y. S. Kim
Dry Pressing
    James S. Reed and Robert B. Runk
Hot Pressing
    M. H. Leipold
Isostatic Pressing
    G. F. Austin and G. D. McTaggart
Slip Casting
    Robert E. Cowan
Doctor-Blade Process
    J. C. Williams
Firing
    Thomas Reynolds III
Ceramic Machining and Surface Finishing
    Paul F. Becher
Surface Treatments
    Minoru Tomozawa
Mechanical Behavior
    R. Nathan Katz and E. M. Lenoe
Methods of Measuring Surface Texture
    W. C. Lo
Crystal Growth
    Chandra P. Khattak
Controlled Solidification in Ceramic Eutectic Systems
    Kedar P. Gupta
Controlled Grain Growth
    R. J. Brook
SUBJECT INDEX

VOLUME 10: PROPERTIES OF SOLID POLYMERIC MATERIALS
    Part A

Morphogenesis of Solid Polymer Microstructures
    J. H. Magill

Molecular Aspects of Rubber Elasticity
  Thor L. Smith
INDEX

  Part B

Anisotropic Elastic Behavior of Crystalline Polymers
  R. L. McCullough
Mechanical Properties of Glassy Polymers
  S. S. Sternstein
Fatigue Behavior of Engineering Polymers
  J. M. Schultz
Electronic Properties of Polymers
  R. Glen Kepler
Electric Breakdown in Polymers
  R. A. Fava
Environmental Degradation
  F. H. Winslow
INDEX

VOLUME 11: PROPERTIES AND MICROSTRUCTURE

Direct Observation of Defects
  R. Sinclair
Crystal Defects in Integrated Circuits
  C. M. Melliar-Smith
Microstructure of Glass
  L. D. Pye
Microstructure Dependence of Mechanical Behavior
  Roy W. Rice
Microstructure and Ferrites
  G. P. Rodrigue
INDEX

VOLUME 12: GLASS I: INTERACTION WITH ELECTROMAGNETIC RADIATION

Introduction
  Robert H. Doremus
Optical Absorption of Glasses
  George H. Sigel, Jr.
Photochromic Glass
  Roger J. Araujo
Anomalous Birefringence in Oxide Glasses
  Takeshi Takamori and Minoru Tomozawa
Light Scattering of Glass
  John Schroeder
Resonance Effects in Glasses
  P. Craig Taylor
Dielectric Characteristics of Glass
  Minoru Tomozawa
INDEX

VOLUME 13: WEAR

Theories of Wear and Their Significance for Engineering Practice
    F. T. Barwell
The Wear of Polymers
    D. C. Evans and J. K. Lancaster
The Wear of Carbons and Graphites
    J. K. Lancaster
Scuffing
    A. Dyson
Abrasive Wear
    Martin A. Moore
Fretting
    R. B. Waterhouse
Erosion Caused by Impact of Solid Particles
    G. P. Tilly
Rolling Contact Fatigue
    D. Scott
Wear Resistance of Metals
    T. S. Eyre
Wear of Metal-Cutting Tools
    E. M. Trent
INDEX

Introduction to Superconducting Materials

DAVID DEW-HUGHES

Department of Energy and Environment
Brookhaven National Laboratory
Upton, New York

I. Phenomena and Applications of Superconductivity 1
   A. Superconductivity Phenomena 1
   B. Applications of Superconductivity 7
   C. Materials Requirements for Superconducting Devices 9
II. Elementary Theories of Superconductivity 10
   A. The London Theory 10
   B. Ginzburg–Landau (G–L) Theory 13
   C. The Upper Critical Field 16
   D. The Critical Temperature 22
III. Irreversible Properties of Type II Superconductors 27
   A. Flux Pinning and the Critical State 27
   B. Theories of Flux Pinning 31
   C. Flux Pinning and Microstructure 37
   D. Instability 39
   References 44

I.   Phenomena and Applications of Superconductivity

A.   *Superconductivity Phenomena*

Superconductivity, discovered by Onnes in 1911 [1], is the complete loss† of electrical resistance at some finite, but low, temperature. It is a property of the metallic state, in that all known superconductors are metallic under the conditions that cause them to superconduct.‡ The phe-

---

† It is not possible to determine the complete absence of any quantity, it is only possible to place an upper limit, depending on the sensitivity of the measuring instruments, on the quantity. Superconductivity is only presumed to be the complete loss of resistance; the known upper limit on conductivity is 10^{14} times that of the purest copper at close to absolute zero.

‡ A few, normally nonmetallic, materials can be made to superconduct under, for example, very high pressure. The pressure, however, converts them to metals before they become superconductors.

1

nomenon is not a particularly rare one; a quarter of the elements (Fig. 1) and over 1000 alloys and compounds are superconductors [2]. Metals that are *not* superconductors are usually either good electrical conductors at normal temperatures, such as the alkali and noble metals, or transition metals with strong magnetic moments. Superconductivity appears below a temperature that is characteristic of each superconductor; this is called the critical temperature T_c. Above T_c, superconductors behave exactly as normal metals, their resistivity increasing with decreasing temperature. At T_c, the resistance drops sharply to zero and remains zero at all temperatures below T_c. Figure 2 shows a curve of resistance versus temperature for thin films of vacuum-evaporated Nb_3Sn and Nb_3Ge. Critical temperatures are of the order of a few kelvins; the element with the highest T_c is niobium (9.2 K) and the highest known T_c is ~23 K for Nb_3Ge.

Almost immediately after the discovery of superconductivity, Onnes found that superconductivity was destroyed by the passage of a transport current greater than a certain critical value I_c [3] or by the application of an external magnetic field greater than a critical value H_c [4]. The critical field is, like the critical temperature, a characteristic of the material. It is a function of temperature, falling from a value $H_c(0)$, usually 10^{-2}–10^{-1} T, at 0 K, to zero at the critical temperature. This temperature dependence can be approximated by

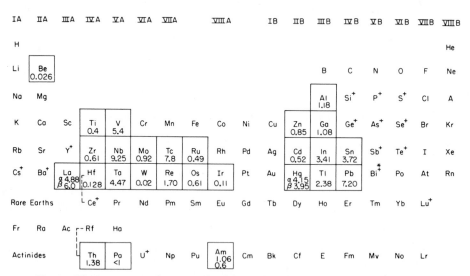

Fig. 1. The occurrence of superconductivity among the elements. Each superconducting element is boxed and its critical temperature in kelvins is given. The elements marked with an asterisk are superconducting only in the amorphous state (bismuth is also a superconductor under high pressure). (Data from Roberts [2].)

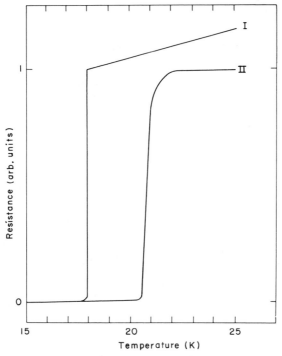

Fig. 2. Resistance (in arbitrary units) versus temperature for evaporated thin films of Nb_3Sn (courtesy M. Gurvitch) and Nb_3Ge (courtesy H. Wiesmann). The gradual onset of superconductivity in Nb_3Ge compared to the sharp transition in Nb_3Sn is indicative of deviations from the ideal stoichiometric ratio in this metastable material (see the chapter by Dew-Hughes on the physical metallurgy of A15 compounds, this volume).

$$H_c(T)/H_c(0) = 1 - (T/T_c)^2$$

and is shown in Fig. 3.

The critical current is a function of temperature, applied magnetic field, and specimen size. Silsbee [5] hypothesized that the critical current was the current that just produced the critical field at the surface of the specimen. The field due to a current in a wire falls off as the reciprocal of the distance from the axis of the wire; thus, for a conductor of radius a,

$$I_c = \tfrac{1}{2}aH_c$$

This relation has been verified for type I superconductors.† In the presence of an applied field, the sum of the applied field and the self-field of the current must not exceed H_c at the surface, and I_c is correspondingly reduced.

† The distinction between type I and type II superconductors is taken up in Section II,C.

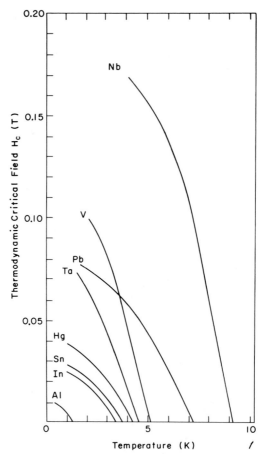

Fig. 3. Critical field versus temperature for some elemental superconductors. (After D. Schoenberg, "Superconductivity." Cambridge Univ. Press, London and New York, 1960.)

A consequence of zero resistivity is that everywhere within a supercon-ductor the electric field E is zero. In 1933 Meissner and Ochsenfeld [6] were able to show that magnetic fields are completely excluded from the body of a superconductor in its superconducting state; i.e., the magnetic induction B is also zero. This is known as the Meissner effect. The mag-netic induction does not fall abruptly to zero at the surface of the super-conductor, but decreases exponentially over a characteristic distance, the penetration depth λ, from the surface (Fig. 4). For a pure metal supercon-ductor, λ is typically 5×10^{-8} m. The flux is excluded by supercurrents that flow in the penetration layer so as to produce a magnetic field within the superconductor that exactly cancels any externally applied field. A

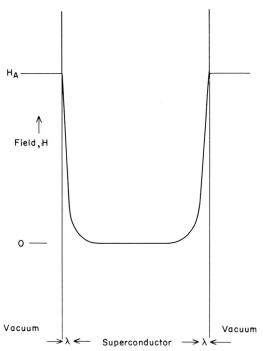

Fig. 4. Penetration of a magnetic field into a superconductor, showing the exponential decay over the penetration depth λ.

similar effect occurs within the skin depth of a normally conducting metal when subjected to a changing magnetic field, but once the field is held constant the normal resistance causes the skin depth current to decay. In a superconductor the lack of resistance allows the current to persist, and flux is permanently excluded from the bulk. This, however, is not the complete explanation of the Meissner effect, since flux is excluded whenever a superconductor goes from the normal to the superconducting state by lowering the temperature in a constant applied magnetic field. Herein lies the difference between a superconductor, in which $E \equiv B \equiv 0$, and a perfect conductor, in which only $E \equiv 0$. The Meissner effect has important consequences for the thermodynamics of the superconducting state, as will be seen in Section II,B. The transition between the superconducting and normal states in a magnetic field is reversible (Fig. 5a). As the external field H_a is raised from 0, the induction B within the superconductor remains at 0 until the applied field equals H_c, at which B rises suddenly to equal $\mu_0 H_c$. For all fields $> H_c$, $B = \mu_0 H_a$. (Any diamagnetic or paramagnetic effects in the normal state are assumed to be negligible. Superconductivity and ferromagnetism are usually incompatible.) As the

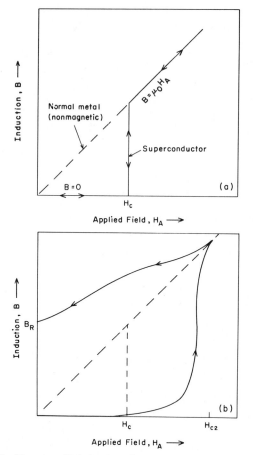

Fig. 5. Magnetic (*B* versus *H*) behavior of superconductors. (a) Ideal, reversible; (b) irreversible, hysteretic.

field is reduced, $B = \mu_0 H_a$ until $H_a = H_c$, at which B suddenly drops to 0, and remains so for all values of $H_a < H_c$.

A few years after its discovery, Mendelssohn and Moore [7] found that some superconducting alloys did not show a complete Meissner effect. These alloys were able to remain superconducting in magnetic fields up to a value H_{c2}, the upper critical field, ~ 1–2 T, considerably higher than their (thermodynamic) critical field H_c. They were also magnetically irreversible (Fig. 5b), retaining trapped flux after removal of the externally applied field. The significance of these materials was not fully understood for nearly 30 years.

The first high-temperature, high-field superconductor, NbN, was discovered in 1941 [8]; it has a critical temperature of 15 K and an upper critical field of 15 T. The impetus for the application of superconductivity came when Hardy and Hulm found that V_3Si, a compound with the A15 structure, had a T_c of 17.1 K [9]. The same year, 1953, Nb–Zr, a ductile bcc alloy, was found with a T_c of 11 K and an H_{c2} of 10 T [10]. The following year Matthias *et al.* achieved a T_c of 18 K in another A15 compound, Nb_3Sn [11]. Further ductile alloys, Mo–Re [12] and Nb–Ti [13], were discovered in 1961, and a new A15 compound, V_3Ga, in 1962 [14]. T_c was not raised for 13 years, until in 1967 Matthias and co-workers prepared a ternary A15 compound, $Nb_3(Al,Ge)$ with T_c slightly greater than 20 K [15], subsequently raised to 20.7 K [16]. This material has an H_{c2} of 40 T. The critical temperature of binary A15 compounds was steadily raised: Nb_3Al, 18.9 K, 1969 [17]; Nb_3Ga, 20.3 K, 1971 [18]; and, finally, Nb_3Ge, 23 K, 1973 [19, 20]. This remains at present the highest T_c, and it is interesting to note that since 1953 the record for T_c has remained with the A15 class of compounds. These compounds are discussed more fully in a later chapter by Dew-Hughes.

A more recent development is that of the ternary lead molybdenum sulfides, $PbMo_5S_6$. For these, T_c is only ~ 15 K but H_{c2} is the highest ever measured at 50–60 T [21]. These materials are referred to in the final chapter (Dew-Hughes and Luhman).

Commercial superconductors are available based on the two ductile alloys, Nb–Zr (25–35 at. % Zr) and Nb–Ti (60–70 at. % Ti), although the former, being more difficult to fabricate and having lower values for H_{c2} and critical current density J_c, has been almost entirely superseded by the latter. The metallurgy of Nb–Ti is discussed in the chapter by McInturff. A15 compounds Nb_3Sn and V_3Ga are also commercially produced as conductors. It has not yet proved possible to fabricate the other A15 compounds into viable conductor configurations. Luhman deals with the problems of A15 conductors later in this volume.

B. Applications of Superconductivity

The applications of superconductivity, which are only briefly listed here, are basically of three types:

(1) Superconductors may be used for the production of large magnetic fields. Large refers not only to the field intensity but also to the volume over which it is generated. Magnets and magnet conductors are dealt with in detail in the chapter by Dahl, this volume.

(2) Superconductors have been proposed for power transmission cables. A superconducting cable cannot compete with overhead transmission lines, except over very large distances (several hundred kilometers).

However, where high-power-density lines are forced underground for amenity or other reasons, a superconducting cable may show some economic advantages. The superconducting material requirements for such a cable are very similar to those for magnet materials, with the additional constraint that for ac lines the ac losses must be minimized. This problem is treated in this volume by Bussière.

(3) Superconductors can be employed in electronic circuitry. Devices were originally based on the idea of using the superconducting–normal transition as a switch. Performance was disappointing compared to that of thin-film transistor developments, and this application has been completely dropped. Devices based on a different property of superconductors, the Josephson effect, are still of great interest. The critical current that flows between two superconductors weakly coupled together through a weak link, which may be a very thin ($\sim$20-Å) oxide barrier, a layer of nonsuperconducting metal, or a narrow bridge of superconductor, varies in an oscillatory fashion depending on the local value of the magnetic field in the weak link. This effect can be used as the basis of highly sensitive ammeters, gauss meters, and voltmeters, as well as logic and memory elements. It is being developed as a voltage standard. The materials problems here are quite different from those associated with applications (1) and (2), and are basically those of thin-film circuitry. These are interesting in their own right, but they will not be dealt with in this volume.

The most important application, in terms of quantity of material employed, is, and will be for many years to come, the production of magnetic fields. First, magnetic fields are used in physics laboratories for research purposes, and small superconducting magnets are now the standard way of producing fields above a few kilo-oersteds. Highly stable magnetic fields for NMR and for high-resolution, high-voltage electron microscopy are provided by superconducting magnets. Large magnetic fields are also required in high-energy nuclear physics, and superconducting magnets have been built for beam bending and focusing, as well as for large bubble chambers.

Power generation by magnetohydrodynamic methods or thermonuclear fusion will require large fields that can only be provided by superconducting magnets. Magnetic ore separation, sewage treatment, and water purification are all areas in which superconducting magnets are expected to make an impact over the next few years. Superconducting motors and generators, which are now under development, are devices in which the superconductor is used to generate high magnetic fields. Superconducting magnets have also been proposed for the levitation of high-speed tracked vehicles. Magnets have also been designed to assist in brain surgery and abdominal operations.

The first successful superconducting magnet was constructed by Yntema in 1955 by winding cold-worked niobium wire around an iron core. This magnet achieved a field of 0.7 T [22]. Commercial magnets, using Nb–Zr wire, were first made available by the Oxford Instrument Co. in 1962–1963. At the same time, laboratory magnets capable of producing 10 T were wound from Nb_3Sn [23]. The development in the past 13 years has been rapid, with maximum fields of 17 T now available, and field volumes, at lower inductions, of cubic meters.

C. Materials Requirements for Superconducting Devices

Superconductors are employed in large-scale electrical devices mainly because they can be constructed with extremely high overall current densities. The critical current density of the superconductor itself lies in the range 10^9–10^{12} A/m². These figures are reduced by a factor of up to 10 for the overall current density in the conductor because of the necessity of providing a normal conductor (Al or Cu) for stability, and reinforcement (stainless steel) for mechanical strength. This density compares very favorably with the maximum allowable current density in uncooled copper of 2×10^6 A/m². Forced cooling cannot raise this figure by much more than a factor of 10.

These high current densities allow the construction of smaller, lighter, and more compact machines. Compactness is aided by the fact that the lack of power dissipation consequent upon the absence of electrical resistivity reduces the necessity for internal cooling. It is the reduction in size that largely accounts for the economic attractiveness of superconducting machinery, lowering capital, installation, maintenance, and running costs. The reduced power requirements are an added attraction, though they must be balanced against the power consumed by refrigeration.

It is obvious from the foregoing that the most important requirement of a conductor is that it be able to carry the highest possible current, and therefore the superconductor from which it is made should have as high a critical current density J_c as possible. The principles that determine J_c are discussed in Section III of this chapter.

Because the superconductor invariably operates in a high-field environment, due to the self-field of the current it is carrying if to nothing else, a high value of the upper critical field H_{c2} is essential. The basic theory of high-field, type II superconductors is developed in Section II.

If the conductor is to experience alternating currents, as it will in some of the transmission line proposals, or ripple fields, as may happen in some of the alternator designs, then ac losses must be low. The production of superconductors with low ac losses is dealt with in the chapter by

Bussière in this volume. The relation between surface properties and high-frequency behavior is treated in the chapter by Strongin, Varmazis, and Joshi in this volume. Resistance to radiation damage may be a requirement for superconductors in controlled thermonuclear reactor (CTR) fusion magnet applications. Radiation effects on superconductors are covered in the chapter by Sweedler, Snead, and Cox, this volume. The foregoing are properties of the superconductor itself. The conductor into which the superconductor is incorporated must be mechanically strong and able to withstand forces imposed by current and field (Lorentz forces). It must be stable against flux jumps and sudden changes in current, field, or temperature. This requirement necessitates that the conductor be small in at least one dimension, that parallel to the direction of flux motion, and be associated with good normal conductor (aluminum or copper). A multifilamentary configuration is most effective in promoting stability.

It must also be possible to fabricate the conductor in long (i.e., greater than 1 km) lengths. This means that at some stage during processing the conductor must be ductile. The availability of constituents must be good. The construction of a device that would deplete a considerable portion of the known world supply of a particular element is neither sensible nor desirable. The final, and perhaps most important, consideration is that the cost must be kept as low as possible. The introduction of a new technology into an industrial environment can only be encouraged if substantial, not merely marginal, economic advantages can be demonstrated.

It is the aim of the rest of this volume to examine in greater detail most of the foregoing points, and to describe how the various metallurgical problems involved with developing conductors are overcome.

II. Elementary Theories of Superconductivity

A. The London Theory

In any conductor, the equation of motion for an electron (of charge e and mass m) is

$$\frac{d\mathbf{v}}{dt} = \frac{e\mathbf{E}}{m} - \frac{\mathbf{v}}{\tau} \tag{1}$$

where the first term on the right-hand side represents the acceleration of the electron due to the electric field $\mathbf{E}$, and the second term is its deceleration due to scattering (τ is the mean time between scattering events). The average velocity $\bar{\mathbf{v}}$ of the electrons is given by

$$\bar{v} = J/ne \qquad (2)$$

where J is the current density and n is the number of electrons per unit volume. Substituting for v in (1) gives

$$\frac{dJ}{dt} = \frac{ne^2E}{m} - \frac{J}{\tau} \qquad (3)$$

In a superconductor, the scattering processes are absent, and (3) can be written

$$\frac{dJ_s}{dt} = \frac{n_s e^2 E}{m} \qquad (4)$$

where the subscript s refers the quantity to the superelectrons only. Thus, an electric field can only appear in a superconductor when the supercurrent density J_s is changing; when J_s is constant $E = 0$.

Taking the curl $[\nabla \times$, where $\nabla = (\partial/\partial x)\mathbf{i} + (\partial/\partial y)\mathbf{j} + (\partial/\partial z)\mathbf{k}]$ of both sides of Eq. (4) and remembering from Maxwell's equations that $\nabla \times E = -dB/dt$ yields

$$\nabla \times \frac{dJ_s}{dt} = \frac{n_s e^2}{m} \nabla \times E = -\frac{n_s e^2}{m} \frac{dB}{dt} \qquad (5)$$

If it can be assumed that the space and time variables are independent of one another, Eq. (5) can then be rewritten

$$\frac{d}{dt} \nabla \times J_s = \frac{d}{dt} \left(-\frac{n_s e^2}{m} B \right)$$

from which it follows that

$$\nabla \times J_s = -(n_s e^2/m)B \qquad (6)$$

In 1935, F. London and H. London suggested that this equation did describe the electrodynamics of superconductors [24]. Defining a quantity Λ (with the dimensions time^2) $= m/n_s e^2$, they proposed that the following two equations, derived from (4) and (6), were applicable to the superelectrons:

$$\frac{d}{dt} (\Lambda J_s) = E \qquad (7)$$

and

$$\nabla \times (\Lambda J_s) = -B \qquad (8)$$

Using these two equations they were able to develop a phenomenological theory of superconductivity.

The Meissner effect is derived by considering another of Maxwell's equations, $\nabla \times \mathbf{B} = \mu_0 \mathbf{J}$. Taking the curl of both sides, gives

$$\nabla \times \nabla \times \mathbf{B} = \mu_0 \, \nabla \times \mathbf{J}_s = (\mu_0/\Lambda)\mathbf{B}$$

Now $\nabla \times \nabla \times \mathbf{B} \equiv \nabla(\nabla \cdot \mathbf{B}) - \nabla^2 \mathbf{B}$, and since a third Maxwell equation states $\nabla \cdot \mathbf{B} = 0$, $\nabla \times \nabla \times \mathbf{B} = -\nabla^2 \mathbf{B}$ and

$$\nabla^2 \mathbf{B} = (\mu_0/\Lambda)\mathbf{B} \tag{9}$$

The exact solution of this equation depends upon specimen geometry, but is of the form [25]

$$B(x) = B(0) \exp[-x(\mu_0/\Lambda)^{1/2}] \tag{10}$$

$B(0)$ is the value of the magnetic induction at the surface of the superconductor, and $B(x)$ its value at a distance x within the superconductor. $B(x)$ is eventually zero for distances $x > (\Lambda/\mu_0)^{1/2}$; i.e., this represents the London penetration depth λ_L, and thus

$$\lambda_L = (m/\mu_0 n_s e^2)^{1/2} \tag{11}$$

The London theory is a "local" theory in that the current density $\mathbf{J}_s(r)$ at the point r is determined by $\mathbf{A}(r)$, the value of the vector potential at the same point r. Pippard [26] argued that an electron traveling from a normal region of a metal into a superconducting region cannot change its wave function from that of the normal state to that of the superconducting state abruptly, but that this change must take place over a distance ξ, the "range of coherence." Pippard estimated that $\xi \approx 1 \ \mu m$ for pure metals. He then proposed a "nonlocal" variation of the London theory, in which $\mathbf{J}_s(r)$ is now determined by the value of $\mathbf{A}$ averaged over a sphere of radius ξ centered at r. A similar situation exists for normal metals, e.g., in the anomalous skin effect, in which averaging must take place over the normal electron mean free path l. In fact, later work on superconductors showed that λ increased and ξ decreased in a superconductor as the normal electron mean free path decreased.

The Pippard theory yields specific results in two limiting cases. For "dirty" superconductors, in which $l \ll \xi_0$ (the value of ξ for the pure metal)

$$\lambda = \lambda_L(\xi_0/l)^{1/2} \tag{12}$$

This is known as the London limit since in this limit the short mean free path leads to a local relation between $\mathbf{J}_s$ and $\mathbf{A}$. For "clean" superconductors, in which $l \gg \xi_0$

$$\lambda_\infty = [(\sqrt{3}/2\pi)\xi_0\lambda_L^2]^{1/3} \tag{13}$$

known as the Pippard limit. In the "dirty" limit the range of coherence is reduced to

$$\xi_d \approx (\xi_0 l)^{1/2} \qquad (14)$$

B. Ginzburg–Landau (G–L) Theory

The Gibbs function of a magnetic system is given by

$$G_s(H) = G_s(0) - \mu \int_0^H M \, dH$$

For a type I pure metal superconductor, between $H = 0$ and $H = H_c$, $B = 0$ and $M = -H$. Thus,

$$G_s(H) = G_s(0) + \tfrac{1}{2}\mu_0 H^2$$

In the normal state, ignoring small diamagnetic or paramagnetic effects,

$$G_n(H) = G_n(0)$$

and at H_c, the Gibbs functions of the normal and superconducting states are equal:

$$G_n(0) = G_n(H_c) = G_s(H_c) = G_s(0) + \tfrac{1}{2}\mu_0 H_c^2$$

or

$$\Delta G_{n-s} = G_n(0) - G_s(0) = \tfrac{1}{2}\mu_0 H_c^2 \qquad (15)$$

The Gibbs function of the superconducting state at zero field is lower than that of the normal state by an amount $\tfrac{1}{2}\mu_0 H_c^2$ per unit volume. This quantity is often referred to as the superconducting condensation energy.

Equation (15) forms the basis for the thermodynamics of superconductivity. Pressure and stress effects on the transition can be derived by appropriate differentiation of this equation.

The Ginzburg–Landau theory was developed in 1950 [27] to give a more adequate description of the behavior of a superconductor in a magnetic field. The normal-to-superconducting transition was thought of as an ordering process, describable in terms of an order parameter ψ. In the normal state, $\psi = 0$; in the superconducting state ψ has a value between 0 and 1. $|\psi|^2$ represents the fraction of conduction electrons that are in the superconducting momentum state, and are described by the single superconducting wave function. In the fully superconducting state $|\psi|^2 = \psi = 1$. In the presence of an externally applied magnetic field, $\mathbf{H}_a = \mathbf{B}_a/\mu_0$, two extra terms are required. The first represents the energy arising from the work done in expelling the magnetic field from the super-conductor. If the Meissner effect is not complete, and there is an internal induction B, then the Gibbs function is increased by

$$(1/2\mu_0)(\mathbf{B}_a - \mathbf{B})^2 = (1/2\mu_0)(\mathbf{B}_a - \mathbf{\nabla}\times \mathbf{A})^2$$

In addition, if there is a spatial variation in ψ within the superconductor, i.e., $\nabla\psi \neq 0$, then the superelectrons are given a momentum $-i\hbar\nabla\psi - e\mathbf{A}\psi$ and an increase in kinetic energy $(1/2m)|-i\hbar\nabla\psi - 2e\mathbf{A}\psi|^2$.

The Gibbs function for a superconductor in an externally applied magnetic field now becomes

$$G_s(\mathbf{B}_a) = G_0 + \alpha|\psi|^2 + \frac{\beta}{2}|\psi|^4$$

$$+ \frac{1}{2\mu_0}(\mathbf{B}_a - \mathbf{\nabla}\times \mathbf{A})^2 + \frac{1}{2m}|-i\hbar\nabla\psi - 2e\mathbf{A}\psi|^2 \qquad (16)$$

This can be minimized by differentiation to give

$$(1/2m)(-i\hbar\nabla - 2e\mathbf{A})^2\psi + \alpha\psi + \beta|\psi|^2\psi = 0 \qquad (17)$$

which allows the order parameter to be calculated in the presence of the field, and

$$\mathbf{\nabla}\times \mathbf{B} = \mathbf{J}_s = \frac{e\hbar}{im}(\psi^*\nabla\psi - \psi\nabla\psi^*) - \frac{4e^2}{m}\psi^*\psi\mathbf{A} \qquad (18)$$

which gives the distribution of supercurrent. These are the Ginzburg–Landau equations. These equations are not easy to solve, even for the simplest geometries. For the case of a superconductor that fills all of space to one side of the $x = 0$ plane, with a magnetic field along the z direction, independent of y and z coordinates, the problem is reduced to one of a single dimension, quantities B and ψ varying only in the x direction.

$$\mathbf{B} = \mathbf{\nabla}\times \mathbf{A}; \quad B_x = B_y = 0, \quad B_z = dA_y/dx, \quad \text{and} \quad A_x = A_z = 0$$

The only variation of ψ is in the x direction, and thus $\mathbf{A}\cdot\nabla\psi = A_y \cdot d\psi/dx = 0$ and the equation for the Gibbs function becomes

$$G_s(\mathbf{B}) = G_n + \alpha|\psi|^2 + \frac{\beta}{2}|\psi|^4 + \frac{1}{2\mu_0}\left(\mathbf{B}_a - \frac{d\mathbf{A}}{dx}\right)^2$$

$$+ \frac{1}{2m}\left[\hbar^2\left(\frac{\partial\psi}{\partial x}\right)^2 + 4e^2\mathbf{A}^2|\psi|^2\right] \qquad (19)$$

Minimizing the Gibbs function with respect to $\mathbf{A}$ leads to a relation

$$\frac{d^2\mathbf{A}}{dx^2} = \frac{4|\psi|^2\mu_0 e^2\mathbf{A}}{m}$$

which has a solution of the form

$$A(x) = A(0)e^{-x/\lambda}$$

from which

$$B(x) = B_a e^{-x/\lambda}$$

where $\lambda = (m/4|\psi|^2\mu_0 e^2)^{1/2}$. This is identical with Eq. (10), with λ equal to the London penetration depth, Eq. (11), and $n_s = 4|\psi|^2$. Thus, the Ginzburg–Landau equations also predict the Meissner effect. Ginzburg and Landau introduce a new parameter:

$$\kappa = \lambda/\xi = 2\sqrt{2}\lambda^2 e\mu_0 H_c/\hbar \tag{20}$$

Superconducting and normal regions can exist side by side in the same piece of metal. This happens, for instance, in a specimen with a nonzero demagnetization coefficient at a field below H_c, when the superconductor breaks up into a mixture of superconducting and normal regions called the intermediate state. An extra energy term must be associated with the boundaries between the normal and superconducting states. This boundary energy, α_{ns} per unit area, can be estimated as follows: Imagine two pieces of identical superconductor, each with a large plane boundary, held some distance apart. One is supposed to be in the normal state and penetrated by a magnetic field $H_a < H_c$; the other is in the superconducting state with $H = 0$. The two pieces are brought together so that their plane faces are in contact along the plane $x = 0$. Because ψ cannot change abruptly, but now does so over a distance equal to ξ, this is equivalent to a volume (superconductor area) $\times$ ξ becoming normal, and thus there is an increase in energy of $\frac{1}{2}\mu_0 H_c^2\xi \times$ area. At the same time the magnetic field penetrates into the superconductor over a distance λ, lowering the energy by an amount $\frac{1}{2}\mu_0 H_a^2 \times$ area (Fig. 6). The surface energy is therefore

$$\alpha_{ns} = \frac{1}{2}\mu_0(H_c^2\xi - H_a^2\lambda) \tag{21}$$

which, since for pure metals $\xi \gg \lambda$, is often written

$$\alpha_{ns} = \frac{1}{2}\mu_0 \Delta H_c^2$$

A more accurate result, using the Ginzburg–Landau equations, gives for $\kappa \ll 1$ (i.e., $\xi \gg \lambda$) $\Delta = 1.89\xi$. Much more interesting is the result that for $\kappa \gg 1$, $\Delta \approx -\lambda$; i.e., the surface energy is negative! The possibility of

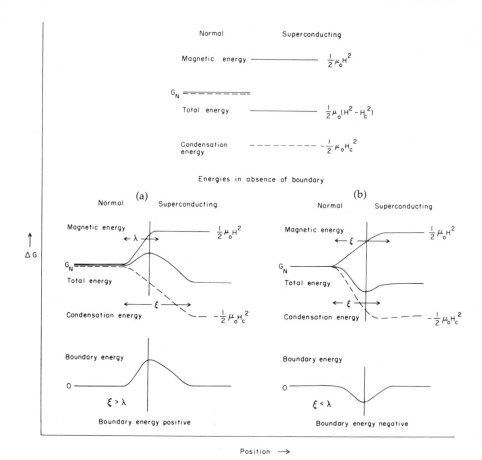

Fig. 6. Origin of the superconducting–normal boundary surface energy. (a) $\xi > \lambda$, positive surface energy; (b) $\xi < \lambda$, negative surface energy.

negative surface energy also follows from Eq. (21) when $\lambda > \xi$. Since α is positive for small κ, and negative for large κ, there must be some intermediate value of κ at which $\alpha = 0$. This is found, in the Ginzburg–Landau theory, to be when $\kappa = 1/\sqrt{2}$.

Ginzburg and Landau chose to ignore the implications of negative boundary energy, believing that since for most known superconductors $\kappa \sim 0.1$, large κ was physically unrealistic.

C. The Upper Critical Field

In 1957, Abrikosov [28] published his theoretical paper that considered the consequences of a superconductor with $\kappa > 1/\sqrt{2}$. This paper was not immediately available in translation, and its importance was not fully appreciated until the Colgate Conference in 1963 [29]. He showed that for large κ superconductivity could exist up to a field, called the upper critical field, given by

$$H_{c2} = \sqrt{2}\kappa H_c \qquad (22)$$

Referring back to Eq. (21), it can be seen that for $\lambda > \xi$ ($\kappa > 1$), the surface energy is still positive at low values of the external applied field H_a, but goes negative as H_a exceeds $H_c(\xi/\lambda)^{1/2} = H_c/(\kappa)^{1/2}$, a value known as the lower critical field H_{c1}. A more accurate estimate by Abrikosov, for $\kappa \gg 1$, is

$$H_{c1} = (H_c/\sqrt{2}\kappa)(\ln \kappa + 0.08)$$

Numerical calculations show this to be valid for $\kappa > 20$.

When a superconductor with $\kappa > 1/\sqrt{2}$ is subjected to an externally applied field, it shows a Meissner effect up to H_{c1}. At this field normal regions are nucleated at the surface of the superconductor. These normal regions carry magnetic flux, and initially move freely into the body of the superconductor. The superconductor is now said to be in the mixed state. As the external field is raised, more of these normal regions are nucleated; the induction B within the superconductor rises, until their mutual repulsion makes it increasingly difficult for more normal regions to enter. At some sufficiently high field the normal regions are so densely packed within the superconductor that they begin to overlap and the normal state is achieved. This field is the upper critical field H_{c2} given by Eq. (22). Provided that the motion of these normal regions is in no way hindered, reduction of the magnetic field allows them to move back out of the superconductor. The magnetization and induction curves, shown in Fig. 7, are reversible. Superconductors that show this behavior, with $\kappa > 1/\sqrt{2}$, are known as type II, in contrast to type I with $\kappa < 1/\sqrt{2}$.

The normal regions are continuous in the direction parallel to the field since flux must be continuous, and these regions may be thought of as

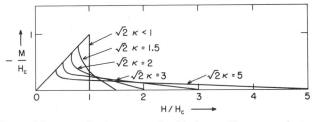

Fig. 7. Reversible magnetization curves for ideal type II superconductors, for some different values of κ. (After Goodman [30].)

flexible rods of normal material. The flux within one of these normal regions is quantized, as first suggested by London, into units of the flux quantum $\phi_0 = (h/2e)$ ($= 2 \times 10^{-15}$ Wb). It can be shown that the energy of one of these normal regions, or flux lines, is proportional to $\Phi_c^2 = n^2 \phi_0^2$, and it is therefore energetically favorable for a multiquanta flux line of $n\phi_0$ to break up into n single-quantum flux lines. The mixed state of a type II superconductor consists of a lattice of single-quantum flux lines. Abrikosov originally postulated a square array; it is now known from both theory and direct observation that the flux-line lattice is a triangular array (Fig. 8). Each flux line may be regarded as having a core of normal electrons, ξ in diameter, the flux quantum being supported by circulating supercurrents that decay over a radius λ from the core (Fig. 9). The magnetic induction in the mixed state B is $n\phi_0$, where n is the area density (number crossing unit area normal to B) of flux lines.

It is possible to express H_c and H_{c2} in terms of ϕ_0:

$$H_c = \frac{\phi_0}{2\pi\sqrt{2}\,\mu_0 \xi \lambda}$$

$$H_{c2} = \sqrt{2}\kappa H_c = \sqrt{2}\,\frac{\lambda}{\xi}\,\frac{\phi_0}{2\pi\sqrt{2}\,\mu_0 \xi \lambda} = \frac{\phi_0}{2\pi\mu_0 \xi^2} \qquad (23)$$

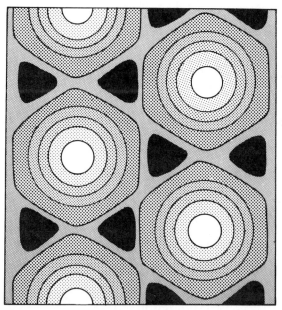

Fig. 8. The spatial configuration of $|\psi|^2$ near H_{c2} for a triangular flux array in a type II superconductor. The area density of the dots is approximately proportional to the square of the reduced order parameter. (After Kleiner *et al.* [30a].)

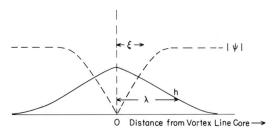

O   Distance from Vortex Line Core →

Fig. 9. The variation of the local magnetic field h and the order parameter $|\psi|$ across an isolated flux line in a type II superconductor.

It has already been mentioned that both λ and ξ are related to the normal electron mean free path l when $l < \xi_0$ [Eqs. (12) and (14)]. Thus both κ and H_{c2} can be expressed in terms of l and the normal state resistivity ρ_n.

$$\kappa = \frac{\lambda}{\xi} = \lambda_L \left(\frac{\xi_0}{l}\right)^{1/2} \frac{1}{(\xi_0 l)^{1/2}} = \frac{\lambda_L}{l}$$

and

$$\kappa = \kappa_0 + 2.37 \times 10^6 \gamma^{1/2} \rho_n \quad [30] \tag{24}$$

$$H_{c2}(0) = \frac{\phi_0}{2\pi\mu_0\xi_0 l} = 3.11 \times 10^3 \gamma \rho_n T_c \quad \text{for} \quad \kappa \gg 1 \quad [31] \tag{25}$$

where γ is the Sommerfeld constant, the temperature coefficient of the electronic specific heat (all quantities are in SI units). Thus an increase in normal state resistivity will increase κ and H_{c2}. As defined in Eq. (20) and as used above, κ is the Ginzburg–Landau κ, κ_{GL}. Maki [32] has defined three other kappas:

$$H_{c2} = \sqrt{2}\kappa_1 H_c \quad \text{defines} \quad \kappa_1 \tag{26}$$

$$-\left(\frac{dM}{dH}\right)_{H \approx H_{c2}} = \frac{1}{1.16(2\kappa_2{}^2 - 1)} \text{defines} \; \kappa_2 \tag{27}$$

and

$$H_{c1} = \frac{H_c}{\sqrt{2}} \frac{\ln \kappa_3}{\kappa_3} \quad \text{defines} \quad \kappa_3 \tag{28}$$

Near T_c, $\kappa_1 = \kappa_2 = \kappa_3 = \kappa_{GL}$, but as the temperature is lowered κ_1, κ_2, and κ_3 all increase. For $\kappa_{GL} \gg 1$ ($\xi_0 \gg l$), the temperature dependences of κ_1 and κ_2 are similar, and both increase to $1.2\kappa_{GL}$ at $T = 0$. κ_3 rises more rapidly, $\kappa_3(0) \cong 1.5\kappa_{GL}$. For lower values of κ_{GL} the temperature dependence of the other kappas is different [33].

Values of $H_{c2}(0)$ can be very large; the largest known to date is 60 T for

PbMo$_{5.1}$S$_6$ [21] and is difficult to determine directly. However, $H_{c2}(0)$ can be derived from the slope of the H_{c2} versus T curve near T_c [34]:

$$H_{c2}(0) = -0.693 T_c \left[\frac{dH_{c2}(T)}{dT} \right]_{T \approx T_c} \tag{29}$$

Table I lists, for a variety of type II superconductors, T_c, γ, $-[dH_{c2}(T)/dT]_{T \approx T_c}$, and $H_{c2}(0)$ calculated from Eq. (29).

So far, the magnetic properties of the normal state have been ignored, it being assumed that the free energy of the normal state is unaffected by any magnetic field. This assumption is reasonable at low fields since superconductors in the normal state are either diamagnetic or paramagnetic, with very small susceptibilities. These small susceptibilities, however, give rise to a very large effect as the high critical fields, of the order of tens of teslas, found in some type II superconductors are approached.

If the normal state has a susceptibility χ, then in a magnetic field H the Gibbs function per unit volume is changed by $-\frac{1}{2}\chi H^2$. The high-temperature, high-field superconductors are based on transition metal alloys or compounds, with a noneven number of electrons per atom. They will, therefore, be paramagnetic rather than diamagnetic, and their susceptibility χ will be positive. In a magnetic field the Gibbs function of the normal state will be lowered and there will be a field H_p given by

$$\tfrac{1}{2}\chi H_p^2 = \Delta G_{ns}$$

above which the normal state always has a lower Gibbs function than the superconducting state in zero field. The Gibbs function of the superconducting state increases as a field is applied, and H_p thus represents the maximum field at which superconductivity will persist; the actual upper critical field is expected to be below this. If the normal state susceptibility is due solely to Pauli paramagnetism, then [35, 36].

$$H_p(0) = 1.84 T_c \tag{30}$$

The experimental value of the upper critical field is close to either H_{c2} given by Eq. (25) or H_p given by Eq. (30), whichever is the least, provided that they are widely different. If they are similar in magnitude, then the effect of the magnetic field on the superconducting state must also be taken into account. Values of $H_p(0)$ are also given in Table I.

Maki [32] has defined a parameter $\alpha = \sqrt{2} H_{c2}(0)/H_p(0)$ where $H_{c2}(0)$ is given by Eq. (25) and $H_p(0)$ by Eq. (30). The actual paramagnetically limited critical field $H_{c2}(0)^*$ should be related to $H_{c2}(0)$ by

$$H_{c2}(0)^* = H_{c2}(0)(1 + \alpha^2)^{-1/2} \tag{31}$$

TABLE I

Upper Critical Fields for High-Field Superconductors[a]

Superconductors	T_c (K)	γ (kJ m⁻³ K⁻²)	$-\left[\dfrac{dH_{c2}(T)}{dt}\right]_{T=T_c}$ (T K⁻¹)	$H_{c2}(0)$ (T)	$H_p(0)$ (T)	α	$H_{c2}(0)^*$ (T)	$H_{c2}(0)_{expt}$ (T)
bcc alloys								
V 40 at. % Ti	7.0	1.1	—	24.0	12.9	2.6	8.6	11
Nb 37 at. % Ti	9.2	—	1.59	10.1	16.9	0.85	7.7	9.6
Nb 56 at. % Ti	9.0	—	2.52	15.7	16.6	1.3	9.6	14.1
Nb 70 at. % Ti	7.2	1.0	—	17.5	13.4	1.8	8.5	14
Nb 25 at. % Zr	10.8	0.79	1.15	8.6	19.9	0.6	7.4	9.2
B1 (NaCl) compounds								
NbN	15.7	0.25	1.25	13.5	29.2	0.66	11.6	15.3
Nb(C₀.₃N₀.₇)	17.4	0.21	—	—	32.5	—	—	11
A15 (Cr₃Si) compounds								
V₃Ga	14.8	3.04	3.4	34.9	27.2	1.82	16.8	25
V₃Si	16.9	2.50	2.9	34.0	31.1	1.55	18.4	24
Nb₃Sn	18.0	1.42	2.4	29.6	33.1	1.25	18.0	28
Nb₃Al	18.7	—	2.52	32.7	34.7	1.34	19.6	33
Nb₃Ga	20.2	—	2.43	34.1	37.2	1.30	20.8	34
Nb₃Ge	22.5	—	2.38	37.1	41.3	1.27	23.0	38
Nb₃(Al₀.₇Ge₀.₃)	20.7	—	3.1	44.5	38.0	1.65	23.1	43.5
Ternary sulfides								
PbMo₅.₁S₆	14.4	—	6.0	59.9	26.5	3.2	17.9	60

[a] Data from references 21, 38, 39.

From (25), (29), and (31), α is found to be given by [37]

$$\alpha = 2.39 \times 10^3 \gamma \rho_N = -0.533 \left[\frac{dH_{c2}(T)}{dT}\right]_{T \approx T_c} \tag{32}$$

and substituting from (25) and (32) into (31), the paramagnetically limited critical field $H_{c2}(0)^*$ is related to the critical temperature by [38]

$$H_{c2}(0)^* = 1.3 T_c \alpha (1 + \alpha^2)^{-1/2} \tag{33}$$

Table I also lists values of α derived from (32) and of $H_{c2}(0)^*$ from (33). These are compared with values of $H_{c2}(0)_{exp}$, which have been extrapolated from experimental results at 4.2 K. (These data have been abstracted from references 21, 38, and 39.) There are several interesting conclusions to be drawn from this table. With the exception of the vanadium based A15 compounds and some of the bcc alloys, the upper critical fields derived from the initial slopes of the magnetization curves are very close to the experimental values. These materials are therefore not paramagnetically limited. The vanadium-based A15 compounds, and the V 40 at. % Ti and Nb 70 at. % Ti bcc alloys, whose theoretical estimates of $H_{c2}(0)$ are derived from normal state resistivity, very clearly do suffer a paramagnetic limitation. The B1 (NaCl) compounds have very low values of $H_{c2}(0)$, due to their low electronic specific heats, and thus paramagnetic effects are not important. Two compounds, $Nb_3(Al, Ge)$ and $PbMo_{5.1}S_6$, have experimental values of H_{c2} greater than $H_p(0)$. These and the other Nb-based A15 compounds are able to escape the paramagnetic limitation because of spin–orbit coupling induced electronic spin–flip scattering. This effectively increases the paramagnetism of the superconducting state, reducing the effect of the magnetic field on $G_s(H)$. The paramagnetic critical field is increased and in the limit of strong spin–flip scattering becomes [40]

$$H_{so}(0) = 1.33 \sqrt{\lambda_{so}} \, H_p(0) \tag{34}$$

where λ_{so} is the spin–flip scattering frequency parameter. Spin–flip scattering is most effective in materials of high resistivity with atoms of large atomic mass.

The origins of the various critical fields are depicted schematically in Fig. 10.

D.   The Critical Temperature

Superconductivity is most unusual and physically interesting because it is a macroscopic quantum state (the only other example is superfluidity in

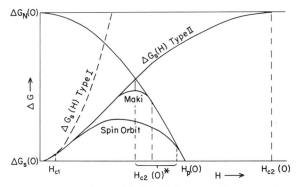

Fig. 10. The origin of the various critical fields in a type II superconductor is shown by plotting the Gibbs function ΔG against applied magnetic field for the superconducting, normal, and mixed states (schematic).

liquid helium II). Below the critical temperature T_c an attractive interaction causes some of the valence electrons to condense into pairs; these pairs, called Cooper pairs, obey Bose–Einstein rather than Fermi–Dirac statistics. The loss of resistance is not due to a disappearance of the normal scattering processes but to the way the electron pairs react to scattering. If one member of a pair is scattered and suffers a change in momentum, the interaction that binds the members of a pair is such that the other member changes its momentum so as to keep the total momentum of the pair constant. Thus, despite scattering, the net electron momentum remains constant, and current flows without resistance.

In most superconductors, the attractive interaction between the members of a pair is via the intermediary of a phonon, the quantum of lattice vibration. Superconductors are those materials in which the electron–photon interaction is strong; it is for this reason that superconductors have lower electrical conductivity in the normal state than most nonsuperconductors since the largest contribution to normal resistivity is from scattering by phonons. The electron–phonon interaction may be described in a very elementary fashion as follows: A negatively charged electron moving through a lattice of positive ions will attract these ions toward it, and the lattice may relax inward, giving rise to a concentration of positive charge (Fig. 11). This region of positive charge will follow the electron through the lattice, lagging slightly behind due to the inertia of the much heavier ions. A pair is formed when a second electron is attracted to this region of positive charge, and therefore follows the first electron.

The Bardeen, Cooper, and Schrieffer (BCS) theory [41], now the accepted theory of superconductivity, is able to treat the formation and behavior of Copper pairs in quantum mechanical formulation. The theory,

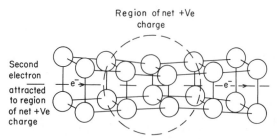

Fig. 11. The origin of the electron–phonon interaction. An inward relaxation of the positive ions of the crystal lattice toward the path of an electron creates a concentration of positive charge, which attracts a second electron.

for which the authors received the 1972 Nobel prize in physics, assumes that, with the exception of the attractive interaction, there is no difference between normal and superconducting states, and that the interaction potential V is constant and isotropic for all electrons within the Cooper shell and zero for all other electrons. The criterion for superconductivity is that V be negative.

The condensation of electrons into Cooper pairs causes an alteration of the energy levels near the Fermi surface in the superconducting state, and an energy gap appears (Fig. 12a). At 0 K, the size of the gap is

$$2\Delta = 4\hbar\omega_{ph} \exp[1/N(0)V] \approx 3.5kT_c \tag{35}$$

This represents the energy required to depair the Cooper electrons; it varies with temperature as does $H_c(0)$ (Fig. 12b). Equation (35) has been verified experimentally for BCS, or weak-coupling, superconductors.

It is possible, with the BCS theory, to make a quite accurate calculation

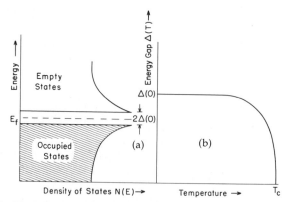

Fig. 12. (a) The BCS energy gap in the excitation spectrum of a superconductor at the Fermi level. (b) The variation of the gap energy with temperature.

of the critical temperature of a superconductor, provided that the normal state properties of the material are well understood. It is unfortunate that the metals that are high-T_c superconductors are least understood in the normal state. In these materials it is impossible to make an accurate estimate of V from first principles without knowing T_c from experiment. Predictions of high values of critical temperature follow from empirical rules.

From an examination of all the available experimental data, Matthias [42] systematized the occurrence of superconductivity and formulated empirical rules. The important rules are

(1) Superconductivity occurs only in metallic systems, and never if the system exhibits ferro- or antiferromagnetism.†

(2) Superconductivity occurs when the electron-to-atom ratio e/a lies between 2 and 8. T_c varies with e/a for metals, and alloys between metals, in the same period of the periodic table. Nontransition metals show T_c increasing as e/a increases from 2 to 6 (beyond 6 the nontransition elements are nonmetallic). The transition metals show a much more complicated behavior, with peaks at $e/a = 4.7$ and 6.5, and a sharp minimum in between (Fig. 13).

(3) Certain crystal structures are particularly favorable for superconductivity. The highest critical temperatures are found in the Cr_3Si (A15) and NaCl (B1) structures, with $e/a = 4.7$. Sigma phases ($D8_b$), α-Mn (A12), and Laves phases are crystal classes that produce compounds with $e/a \cong 6.5$ and critical temperatures up to ~ 10 K.

The first rule needs no further comment here. The dependence of T_c

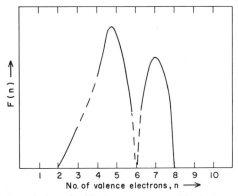

Fig. 13.   Matthias's variation of critical temperature with electron-to-atom ratio for the superconducting transition metals, their alloys, and their compounds.

† Materials that exhibit both ferromagnetism and superconductivity have recently been discovered.

upon the electron-to-atom ratio is related to the effect of the density of states $N(0)$. Both the Sommerfeld constant γ and the T_c for transition metals show maxima at 4.7 and 6.5 electrons/atom and minima at 4 and 6 electrons/atom. The value of V would appear to be fairly constant between about 4 and 8 electrons/atom. Beyond 8 electrons/atom the paramagnetic spin susceptibility rises rapidly, which explains the decrease in T_c in this region, despite a third peak in γ at $e/a \cong 10$.

The effect of crystal structure is least well understood. The A15 compounds with the highest critical temperatures are those formed between either Nb or V and group IIIA or IVA elements. The stoichiometric composition is A_3B, with A representing the group VB elements, and $e/a = 4.5$ or 4.75. The critical temperature is a very sensitive function of stoichiometry; deviations from the ideal composition cause a sharp decrease in T_c, as discussed in the chapter by Dew-Hughes on the physical metallurgy of A15 compounds, this volume.

The high-T_c B1 compounds are carbides and nitrides of transition elements, notably Mo, Nb, Ta, and Ti, or mixtures of these, with 4.5 or 5.0 electrons/atom. The highest critical temperatures in a B1 structure (~ 18 K) are found in mixtures of NbN with either NbC′ or TiC that give an e/a ratio very close to the ideal value of 4.7. Thus, similarly to transition metals, the superconducting behavior of the B1 compounds appears to be dominated by the electron density of states. T_c values decline with departure from stoichiometry, though less drastically than for the A15 compounds.

The A12, D8$_b$, and Laves phases with high transition temperatures all have e/a ratios close to 6.5, the second favorable value. It would appear that the primary role of crystal structure lies in allowing the formation of compounds with a favorable electron-to-atom ratio, and other effects on superconductivity are secondary.

The BCS theory has been extended by McMillan [43] to the case of strong-coupling superconductors, with V large. His expression for T_c is

$$T_c = \frac{\theta_D}{1.45} \exp\left[-\frac{1.04(1 + \lambda)}{\lambda - \mu^*(1 + 0.62\lambda)} \right] \tag{36}$$

where $N(0)V = \lambda - \mu^*$; λ is the (attractive) coupling constant, and μ^* the Coulomb repulsion between electrons. McMillan then assumes $\lambda \approx \eta/M \langle \omega_{ph}^2 \rangle$. η is a constant for a particular class of materials and M is the atomic mass. In the strong-coupling limit, $\lambda \gg \mu^*$ and

$$T_c \approx \langle \omega_{ph} \rangle \exp[-(1 + \lambda)/\lambda] = \langle \omega_{ph} \rangle (-M \langle \omega_{ph}^2 \rangle / \eta - 1)$$

which has a maximum when $\langle \omega_{ph} \rangle = (\eta/2M)^{1/2}$,

$$T_{c\,\max} = (\eta/2M)^{1/2} e^{-3/2}$$

McMillan uses this equation to predict maximum T_c's for series of metals and compounds for which η is known.

Allen [44] has reexamined McMillan's theory and shown that it is valid only for $\lambda < 1.5$, and that for larger values of λ T_c can be considerably higher than the McMillan limit. A study of the available data suggest that η is not necessarily constant for a given series, and that high T_c is more often a direct consequence of a large value of η. Hopfield [45] has proposed that a strong admixture of p- and d-wave components of the wave functions at the Fermi surface favor large values of η. Such a situation also produces a tendency toward covalent bonding. It would seem that what is needed for high critical temperatures is a structure with both a high density of states $N(0)$ and a strong tendency to covalent bonding. The success of the A15 structure in providing the highest critical temperatures may be that it represents the best compromise between these two requirements, which are to some extent contradictory.

The role of the density of states is ambiguous. There is much evidence (see for example the chapter by Dew-Hughes on the physical metallurgy of A15 compounds, this volume) that in a given material anything that enhances or degrades $N(0)$ will likewise raise or lower T_c, but between two similar materials it is not true that the one with the highest $N(0)$ will necessarily have the highest T_c. For example, γ for V_3Ga is twice that for Nb_3Sn, but the T_c of the latter is 3 K higher than that of the former.

The prediction of a value of T_c for a new member of an already well-known class of materials can only be carried out by extrapolation of empirical correlations between experimental T_c and material properties, such as has been done for A15 compounds by the author [39], or by Poon and Johnson, who relate T_c to the melting point [46]. It is almost impossible to make any sensible prediction about the superconducting behavior of an entirely new class of materials.

III. Irreversible Properties of Type II Superconductors

A. Flux Pinning and the Critical State

As mentioned in the introduction to this chapter (Section I,A), Silsbee hypothesized that the critical current in a type I superconductor is that current which just produces the critical field at the surface of the superconductor. The critical current is reduced by the application of an external field since the field it now has to produce is the difference between the critical field and the applied field. The critical currents of type II superconductors below H_{c1} agree with the Silsbee hypothesis, with H_c replaced by H_{c1}, but above H_{c1} they are much lower than would be expected by inserting H_{c2} in the Silsbee rule, and the variation with an applied field can be quite complicated. For fully reversible type II superconductors the critical current above H_{c1} is practically zero.

Flux lines in the mixed state of a type II superconductor experience a Lorentz force F_L whenever a current flows in the superconductor, given by

$$\mathbf{F}_{L(v)} = \mathbf{J} \times \mathbf{B} \quad \text{per unit volume of superconductor} \tag{37}$$

where $\mathbf{J}$ is the current density and $\mathbf{B}$ the flux density ($= n\phi_0$, where n is the number of flux lines per unit area). The force acts in a direction normal to both the flux lines and the current. Unless otherwise prevented, the flux lines will move in the direction of this force, and in so doing induce an electric field $\mathbf{E} = \mathbf{V} \times \mathbf{B}$ where $\mathbf{V}$ is the velocity of the flux lines; the superconductor now shows an induced resistance whose value approaches that of the normal state resistance as B gets close to $\mu_0 H_{c2}$.

The critical current is that current which just produces a detectable voltage across the specimen, and is therefore that current which first causes the flux lines to move. If there is no hindrance to the motion of flux lines, then the critical current above H_{c1} is zero and the superconductor is "reversible." It is possible to "pin" flux lines and prevent them from moving by interaction with microstructural features of the material, and a pinning force $\mathbf{F}_p$ is exerted on the flux lines that opposes the Lorentz force. The critical current density $\mathbf{J}_c$ is determined by the magnitude of the pinning force:

$$\mathbf{J}_c \times \mathbf{B} = \mathbf{F}_{p(v)} \quad \text{per unit volume} \tag{38a}$$

or

$$\mathbf{J}_c \times \phi_0 = -\mathbf{F}_{p(l)} \quad \text{per unit length of flux line} \tag{38b}$$

Pinning is due to crystal lattice defects, such as dislocations found in heavily cold-worked materials, impurities, or precipitates of a second phase. Strong pinning materials are metallurgically dirty and by analogy with mechanically strong or magnetically hard materials, irreversible superconductors are called hard superconductors. The critical current is not a property of a particular composition, but of a particular sample of superconductor, and is strongly influenced by the sample's metallurgical history; just as for mechanical strength, its value cannot be predicted with any accuracy.

When flux is pinned, it is reluctant to enter the specimen in an increasing field, and the entry of appreciable quantities of flux may not occur until a field several times H_{c1} has been applied. When the flux does penetrate, it does so slowly and penetration is not complete until the applied field is equal to H_{c2}. On reducing the field the flux is reluctant to leave the specimen, and there is remanent flux when the external field is zero. Reversal of the field causes the specimen to describe a complete magnetic hysteresis loop. The area of the loop is greater the stronger the flux pinning, and may be related directly to the critical current of the material. Minor loops are traversed if the field is reversed below H_{c2}. Schematic magnetization and critical current curves for a material in which the pinning has been progressively increased are shown in Fig. 14.

The relation between magnetization and critical current is established using the concept of the "critical state" [47, 48]. It is assumed that the superconductor everywhere carries either the critical current or no current at all. This model requires that everywhere within the superconductor flux lines be in equilibrium. Any force acting so as to try to move a flux line is just, and only just, opposed by a pinning force. Any imposed disturbance, by changing either a transport current or an external magnetic field, results in a redistribution of flux until the critical state is restored.

The total current density $\mathbf{J} = \nabla \times \mathbf{H}$. The critical state equation can be written [49]

$$\mathbf{J} \times \mathbf{B} = -\mathbf{F}_{p(v)}(B) = \nabla \times \mathbf{H}(B) \times \mathbf{B} \qquad (39a)$$

or

$$\mathbf{J} \times \boldsymbol{\phi}_0 = -\mathbf{F}_{p(l)}(B) = \nabla \times \mathbf{H}(B) \times \boldsymbol{\phi}_0 \qquad (39b)$$

where the pinning force is now recognized to be a function of the local value of the induction B. The resulting flux distribution is derived from

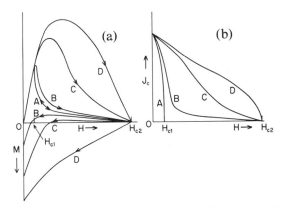

Fig. 14. (a) Magnetization (M) and (b) critical current (J_c) versus applied magnetic field for four samples of the same material with different levels of flux pinning. Sample A, reversible, no flux pinning; sample B, very weak pinning; sample C, moderate pinning; sample D, strong pinning.

$$\nabla \times \mathbf{H}(B) = \frac{\partial H}{\partial B} \nabla \times \mathbf{B} = \mathbf{J}_c(B) = \frac{F_{p(v)}(B)}{B} = \frac{F_{p(l)}(B)}{\phi_0} \qquad (40)$$

where $\mathbf{H}(B)$ is the external field that would be in equilibrium with the internal induction $\mathbf{B}$, and $\partial H/\partial B$ is the slope of the ideal reversible magnetization curve for the material. Integration of (40) with appropriate boundary conditions gives the actual flux distribution within the superconductor and hence the magnetization, if it is known how either J_c or F_p varies with B. This information is inferred from measured magnetization or current curves. The flux distribution in a cylinder of type II material in a magnetic field parallel to the axis is shown schematically in Fig. 15.

Many empirical expressions have been proposed to describe the relation between F_p and B. A better procedure is to derive expressions for particular models of the primary interaction between flux lines and the microstructure, and compare these with experimental results of J_c versus H. It is found that both theoretical expressions and experimental results can often be expressed in the general form [50]

$$F_p(b) = \text{const}(H_{c2})^n fn(h) \qquad (41)$$

where h is the reduced field, $-H/H_{c2}$, and $b = B/\mu_0 H_{c2}$. Plots of $J_c b$ versus h show a maximum, the position of which is characteristic of a particular pinning mechanism. Figure 16 shows a series of $J_c H$ versus H curves, all at 4.2 K, for an Nb 65 at. % Ti alloy after various heat treatments. The changing shape of the curves is indicative of the changing mechanisms of flux pinning as a dislocation cell structure is replaced by precipitates of titanium oxides and suboxides. Figure 17 shows, on a similar plot, data for a variety of A15 superconductors.

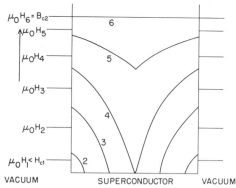

Fig. 15.   Flux distribution across a cylinder of irreversible type II superconductor as the external magnetic field is progressively increased.

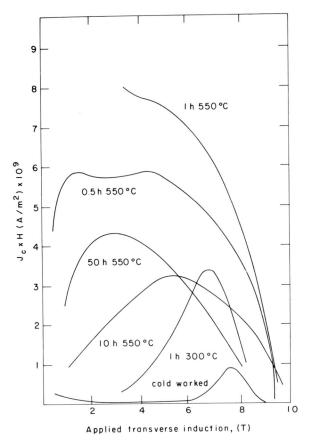

Fig. 16. Lorentz force ($J_c \times H$) versus applied field H for a Nb–Ti alloy annealed in vacuum. (After Witcomb and Dew-Hughes[51].)

B. Theories of Flux Pinning

High values of the critical current result from strong interactions that prevent the motion of flux lines. The metallurgical problems are similar to those of producing mechanically strong or magnetically hard materials. In strong materials the movement of crystal dislocations is hindered by regions whose elastic properties differ, and in hard magnetic materials domain boundaries are obstructed from moving by regions whose magnetic properties differ, from those of the bulk of the material. It is expected that in high-current superconductors regions whose superconducting properties show a difference from those of the bulk of the material will interfere with the motion of flux lines. The nature and origin of these regions are di-

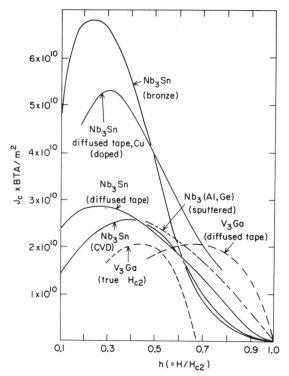

Fig. 17. Lorentz force ($J_c \times H$) versus reduced field h ($= H/H_{c2}$) for various A15 superconductors. (After Dew-Hughes [39].)

verse, but all are "imperfections" in the regularity of the crystal structure of the matrix.

Any theory of flux pinning consists of two parts, the local interaction between one flux line and one pinning center, and the summation of this local force over the entire lattice. These two factors are not necessarily independent, and further complications are introduced by their dependence upon type of pinning center and the possibility of their changing with field and/or temperature, scale of the microstructure, and the presence of more than one type of pinning center in any given sample. Experiments carried out on model systems, such as pure niobium or lead alloys, may give results that are not directly applicable to the high-κ materials of practical interest. The problem of critical currents and flux pinning has been covered in detail in an excellent review by Campbell and Evetts [52].

Flux lines interact with pinning centers because the superconducting properties of the latter differ from those of the bulk of the superconductor. The strength of the interaction is related to the magnitude of this difference. The difference may be small and may manifest itself as a change in critical temperature, critical field, or Ginzburg–Landau κ. The difference may be large, as is the case when the pinning center is nonsuperconducting.

Three types of pinning interaction can be recognized: elastic, core, and magnetic. There is a change in volume ($\sim -10^{-6}$) and in elastic constants ($\sim -10^{-4}$) in going from the normal to the superconducting state. Thus, the normal core of a flux line is both slightly more dense and more stiff than the bulk of the superconductor. The periodic change in density throughout a flux-line lattice produces a stress field of similar periodicity. Any defect, such as a dislocation, that has an associated strain field can interact with the stress field [53–56]; this is called the first-order, parelastic, interaction. There is also a second-order, dielastic, interaction between the square of the strain field and the elastic modulus variation [55–60]. The first-order interaction is believed to predominate over the second-order interaction [61, 62].

The core interaction results from a local change in the superconducting condensation energy at the pinning center. As a flux line passes through a center, the self-energy of the length of line intersecting the center is changed. The normal core of the flux line represents a volume, $\pi\xi^2$ per unit length, of energy per unit volume $\sim \frac{1}{2}\mu_0 H_c^2$. If the pinning center is a normal particle or void of diameter $a > \xi$, the flux line can lower its energy by $\frac{1}{2}\mu_0 H_c^2 \pi\xi^2 a$ if it positions itself along a diameter of the void. The interaction falls off over a distance $\sim \xi$, and the force needed to separate the flux line from the center is [63]

$$f_p = \frac{1}{2}\mu_0 H_c^2 \pi\xi a \tag{42}$$

The difference in energy between the flux-line core and a position midway between the flux lines is reduced as the induction, and hence the density of the flux lines, is increased. Campbell and Evetts take this into account by multiplying expression (42) by $(1 - b)$ [52]. They also point out that for a spherical center the length of interacting flux line continuously decreases from a to 0 as the flux line moves away from a diameter, and thus a is the appropriate interaction distance, and the pinning force becomes

$$f_p = \frac{1}{2}\mu_0 H_c^2 \pi\xi^2 (1 - b) \tag{43}$$

If the diameter $a < \xi$, then

$$f_p = \frac{1}{2}\mu_0 H_c^2 (V/\xi)(1 - b) \tag{44}$$

where V is now the volume of the center. However, in this case, because of its small size, the center must be considered as being capable of producing only a small fluctuation in the superconducting properties. Expressing this as a fluctuation in the Ginzburg–Landau free energy [52], gives

$$f_p = \tfrac{1}{2}\mu_0 H_c^2 \frac{\delta H_{c2}}{H_{c2}} \frac{V}{\xi} (1 - b) \tag{45}$$

An identical result is found if the center itself represents a small change in superconducting properties, as, for example, would be produced by a change in κ, due to a local variation in normal-state resistivity. Substituting for H_c in terms of κ and H_{c2}, gives for the pinning force

$$f_p = \mu_0 \tfrac{1}{4} H_{c2}^2 \frac{\Delta\kappa}{\kappa^3} \frac{V}{\xi} (1 - b) \tag{46}$$

for small ($a < \xi$) centers,

$$f_p = \mu_0 \tfrac{1}{4} H_{c2}^2 \frac{\Delta\kappa}{\kappa^3} \pi\xi^2 (1 - b) \tag{47}$$

for spherical centers, and

$$f_p = \mu_0 \tfrac{1}{4} H_{c2}^2 (\Delta\kappa/\kappa^3) \pi\xi a(1 - b) \tag{48}$$

for large centers.

If the center is larger than the inter-flux-line spacing $d \,[= 1.07(\phi_0/B)^{1/2}]$, then the center will contain a volume of the flux-line lattice, and it is necessary to consider magnetic and inter-flux-line interactions in addition to the core interaction. The Gibbs function per unit volume of a superconductor in the mixed state is [64]

$$G_m(H, T) = G_n(H, T) - \mu_0(H_{c2} - H)^2/2.32(2\kappa_2^2 - 1) \tag{49}$$

where G_n is the Gibbs function of the normal state. The second term on the right-hand side is the Gibbs function of the flux-line lattice per unit volume. The total length of flux line in a unit volume is B/ϕ_0, and G (lattice) per unit length of flux line is [65]

$$g = \mu_0\phi_0(H_{c2} - H)^2/2.32(2\kappa_2^2 - 1)B \tag{50}$$

The order parameter varies sinusoidally with a wavelength of $2d$, and the maximum force, obtained by differentiating with respect to d, is $-\pi g/d$. The pinning force per unit length of flux line for a normal pinning center or void is

$$f_p(\text{per unit length}) = \mu_0\pi\phi_0(H_{c2} - H)^2/2.32(2\kappa_2^2 - 1)Bd \tag{51}$$

If the pinning arises from a change in κ, then the pinning force is derived by differentiating (51) with respect to κ. Again, neglecting the difference between κ_1 and κ_2, for $\kappa \gg 1$

$$f_p(\text{per unit length}) = \mu_0\phi_0(H_{c2} - H) \, \Delta\kappa/2.32\kappa^3 d \tag{52}$$

The magnetic interaction occurs when the pinning center is even larger still, large enough that its dimensions exceed the superconducting penetration depth λ. This distance is the range over which the induction can vary appreciably, and only in regions of dimensions $> \lambda$ can an equilibrium induction, different from that elsewhere within the superconductor, be established. If the microstructure consists of nonsuperconducting precipitates, of a size and separation $> \lambda$, then an equilibrium induction will be established in the precipitate equal to $\mu_0 H(B)_{\text{rev}}$, where $H(B)_{\text{rev}}$ is the magnetic field that would be in equilibrium, at a free surface, with the induction B within the superconductor. $H(B)_{\text{rev}}$ is the reversible curve given by the Abrikosov theory [28]. This difference in flux density may be thought of as being maintained by supercurrents circulating in the boundary between superconductor and precipitate. Flux moving between precipitate and superconductor must cut these currents; the force necessary to achieve this is the pinning force, given by [66]

$$f_p(\text{per unit length}) = \frac{\phi_0(H_{c2} - H)}{1.16(2\kappa_2{}^2 - 1)\lambda} \tag{53}$$

If the pinning center is not a normal precipitate, but a region that can be characterized as having a value of κ slightly different from that of the main body of the superconductor, the pinning force is obtained by differentiating (53) with respect to κ [67]:

$$f_p(\text{per unit length}) = \phi_0(H_{c2} - 2H) \, \Delta\kappa/2.32\kappa^3\lambda \tag{54}$$

This function goes to zero and changes sign at $H = \frac{1}{2}H_{c2}$, leading to a minimum in the J_cB versus H curve. The change in sign indicates that flux prefers to be in the high-κ region at low fields and in the low-κ regions at high fields.

It is now necessary to decide how these individual interaction forces are to be summed over the whole flux-line lattice. It is most straightforward to simply add all of the individual interactions, that is, multiply the foregoing f_p values by the number of interactions per unit volume. This method assumes that the flux lines act individually, and ignores any inter-flux-line forces. The alternative approach, due to Labusch [68], is to assume that the primary forces produce local elastic distortions of the flux-line lattice. The flux lines undergo limited displacement and cannot take up positions of minimum energy. The total pinning force is related to the dissipation of the stored elastic energy in the lattice, and is proportional to $n_v f_p^2/C$, where n_v is the density of pinning sites per unit volume and C is some appropriate lattice elastic constant. This theory has only been worked out in detail for point or line pins, and it has not yet been extended to large surface or volume pins. The current status of the theory, its limitations, and the uncertainty about the current values of the elastic constants are discussed by Kramer [62].

Flux-lattice elasticity can be ignored, and flux lines can be assumed to take up positions of minimum energy, if the lattice is amorphous or if the primary forces are sufficiently strong to cause plastic deformation of the lattice. Unfortunately, there is little direct evidence, but by using the high-resolution Bitter technique, Trauble and Essmann [69] have shown that the flux lattice can become amorphous at high flux gradients. Herring [70] has combined this technique with transmission electron microscopy to observe simultaneously the flux distribution and the underlying defect structure in niobium. The flux lattice is distorted to allow flux lines to coincide with crystal dislocations; where the niobium contains a cellular substructure the flux-line lattice becomes polycrystalline. Defects in the flux line lattice, such as dislocations, will allow displacements in excess of those expected from simple elasticity, and indeed dislocations may well be present in order to maintain a flux gradient [38].

Much of the experimental evidence on samples with strong flux pinning supports the direct summation approach [71]. The total pinning force per unit volume for point pins, described by Eq. (42)–(48), is given by

$$F_p = f_p n_v \tag{55}$$

where n_v is the total number of pinning interactions per unit volume. If L, the average distance between pinning centers, is $\geq d$, the inter-flux-line spacing, then all pins are occupied by flux lines and $n_v \approx 1/L^3$, the pin density. If $L < d$, not all pins intersect with the flux lines at any one time, and $n_v \approx B/L\phi_0$. For large pins, described by Eqs. (51)–(54), the interaction occurs at the interface between pin and matrix, and

$$F_p = f_p(S_v/1.07)(B/\phi_0)^{1/2} \tag{56}$$

where S_v is the area of interface per unit volume perpendicular to the direction of motion of the flux lines.

Finally, Kramer [72] has suggested that the ultimate pinning force that can be exerted must correspond to the shear strength of the flux-line lattice. As the Lorentz force becomes equal to the lattice shear strength, regions of the lattice that are less strongly pinned will shear past more strongly pinned regions. The lattice shear strength, at high fields, is proportional to $h^{1/2}(1 - h)^2$, going to zero at H_{c2} ($h = 1$), as depicted schematically in Fig. 18. As the pinning strength is increased, the peak in

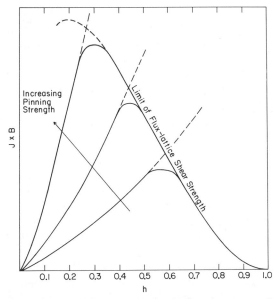

Fig. 18. Kramers's flux-shear theory. The flux-lattice shear strength decreases with increasing external reduced field as $h^{1/2} (1 - h)^2$. The flux-pinning interaction initially increases with h, at a rate depending upon the details of the pinning mechanism. The actual Lorentz force curve is the lower combination of these two curves. As the pinning strength is increased, the lattice shear strength is little altered, and the maximum moves to lower reduced fields.

F_p ($= J_c B$) should occur at lower fields. Such a trend, found for many materials, can be discovered in the data of Fig. 17. This behavior is not, however, universal. If the pinning is isotropic, it is not easy to understand why, on average, any one region of the flux lattice should be less strongly pinned than any other, though statistical fluctuations could become appreciable over sufficiently small volumes.

There would appear to be no single universal theory of flux pinning. The theory appropriate to a particular material will depend upon the microstructure, i.e., the nature of the flux-pinning center, its size, and its distribution.

C. Flux Pinning and Microstructure

Just as crystal imperfections are classified by the number of dimensions that are large compared to interatomic spacings, so superconducting imperfections may similarly be classified by the number of dimensions that are large compared to the flux-line spacing d [66]. Point defects, with no dimension greater than d, are small second-phase particles, small voids, and clusters of crystal defects such as those produced by irradiation with high-energy particles. Pinning by radiation-induced defects is discussed in the chapter by Sweedler, Snead, and Cox, this volume. The size of a point defect is such that it can interact directly with only one flux line at a time.

Line defects, with one dimension greater than d, are represented mainly by crystal dislocations. The orientation between the flux line and the defect is important in determining the strength of the interaction.

Surface defects, with two dimensions greater than d, include grain and twin boundaries, stacking faults, martensitic boundaries, certain configurations of crystal dislocations (such as subgrain and polygonized boundaries), and the surface of the superconductor itself. Again the orientation between the flux line and the plane of the surface can be of importance.

Volume defects, with three dimensions greater than d, are large second-phase particles and voids. Each defect is able to interact with several flux lines. At the high fields available in materials of interest, most second-phase particles are likely to be volume defects.

The following conclusions may be drawn from the vast range of experimental work that has been carried out in relating microstructure to flux pinning [38]:

(a) Strong flux pinning requires regions whose superconducting properties differ from those of the bulk; the greater the difference, the stronger the pinning. The region may be more strongly superconducting than the bulk (i.e., have a higher T_c and H_c), be less superconducting, be nonsuperconducting or be ferromagnetic.

(b) Such regions are most effectively produced by precipitation, powder metallurgy, dislocation cell structures, or neutron irradiation.

(c) For a given volume fraction of these regions, the more finely divided the structure, the stronger the pinning. Quantitative metallography indicates that pinning is, in most cases, directly proportional to the area of interface per unit volume.

(d) In the niobium alloys which form the basis for much commercial material, pinning is due to the dislocation cell structure. Extreme cold work, $\sim 99.995\%$ reduction in area, of ductile niobium-based alloys can result in dislocation cell structures with cells down to ~ 0.1 μm in diameter; J_c is proportional to (cell diameter)$^{-1}$ [73]. Subsequent annealing at $\sim 400°C$ allows of dislocation rearrangement that enhances the difference in superconducting properties between cell and cell wall, increasing pinning strength. This treatment has an additional advantage since the conductor is usually incorporated in a copper matrix, in that the effects of cold work in the copper are reduced, thus decreasing its resistivity. The theory of flux pinning in Nb–Ti is discussed in the chapter by McInturff, this volume.

(e)   Pinning in A15 compounds is produced by a fine grain size and by precipitates. It is difficult to distinguish between the two effects since the presence of precipitates will cause refinement of the grain size. Precipitates have been introduced via the substrate or by the presence of oxygen and nitrogen during the reaction to form the compound. Oxides and nitrides precipitate on a substructure within the grains; the highest critical currents are given by oxide particles ~ 50 Å in diameter at a spacing of 200 Å [74]. Grain boundaries in A15 compounds are expected to be more efficient than those in alloys as flux pinning centers since they represent an interruption in the high degree of order necessary for superconductivity in the compounds. Theories of flux pinning in A15 compounds are covered in more detail in the chapter by Dew-Hughes on the physical metallurgy of A15 compounds, and in the chapter by Luhman, this volume.

D.   Instability

Instabilities in type II superconductors are an inevitable consequence of the critical state. The flux everywhere within the superconductor is in a metastable condition, the pinning force exactly balancing the Lorentz force. Local flux motion may be generated spontaneously by thermal activation (flux creep), or by an external stimulus such as a sudden change in transport current ΔJ, external magnetic field ΔH, or mechanical shock. This flux motion, called a "flux jump," is dissipative and causes a local rise in temperature. The pinning force usually decreases as the temperature increases (dJ_c/dT is negative), more flux is able to move, and a small flux jump can run away and develop into a "flux avalanche." If the avalanche is not halted, the temperature rise will ultimately be large enough to cause the superconductor to exceed its critical temperature and become normal [75]. The normal state can also be directly induced by an external temperature pulse ΔT, which may be generated by friction if the superconductor is able to move under the action of electromagnetic forces, or by the release of elastic energy consequent upon the cracking of brittle components (such as resins in which the superconductor is "potted") as a result of differential thermal contraction as the superconducting device is cooled to cryogenic temperatures.

The redistribution of flux into a superconductor causes a reduction in the flux gradient leading to a change in the Lorentz force ΔF_L and a temperature rise ΔT that in turn causes a change in the pinning force ΔF_p. If $\Delta F_p > \Delta F_L$, then the critical state will have been exceeded and a flux avalanche will be propagated.

The criterion for the suppression of a flux avalanche can be estimated as follows: Consider a thin-strip superconductor of thickness r carrying a current density J in a field B. The current gives rise to a flux gradient in the strip, as shown in Fig. 19a, such that

$$J = 2 \, \Delta B / \mu_0 r$$

Now suppose that the flux gradient, and hence the current, is reduced. This can be achieved notionally by imagining the transfer of a quantity of flux dB from the left-hand side of the strip to the right-hand side (Fig. 19b). The current has been reduced by an amount

$$dJ = 2 \, dB / \mu_0 r$$

The Lorentz force F_L ($= JB$) is similarly reduced:

$$dF_L = B \, dJ = 2B \, dB / \mu_0 r \tag{57}$$

Moving the quantity of flux dB through a distance r across the current results in a dissipation of energy per unit volume $Jr \, dB$, and hence a temperature rise. $\Delta T = Jr \, dB / c$, where c is the specific heat per unit volume. This assumes that all of the energy is unable to diffuse away from the region in which it is generated. This is reasonable for high-field alloys, in which the magnetic diffusivity D_m ($\sim 10^3$–10^4 cm^2 sec^{-1}) in the normal state is very much greater than the thermal diffusivity D_{th} (~ 1 cm^2 sec^{-1}).

This temperature rise results in a change in the pinning force

$$dF_p = B \frac{dJ_c}{dT} \Delta T = \frac{JBr \, dB}{c} \left(\frac{dJ_c}{dT} \right) \tag{58}$$

Since dJ_c/dT is usually negative, the pinning force is reduced.

The superconducting strip is stable, that is, no flux avalanche will occur, provided that the reduction in Lorentz force dF_L is greater than the reduction in the pinning force dF_p. From (57) and (58) this condition is satisfied when

$$\frac{JBr \, dB}{c} \frac{dJ_c}{dT} > - \frac{2B \, dB}{\mu_0 r}$$

The superconductor is therefore stable to currents less than a critical value J_s, given by [38]

$$J_s = - \frac{2c}{\mu_0 r^2} \left(\frac{dJ_c}{dT} \right)^{-1} \tag{59}$$

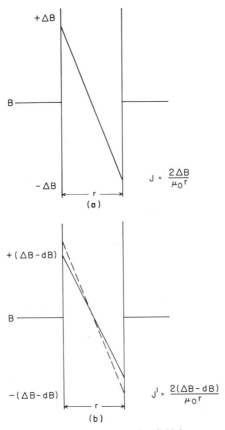

Fig. 19. Distribution of flux across a narrow strip of thickness r, carrying a current density J, in an average induction B. (a) Before a flux jump; (b) after a flux jump that has transferred a quantity of flux dB from the left-hand to the right-hand side of the strip.

This result does not differ significantly from those obtained by other methods [76, 77]. For wire instead of strip samples, the preceding equation is modified by a constant, and r becomes d, the wire diameter.

Methods of stabilization can be understood by reference to Eq. (59) and Fig. 20. The maximum stable current J_s is increased if the specific heat of the material is increased [78]. Specific heats do rise rapidly with temperature at the low temperatures at which superconductors operate, and an instability may be arrested as the temperature and specific heat rise [79].

It can be seen that if dJ_c/dT is positive, the conductor is stable to any current. No practical material has yet been developed that utilizes this concept.

The final, and most important, method of stabilization is to make r as

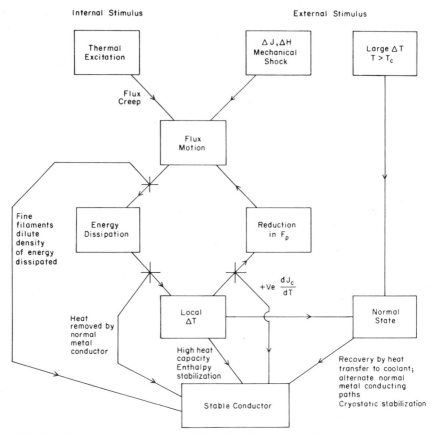

Fig. 20. The origin and propagation of flux disturbances, showing how these can lead to instabilities in superconductors. The various ways in which the effects of these disturbances can be reduced, and stability conferred, are also indicated. (After Dew-Hughes [38].)

small as possible. This is done by fabricating superconductors as fine filaments, of which a number, from 10 to 10,000, are embedded in a nonsuperconducting high-conductivity matrix such as copper or aluminum. D_{th} in copper is $10^3–10^4 \, cm^2 \, sec^{-1}$, and $D_m \sim 1 \, cm^2 \, sec^{-1}$; thus, the copper assists in removing heat from the superconductor, while slowing down the propagation of any magnetic disturbance. The stability conferred by the fine filaments is assisted, and flux jumps in one filament are not transferred to neighboring filaments.

Filamentary stabilization is only effective provided that the filaments are independent of one another. Independence is achieved under conditions of steady current and field, there being no voltage to drive current across the normal matrix separating the superconducting filaments. Every conductor will at some time be subjected to a changing field, either externally imposed or due to a change in the current flowing through it, if only when the device is switched on or off. The changing field induces an emf across the normal matrix, and current flows between neighboring filaments as a result of this emf. The filaments can be decoupled, and stability at high rates of change of field can be achieved by using a high-resistivity matrix or by twisting the filaments [79]. Twisted composites are now common in large magnets. Transposing the filaments by cabling or braiding gives even greater improvement in performance.

If a section of superconductor has been driven normal, and the current is not switched off, then the energy will be dissipated by Joule heating at a rate $J^2 \rho_n \pi R^2 L$ in a length L of conductor of radius R and normal state resistivity ρ_n. The rate at which this energy is transferred to the coolant is $2\pi R L \dot{Q}$, where $\dot{Q}$ is the heat flux from the surface of the conductor into the coolant. The conductor will be able to revert to the superconducting state provided that the rate of energy removal is greater than the rate of energy dissipation. The current density J_R above which recovery will not take place is given by [78, 80, 81]

$$J_R = (2\dot{Q}/\rho_n R)^{1/2} \qquad (60)$$

It must be noted that R is the overall radius of the conductor and that ρ_n is the average resistivity of the conductor in the normal state. Provided that the current density is kept below J_R, the superconductor will always recover from any disturbance.

The crucial quantity is seen here to be $\dot{Q}$. This depends on the nature of the interface, the nature of the coolant, and the temperature difference across the interface. The usual coolant for superconductors is liquid helium and the heat flux $\dot{Q}_f$ at which the evaporation of helium changes from nucleate boiling to film boiling is critical. Once film boiling has begun, the surface of the conductor is covered with an insulating blanket of gaseous helium and the rate of heat transfer drops abruptly. $\dot{Q}_f$ depends on the nature and geometry of the surface, but lies in the range $0.1-1$ W cm^{-2} [82]. Greater heat fluxes can be achieved by cooling with supercritical helium gas at $5-10$ K. Large devices are now usually planned with this type of cooling. In multifilamentary materials the electrical conductivity of the matrix is $\sim 10^3-10^4$ times greater than that of the superconductor in the normal state. The value of ρ_n for the composite conductor, appropriate for Eq. (61), is therefore much lower, and the recovery current much higher, than for a noncomposite material.

By reference to Eqs. (59) and (60) it is possible to design a composite that is completely stable to flux motion and that is able to recover if driven normal by other disturbances. J_s is the current density in each filament, and ideally the filament diameter is chosen so that J_s exceeds the value of J_c at the highest field at which the conductor is to operate. The overall diameter of the composite and the ratio of superconductor to normal material, and hence the number of filaments, are chosen so that the overall current density of the composite conductor is equal to J_R.

With the introduction of filamentary composites, instabilities no longer limit the performance of high-current superconductors. The correct choice of such factors as filament diameter, number of filaments, packing fraction, resistivity of matrix, pitch of twist, transposition, and cabling make it possible to attain any reasonable level of desired stability. Rise time of large coils can still, however, be a problem.

References

1. H. K. Onnes, *Commun. Phys. Lab. Univ. Leiden* Nos. 119b, 120b, 122b (1911).
2. B. W. Roberts, *J. Phys. Chem. Ref. Data* **5**, 581 (1976); B. W. Roberts, Natl. Bur. Stand. (U.S.) Tech. Note 983 (1978).
3. H. K. Onnes, *Commun. Phys. Lab. Univ. Leiden Suppl.* **34**, No. 133d (1913).
4. H. K. Onnes, *Commun. Phys. Lab. Univ. Leiden* No. 139f (1914).
5. F. B. Silsbee, *J. Wash. Acad. Sci.* **6**, 597 (1916).
6. W. Meissner and R. Ochsenfeld, *Naturwissenschaften* **21**, 787 (1933).
7. K. Mendelssohn and J. R. Moore, *Nature (London)* **135**, 836 (1935).
8. G. Ascherman, E. Frederik, E. Just, and J. Kramer, *Phys. Z.* **42**, 349 (1941).
9. G. F. Hardy and J. K. Hulm, *Phys. Rev.* **87**, 884 (1953).
10. B. T. Matthias, *Phys. Rev.* **92**, 874 (1953).
11. B. T. Matthias, T. H. Geballe, S. Geller, and E. Corenzwit, *Phys. Rev.* **95**, 1435 (1954).
12. J. E. Kunzler, E. Buehler, F. S. Hsu, B. T. Matthias, and C. Wahl, *J. Appl. Phys.* **32**, 325 (1961).
13. J. E. Kunzler, *Rev. Mod. Phys.* **33**, 501 (1961).
14. L. H. Wernick, F. J. Morin, F. S. L. Hsu, J. P. Maita, D. Dorsi, and J. E. Kunzler, *in* "High Magnetic Fields" (H. H. Kolm, B. Lax, F. Bitter, and R. G. Mills, eds.), p. 609. MIT Press, Cambridge, Massachusetts and Wiley, New York, 1962.
15. B. T. Matthias, *et al., Science* **156**, 645 (1967).
16. G. Arrhenius *et al., Proc. Natl. Acad. Sci. U.S.A.* **61**, 261 (1968).
17. P. R. Sahm and T. V. Pruss, *Phys. Lett.* **23A**, 707 (1969).
18. G. W. Webb, L. J. Vieland, R. E. Miller, and A. Wicklund, *Solid State Commun.* **9**, 1769 (1971).
19. J. R. Gavaler, *Appl. Phys. Lett.* **23**, 480 (1973).
20. L. R. Testardi, J. H. Wernick, and W. A. Royer, *Solid State Commun.* **15**, 1 (1974).
21. S. Foner, E. J. McNiff, Jr., and E. J. Alexander, *Appl. Supercond. Conf., Oakbrook 1974; IEEE Trans. on Magn.* **MAG-11**, 155 (1975).
22. G. B. Yntema, *Phys. Rev.* **98**, 1197 (1955).
23. D. L. Martin, M. G. Benz, C. A. Bouch, and C. H. Rosner, *Cryogenics* **3**, 161 (1963).
24. F. London and H. London, *Proc. R. Soc. London Ser. A* **149**, 71 (1935); *Physica* **2**, 963 (1935).
25. M. von Laue, "Theorie der Superleitung," 2nd ed. Springer, New York, 1949.
26. A. B. Pippard, *Proc. R. Soc. London Ser. A* **216**, 547 (1953).

27. V. L. Ginzburg and L. D. Landau, *JETP USSR* **20**, 1064 (1950).
28. A. A. Abrikosov, *Sov. Phys. JETP* **5**, 1174 (1957).
29. *Proc. Conf. Sci. Supercond., Colgate Univ. 1963; Rev. Mod. Phys.* **36** (1964).
30. B. B. Goodman, *IBM J. Res. Dev.* **6**, 63 (1962).
30a. W. H. Kleiner, L. H. Roth, and S. H. Autler, *Phys. Rev.* **133**, A1226 (1964).
31. Y. B. Kim, C. F. Hempstead, and A. R. Strnad, *Phys. Rev.* **139**, 1163 (1965).
32. K. Maki, *Physics* **1**, 21, 127, 201 (1964).
33. G. Eilenberger, *Phys. Rev.* **153**, 584 (1967).
34. R. R. Hake, *Phys. Rev.* **158**, 356 (1967).
35. A. M. Clogston, *Phys. Rev. Lett.* **9**, 266 (1962).
36. B. S. Chandrasekhar, *Appl. Phys. Lett.* **1**, 7 (1962).
37. N. R. Werthamer, E. Helfand, and P. C. Hohenberg, *Phys. Rev.* **147**, 295 (1966).
38. D. Dew-Hughes, *Rep. Prog. Phys.* **34**, 821 (1971).
39. D. Dew-Hughes, *Cryogenics* **15**, 435 (1975).
40. O. Fischer and M. Peter, in "Magnetism" (G. Rado and H. Suhl, eds.), Vol. 5, p. 327. Academic Press, New York, 1973.
41. J. Bardeen, L. N. Cooper, and J. R. Schrieffer, *Phys. Rev.* **108**, 1175 (1957).
42. B. T. Matthias, *Prog. Low Temp. Phys.* **2**, 138 (1957).
43. W. L. McMillan, *Phys. Rev.* **167**, 331 (1968).
44. P. B. Allen, "Physique Sous Champs Magnetiques Intenses," p. 95. CNRS, Paris, 1975.
45. J. Hopfield, *Phys. Rev.* **186**, 443 (1969).
46. S. J. Poon and W. L. Johnson, *Phys. Rev. B* **12**, 4816 (1975).
47. C. P. Bean, *Phys. Rev. Lett.* **8**, 250 (1962).
48. Y. B. Kim, C. F. Kempstead, and A. R. Strnad, *Phys. Rev.* **129**, 528 (1963).
49. J. E. Evetts, A. M. Campbell, and D. Dew-Hughes, *J. Phys. C.* **1**, 715 (1968).
50. W. A. Fietz and W. W. Webb, *Phys. Rev.* **178**, 657 (1969).
51. M. J. Witcomb and D. Dew-Hughes, *J. Mater. Sci.* **8**, 1383 (1973).
52. A. M. Campbell and J. E. Evetts, *Adv. Phys.* **21**, 199 (1972).
53. E. J. Kramer and C. L. Bauer, *Philos. Mag.* **15**, 1189 (1967).
54. V. P. Galaiko, *Sov. Phys-JETP Lett.* **7**, 230 (1968).
55. U. Kammerer, *Z. Phys.* **227**, 125 (1969).
56. H. Kronmuller and A. Seeger, *Phys. Status Solidi* **34**, 781 (1969).
57. R. L. Fleischer, *Phys. Lett.* **3**, 111 (1963).
58. W. W. Webb, *Phys. Rev. Lett.* **11**, 191 (1963).
59. R. Labusch, *Phys. Rev.* **170**, 470 (1968).
60. K. Miyahara, F. Irie, and K. Yamafuji, *J. Phys. Soc. Jpn.* **27**, 290 (1969).
61. F. R. N. Nabarro and A. I. Quintanilha, *Mater. Res. Bull.* **5**, 669 (1970).
62. E. J. Kramer, *J. Nucl. Mater.* **72**, 5 (1978).
63. P. W. Anderson, *Phys. Rev. Lett.* **9**, 309 (1962).
64. B. B. Goodman, *Rep. Prog. Phys.* **29**, 445 (1966).
65. R. G. Hampshire and M. T. Taylor, *J. Phys. F* **2**, 89 (1972).
66. A. M. Campbell, J. E. Evetts, and D. Dew-Hughes, *Philos. Mag.* **18**, 313 (1968).
67. D. Dew-Hughes and M. J. Witcomb, *Philos. Mag.* **26**, 73 (1972).
68. R. Labusch, *Cryst. Lattice Defects* **1**, 1 (1969).
69. H. Trauble and U. Essmann, *J. Appl. Phys.* **39**, 4052 (1968).
70. C. P. Herring, *Phys. Lett.* **47A**, 103, 105 (1974); *J. Phys. F* **6**, 99 (1976).
71. D. Dew-Hughes, *Philos. Mag.* **30**, 293 (1974).
72. E. J. Kramer, *J. Appl. Phys.* **44**, 1360 (1973).
73. D. F. Neal, A. C. Barber, A. Woolcock, and J. A. F. Gidley, *Acta Metall.* **19**, 143 (1971).

74. D. Dew-Hughes, *in* "Superconducting Machines and Devices" (S. Foner and B. B. Schwartz, eds.), p. 87. Plenum Press, New York, 1974.
75. J. E. Evetts, A. M. Campbell, and D. Dew-Hughes, *Philos. Mag.* **10**, 339 (1964).
76. R. Hancox, *Appl. Phys. Lett.* **7**, 138 (1965).
77. P. S. Swartz and C. P. Bean, *J. Appl. Phys.* **39**, 4991 (1968).
78. P. F. Chester, *Rep. Prog. Phys.* **30**, 561 (1967).
79. M. N. Wilson, C. R. Walters, J. D. Lewin, and P. F. Smith, *J. Phys. D* **3**, 1518 (1970).
80. Z. J. J. Stekly, *J. Appl. Phys.* **37**, 324 (1966).
81. D. Dew-Hughes, K. Seshan, and K. A. Wallis, *Cryogenics* **14**, 647 (1974).
82. D. N. Lyon, *Adv. Cryog. Eng.* **10B**, 371 (1965).

Magnets, Motors, and Generators

PER F. DAHL

Isabelle Project, Accelerator Department
Brookhaven National Laboratory
Upton, New York

I. Introduction 47
II. Magnet Design Considerations 49
   A. Properties Affecting Flux Jumping Instabilities 49
   B. Methods of Stabilization 51
   C. Intrinsically Stable Conductors 54
   D. High Current Capacity Macrocomposites 59
   E. Other Magnet Aspects Affecting Conductors 60
   F. Multifilamentary A15 Composites 62
III. Applications 65
   A. General-Purpose Magnets 65
   B. Large Magnets 73
   C. Transverse-Field Magnets 76
   D. Electrical Machines 84
   E. Other Applications 90
   References 96

I. Introduction

The discovery of high-field superconductors during 1961–1962 and the commercial availability shortly thereafter of such material in wire or cable form led to the construction of a number of simple dc high-field laboratory magnets in the early 1960s. With few exceptions these early devices displayed disappointing performance, since they were plagued by instabilities and general failure to live up to design expectations. Only gradually was this frustrating behavior interpreted correctly in terms of "flux jumping," and effective solutions to the problem were realized. This chapter recounts briefly this pioneering phase of superconducting magnet development. Next the various stability criteria and stabilization techniques, together

47

with other superconducting magnet design considerations, are discussed in more detail. Some of these considerations are further treated in later chapters. The chapter then describes different types of magnets, and concludes by summarizing the main areas of present and potential applications of these devices.

The properties of superconductors have already been described in the introductory chapter by Dew-Hughes. Materials suitable for magnet construction are the irreversible type II superconductors, with high upper critical fields and high bulk current densities. The parameters that characterize these conductors, namely, field, temperature, and current density, are interrelated and must all be below certain critical values H_{c2}, T_c, and J_c for the superconducting state to exist. They define a three-dimensional H_{c2}–T_c–J_c critical surface, shown schematically in Fig. 1. For a given temperature and field, for instance, the conductor has a definite critical current $I_c(T, H)$; this current value generally has an inverse dependence on temperature and field.

A few small superconducting magnets were wound with niobium wire and tested in the late 1950s, but they were not very successful. The decisive breakthrough awaited the announcement by Kunzler *et al.* [1] at the Bell Telephone Laboratories in 1961 of the very high-field current capacity of the intermetallic compound Nb_3Sn. This event quickly stimulated various magnet development programs, the construction of numerous small solenoid magnets in many laboratories, and the discovery of several other promising superconductors in the form of alloys [2]. The first of these, Nb–Zr, which was the first high-field superconductor to be produced in long lengths of wire that was copper plated and insulated, was soon supplanted by the more ductile alloy Nb–Ti, which also has superior

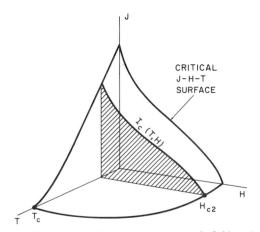

Fig. 1.   Relationship between critical current, magnetic field, and temperature.

superconducting properties. Early magnet conductors were typically in the form of multistrand cables of twisted superconductor and copper strands impregnated with indium; later, composite conductors of filaments in a copper matrix became available. The original hard superconductor, Nb_3Sn, possesses even higher critical field, current density, and temperature than Nb–Ti, but at first its extremely brittle nature prevented it from being used extensively. In its early application, coils were wound with pure Nb and Sn in a suitable form and then heat treated or reacted to form the compound in an irretrievable coil. This technique soon became obsolete (though it has been revived recently with the appearance of filamentary forms of Nb_3Sn) when flexible Nb_3Sn conductors became available in ribbon form as very thin layers of superconductor on a strong and relatively thick substrate, produced by a vapor deposition or diffusion process.

As already noted, these early superconducting coils tended to exhibit irregular or unstable performance, characterized by "degradation" (failure to reach the current or field expected from tests on short samples of the conductor) and "training" (diminishing degradation after each successive transition to the normal state, or quench). The key to a satisfactory magnet lies essentially in the ability of magnet designers to suppress these two related phenomena. The first, premature quenching, is now largely understood and under control. It is in large measure attributed to the process of flux jumping, or sudden discontinuous field changes accompanied by rapid heating as flux penetrates the conductor; this behavior is a consequence of the intrinsic properties of type II superconductors, and has already been considered in some detail in the introductory chapter by Dew-Hughes, Section III,C. An adequate understanding of training has proven more elusive, and essentially all superconducting magnets designed for very high fields (greater than 4 T) still suffer from it to some degree. It appears to be caused by several factors, among them conductor motion due to electromagnetic forces, and relaxation of stresses in the conductor or coil impregnation materials during cooldown or operation. A systematic treatment of this complex subject is well beyond the scope of the present review.

II. Magnetic Design Considerations

A. *Properties Affecting Flux Jumping Instabilities*

Premature quenching may typically occur between 50 and 70% of the expected "short sample" field strength, as illustrated in Fig. 2, which schematically shows critical current versus field for a representative superconducting wire. The curve is the approximately hyperbolic short-

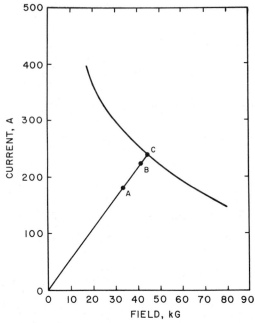

Fig. 2. Dependence of critical current on magnetic field. Point *A*, quench point when a magnet is initially energized; *B*, quench point after repeated training; *C*, "short sample" current.

sample characteristic curve for the wire. Usually this curve is known for a limited field range only. The line 0*C* is the "load line" for a coil wound from this wire, i.e., the expected linear (in the absence of ferromagnetic components in the magnet) relationship between the central field strength in the magnet bore and current in the winding. On the first charge the coil may experience a number of small low-field flux jumps, somewhat erratic in occurrence but proportional to charge rate, and revealed as transients in a coil voltage–transport current plot. The coil finally quenches violently at point *A*, say, with considerable evaporation of liquid helium. Subsequently, it may train during repeated quenches to point *B*, usually still somewhat below the ultimate short-sample point *C*.

The flux jumping process is governed by the magnetization behavior of type II conductors in an external field, and the ability of flux to enter the bulk of these materials [3]. If the superconductor is exposed to a changing external field, local "persistent" diamagnetic current loops are created in the conductor, similar to eddy currents in normal conductors, in agreement with Lenz's law, induced to oppose the changing field. Eventually, as the external field exceeds a critical value, field will penetrate into the interior of the conductor, and from this point on there will exist a local critical current in the interior, determined by the local penetrated flux, and a field gradient is set up between the outside and inside of the material where it is said to be in the "critical state." It is the collapse of these shielding currents and the consequent rapid field change that is responsible for flux jumping. If a perturbation of some kind causes a slight local change in flux, local heat dissipation occurs and, provided the process is sufficiently fast to be adiabatic, the temperature rises. (Mechanical motion may also trigger heat release.) The rise in temperature causes a decrease in the allowed critical current density, in accord with the inverse relationship between current density and temperature (Fig. 1). The drop in current density, in turn, allows more flux to penetrate, with further generation of heat.

This cycle may be regenerative under certain conditions, as was outlined in the introductory chapter by Dew-Hughes, Section III,D— primarily those for which the temperature rise from the energy released exceeds the heat capacity of the material [4]. The energy released is proportional to the square of the critical current density (this being the reason flux jumping occurs predominantly at low fields) as well as the square of the superconductor cross section. Thus, in addition to high volumetric specific heat, thermal conductivity, and critical temperature, the suppression of flux jumping is aided by low critical current density and small conductor size, although not all of these factors are easily controlled. The only parameters available for manipulation are size, thermal environment, and current density. But decreasing the critical current density negates the basic feature of a superconducting magnet and defeats the objective of its design, i.e., to achieve maximum field for a given amount of superconductor. In practice, stabilization against flux jumping is achieved most readily by altering the two other variables, conductor size and thermal environment, as discussed in Sections II,B and II,C.

B. Methods of Stabilization

The simplest approach to the prevention of flux jumping is the use of "fully stabilized" or "cryostatically stable" conductors [5–7]. These consist of superconductors in close thermal and mechanical contact with sufficient normal metal of high electrical conductivity to smooth out local hot spots and to ensure an alternative current path during periods of instability. The effectiveness of the method depends on the interrelationship between the temperature–current characteristics of the superconductor, the resistance of the normal metal, and the rate of heat transfer to the surrounding helium bath, according to Eq. (60) of the introductory chapter by Dew-Hughes, which may also be expressed as

$$I^2\rho = \alpha h \, \Delta T \, AS \qquad (1)$$

Here I is the current, ρ the normal metal resistivity, α a stability parameter, h the heat transfer coefficient (assumed constant), ΔT the liquid–composite temperature difference, and A and S the conductor cross section and surface area, respectively. The composite is stable if $\alpha \leq 1$. Full recovery of the superconductor after the disturbance requires that the metal–liquid interface heat flux $h \, \Delta T$ be sufficiently low that normal nucleate boiling of the helium is maintained at the interface, thereby assuring a tolerable temperature rise there. If the heat flux exceeds a critical value (approximately 0.3 W/cm²), film boiling commences and creates an insulating vapor barrier at the surface and a sharp rise in temperature across it. In this case full recovery to the superconducting state may not be possible (unless the current is reduced sufficiently).

Most stabilized composite conductors consist of NbTi strands embedded in a copper matrix. Copper is used primarily because it is easily codrawn with NbTi, although aluminum is a potential matrix competitor because of its lower resistivity and saturating magnetoresistance effect. It is both lighter and cheaper, but is only useful in very pure form.

Fully stabilized composite conductors are essential for very large devices, such as bubble chamber magnets found in high-energy physics facilities and contemplated energy storage magnets, examples of which are treated later, but their high copper-to-superconductor ratio (10 to 1000) and consequently low overall current density (at most a few kiloamperes per square centimeter) makes them unsuitable for use in more compact high-field magnets where the winding occupies a large fraction of the magnet volume.

The subsequent quest for flux jump stable conductors containing little or no additional stabilizing material has led to the recognition of several other stability criteria. Of these, "adiabatic" or "enthalpy" stabilization [8, 9], dating from 1967, is the most important, and depends on the observation, already noted, that a critical parameter influencing flux jumping is the size of the superconductor. At low temperatures the specific heat of metals, including superconductors, is very low. However, by subdividing the superconductor into many filaments of small cross section, the energy dissipated during an adiabatic flux jump can be absorbed by the material itself with a negligible rise in temperature. The criterion for adiabatic stability [10] is based on Eq. (1-59); it requires that the filament diameter d be limited to

$$d < (1/J_c) \sqrt{3CT_0/\mu_0} \tag{2}$$

where J_c is the critical current density, C the specific heat, and $T_0 = J_c/(dJ_c/dT)$ has the dimension of temperature. For typical materials, $d \approx 25 \ \mu\text{m}$, a size easily achieved in fabricating Nb–Ti conductors.

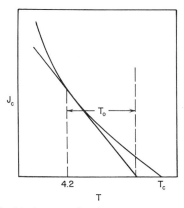

Fig. 3. Dependence of critical current density on temperature. (See text for significance of quantity T_0.)

The significance of the quantity T_0 is apparent in Fig. 3 where critical current versus temperature is plotted for a typical Nb–Ti conductor. Here T_0 is seen to be approximately equal to the difference between the critical temperature and the operating temperature (most superconducting magnets operate in boiling helium at 4.2 K). For a given filament size stability is enhanced by increasing T_c or by lowering J_c, as previously remarked.

Adiabatic stability is further treated in Section II,C, but first a third stabilization technique, known as "dynamic stabilization" [11–14] is noted. It relies on the maintenance of a delicate balance between the rate of movement of magnetic flux in the superconductor, which generates heat, and the rate of removal of this heat, by virtue of the fact that in the superconducting state magnetic flux diffuses through a superconductor much faster than does heat, whereas the reverse is true for pure metals. A dynamic heat balance is achieved by combining the superconductor with a small amount of pure metal, which has the effect of slowing down the motion of flux, and allowing time for the heat to escape to the helium bath by conduction. In this case the expression for the superconductor filament diameter takes the form

$$d < \frac{1}{J_c} \sqrt{\frac{6 K_s T_0}{\rho} \frac{1 - \lambda}{\lambda}} \tag{3}$$

where K_s is the thermal conductivity of the superconductor, λ the space factor of the superconductor within the composite, and T_0 is defined above.

C. *Intrinsically Stable Conductors*

Although all of the various stabilization techniques of Section II,B are still utilized to some extent depending on the particular magnet application, they have largely been supplanted by the appearance of "intrinsically stable" multifilamentary conductors [15], a step beyond adiabatic stabilization. First introduced commercially in 1968, intrinsically stable conductors consist of a twisted composite wire containing many superconducting filaments (mainly Nb–Ti at present for reasons of ductility) embedded in a high-conductivity matrix of normal metal. The filament diameter, however, is chosen not solely to guarantee flux jump stability but is chosen rather to achieve low superconductor power losses under ac or pulsed conditions, which places a more stringent demand on filament size. Although all superconducting magnets must be charged and discharged and thus are never truly dc devices, interest in the ac application of superconductivity was enormously stimulated by proposed high-energy particle accelerators utilizing pulsed superconducting beam guiding and focusing magnets, which were first contemplated in the late 1960s. More recently, the potential for application of superconductors in electrical machinery, power transmission, and, ultimately, plasma confinement devices has emphasized the timely importance of this subject.

Alternating current losses in superconductors are hysteretic in nature [15] and again a consequence of the magnetization behavior of type II superconductors. Essentially, when such a conductor is exposed to a cyclic external field the work done on the diamagnetic circulating currents by the changing field appears as heat, and the integral of this energy over the conductor volume and a field cycle constitutes the loss. It is treated in considerable detail in the chapter by Bussière, this volume. In summary, if a cylindrical superconductor of diameter d is in the presence of an external field cycling between a minimum value, not less than the field level B_s sufficient for complete penetration of the conductor, and some peak value B_p, and if the rate of change of the field does not exceed a certain critical value, then the loss per cycle and per unit length of conductor is independent of frequency and given by

$$W = \frac{d^3}{3} \int_{B_s}^{B_p} J_c \, dB \quad \text{joules/meter} \qquad (4)$$

where $B_s = \mu_0 \, dJ_c$ in mks units. (If the change in field is less than B_s the expression is more complicated, and the loss has a cubic dependence on field.) A more accurate expression would include the effect of a transport current in the conductor; however, this effect is usually quite small. In evaluating Eq. (4), the dependence of critical current density on field may be approximated by the formula of Kim *et al.* [16]

$$J_c = J_0 B_0 / (B + B_0) \tag{5}$$

where J_0 is the current density at zero field and B_0 the field at $\frac{1}{2}J_0$, or by the use of measured integrals over $J_c \, dB$ in the form of tables or graphs. Equation (4) must also be integrated over the coil winding in a magnet, although often the approximation $\overline{B} \approx \frac{1}{2}B_p$ is sufficiently accurate. Typical design considerations for pulsed magnets dictate a filament size of 5–10 μm for an acceptable loss level, and even this is not difficult to obtain with Nb–Ti.

Small filament size, however, is not the only requirement for minimizing the loss. The need for also twisting the filaments in these multifilamentary conductors arises from the fact that because they must be assembled in a metal matrix for mechanical support they will be electrically coupled in the presence of a changing magnetic field by transverse eddy currents induced to flow in the conducting matrix [17]. These transverse currents occupy a length L_c given by

$$L_c{}^2 = \lambda \rho J_c \, da / \dot{B}_c (a + d) \tag{6}$$

where a is the filament separation and ρ the matrix resistivity. The factor λ is dimensionless and approximately unity, and varies with the spatial distribution of filaments. Under these conditions the filaments do not behave as individual filaments and the composite loss exceeds the combined filament loss. If the conductor length L is less than L_c (the "critical length"), however, then only a fraction $L^2/L_c{}^2$ of the currents will cross the matrix. This is effectively achieved in practice by axially twisting the filament bundle with a pitch $4L_c$, thereby decoupling the filaments from each other.

The critical rate of field rise below which Eq. (4) is valid is determined by $\dot{B}_c$ in Eq. (6). Note, however, that even though the filaments are essentially decoupled by twisting, *within* the twist pitch currents still flow through the matrix, and give rise to a rate-dependent normal eddy current loss for $\dot{B} < \dot{B}_c$, in addition to the hysteretic loss. The expected loss per unit length from a circular bundle of filaments of overall diameter D (approximately the composite diameter) in a simple matrix of resistivity ρ may be calculated from

$$W_m = 2D^2 L_c{}^2 \dot{B} B_p / \pi \rho_e \tag{7}$$

Here ρ_e is an effective resistivity, approximately the matrix resistivity but more accurately $\rho_e \approx \rho(a - d)/d$ where a is the filament spacing as before.

In a magnet the matrix loss should be small compared to the filament loss since otherwise there would be no reason for subdividing the superconductor (as far as the losses are concerned). Because the maximum achievable twist pitch is proportional to the composite diameter (the practical limit is about one twist for every five wire diameters), it can be shown that the rate of field change for which these two loss mechanisms are equal is inversely proportional to the number of filaments. [18]. Thus, for composites with very many filaments (several thousand are now quite common) resistive barriers of cupronickel between the filaments have been introduced to limit the matrix loss. Unfortunately the increased matrix resistivity (about 2000 times that of copper) is detrimental with respect to conductor stability (even for filament sizes below the adiabatic stability limit) and magnet quench protection, both of which demand good electrical conductivity. These conflicting requirements have led to commercial composites of increasingly sophisticated form containing both copper and cupronickel in the matrix. In Fig. 4 is shown a cross section of such a three-component composite wire of 3.5-mm diameter, containing 14,701 individual 18-μm filaments.

One important characteristic of the finely subdivided superconductors is associated with the so-called premature onset of resistivity in these materials. The transition to the normal state in wire with superconducting filaments of moderate size, monitored by measuring the voltage per unit length developed across a short sample of the wire in a given external field as the sample current is increased, occurs abruptly at a certain critical current, as shown in curve a in Fig. 5. For such a material there exists an unambiguous curve of critical current versus field. However, as the filament diameter is reduced, a more gradual resistive transition, as indicated by curve b, is found. In this case curves of "critical current" versus field can be drawn for different overall resistivities, as in Fig. 6. Here I_c is the current at which the sample voltage is no longer stable with time, or the quench current. This phenomenon, to some extent characteristic of all composites with very fine (<10 μm) filaments, although not as pronounced in more recent materials, renders the precise meaning of short-sample performance somewhat vague. It is attributed to the existence of narrow regions along the filaments introduced during the drawing down of the copper–superconductor billet, and if not controlled carefully by manufacturers it can result in magnet dissipation under dc conditions. It has become accepted practice to use an effective resistivity of 10^{-12} Ω cm for specifying short-sample critical current in these wires [10].

Note that even for magnet applications where ac losses may not be important, filaments smaller than those required simply for adiabatic stability may be desirable to minimize the perturbing effects of magnetization currents on the field distribution in high-precision magnets, and possibly

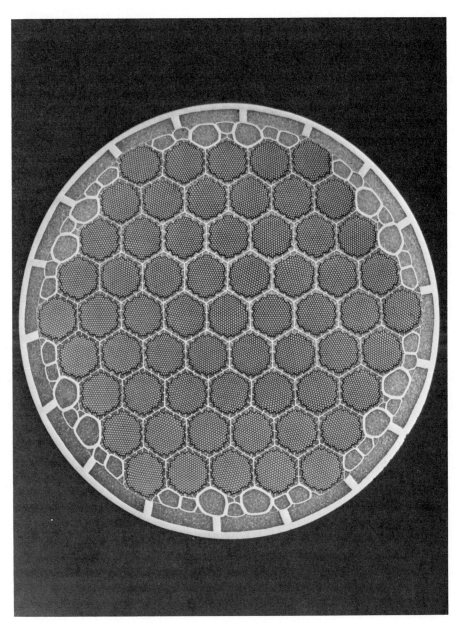

Fig. 4. Cross section of a composite wire containing 14,701 filaments of niobium–titanium. Each filament is surrounded by copper and then cupronickel, and groups of 421 filaments assembled into a "spider can" of cupronickel with cupronickel radial spokes separating copper regions. (Courtesy Imperial Metal Industries Ltd.)

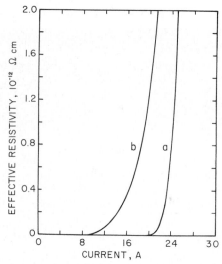

Fig. 5. Transition to the normal state for two composite wires of different filament sizes.

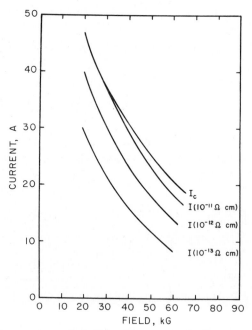

Fig. 6. Current versus magnetic field for different overall effective resistivities in a composite sample containing very fine filaments.

remanent fields. Both of these effects are proportional to the filament diameter.

D. High Current Capacity Macrocomposites

In superconducting magnets a conductor of relatively high current-carrying capacity is usually desirable, which for a given attainable current density in the superconductor dictates a conductor of fairly large cross section and low density of turns in the coil-winding structure. Aside from lessening the winding labor, the consequent simplification in coil structure from limiting the number of turns aids conductor cooling (a benefit not only during operation but also during cooldown by minimizing temperature gradients). Most important, a large conductor limits both the coil inductance and the voltage required for excitation and, conversely, the voltage developed if the magnet is discharged rapidly, as during a quench. (Stabilization by cladding with copper or metallic interleaving in the winding also helps cooling and limits quench transients.) The chief drawback of very high operating currents is the considerable heat flow into the system along the high-current leads to the magnet cryostat. For this reason even large fully stabilized magnets are rarely designed to carry more than about 5000 Amp.

The fully or partially stabilized conductors of Section II,B are usually fabricated in the form of round or rectangular composites, which contain one or many superconducting Nb–Ti rods or filaments in a copper or copper alloy matrix, or as cables of composite strands. They are often shaped by compaction to conform to a specific coil configuration and to obtain high overall current density. Cabling has the advantage that the conductor can be more readily fabricated in long unit lengths, dictated by the feasible production length of the individual constituent wire strands. Such a cable may be a simple twisted conductor, such as a seven-strand cable made by twisting six composite strands around a central member, usually copper but sometimes nonmagnetic steel for strength. A number of strands may also be twisted into a multilayer cable or rope, with the direction of twist altering from layer to layer. The "filling factor" of such a twisted cable is good, but it is prone to degradation and unequal current sharing between the various layers because each strand layer occupies different radial positions and is therefore exposed to a nonuniform field. Such exposure is avoided in a fully transposed cable. In the most elementary form of transposition, which is a simple hexagonal stranding sequence ("Litz" wire), six cabled strands, each consisting of six elementary strands or wires wound around a central core as before, are wound around a larger core; six of the resulting bundles are wound around a third core, and so on. In this geometrical arrangement of cabled strands each strand alternates regularly in position between the inside and outside of the cable along the cable length, and thus has equal inductance. Transposed cables suffer from rather low filling factors ($\sim 50\%$), however, and are sensitive to wire damage if heavily compacted.

For pulsed application, as noted in the last section, intrinsically stable conductors with filament sizes ranging down to about 5 μm are required. Such a filament has very limited current capacity (typically 30 mA at 5 T), so as many as 10^5 filaments must be connected in parallel to carry a realistic operating current, which may be a few thousand amperes. An example of a "double composite" was shown in Fig. 4, although in such a composite there are unresolved possible theoretical limits, arising from the lack of true transposition of the filaments [19], as well as more practical limits on the number of filaments feasible in a single wire of this type. Instead, several hundred simple twisted composite strands, each containing up to about 1000 filaments, may again be formed into transposed cables. However, these suffer from the several aforementioned difficulties. One superior configuration, developed at the Rutherford Laboratory, is a flat twisted cable that consists of a hollow tube of helically twisted wires that is rolled flat [18, p. 496]. This cable is fully transposed, and the filling factor is 80% or better, but the aspect ratio (ratio of width to thickness) is limited to about 5:1. This ratio places a limit on the number of strands and consequently requires wires with many filaments. Another approach pursued at Brookhaven is the use of a flat braid [20], in which the number of wires and aspect ratio (aspect ratio is an important design consideration for certain precision saddle-type magnets, discussed in Section III,C) are less restricted. Braids are particularly promising for A15-based conductor composites. They are less rigid than flattened cables, but the "packing factor" is not as good.

The importance of these macrocomposites depends also on certain additional magnet performance considerations, which are considered next.

E. *Other Magnet Aspects Affecting Conductors*

As noted at the outset, degradation and training have been the principal obstacles to reliable magnet performance. The preceding sections have considered various methods of stabilization and the choice of operating current. These factors have led to the use of transposed cables or braids made from intrinsically stable composite wires. Nevertheless, degradation and training have not yet been altogether eliminated; this behavior continues to bother all but fully stabilized devices to some degree, and especially magnets with complex winding configurations and those designed for very high fields and high rates of field change.

The source of the problem lies mainly in conductor motion, since even magnets wound from filamentary wire composites will be prone to dissipation and generally behave in an unstable manner if the wire is allowed to move under the influence of magnetic forces. Furthermore, in all but the simplest solenoids it is difficult to adequately restrain the individual wires in a braid or cable by winding tension alone. A partial remedy is to insulate the individual strands with a thermosetting organic coating that is heat treated after winding to form a strong but relatively porous structure. Magnets utilizing organically insulated wires immobilized in this manner do exhibit reduced ac losses but still suffer from considerable training, especially in the case of fully "potted" epoxy coil structures and if the superconductor-to-copper composite ratio is high.

A much better remedy has been found in the exceptionally stable magnet performance realized by bonding the individual uninsulated twisted strands together in a cable or braid that is filled with a secondary matrix based on a soft metal, such as indium or tin [21]. Suitable indium-based fillers are In–Tl and In–Pb; examples of fillers based on tin are Sn–Ag or Sn–Bi. The significant reduction in degradation and training in these magnets can be ascribed to several thermal and mechanical properties of this type of conductor: enhanced heat capacity and especially thermal conductivity, excellent mechanical stability, and elimination of wire motion. (Conductor motion as a whole, however, must be arrested by a properly designed coil support structure; see Section III,C.) Such a conductor may be thought of as an "intrinsically stable cable" analogous to the multifilamentary wire, where now the transposed strands in a soft metal matrix play the role of individual filaments in a pure metal or metal alloy matrix. The analogy, however, also points to a complication; namely, considerable electrical coupling between the wires of the conductor for rapid field changes, giving rise to additional rate-dependent magnetization effects and losses similar to those described by Eq. (7), but governed by the effective resistivity of the filler metal and the cable or braid transposition length. There are limits on how much the transposition pitch length can be shortened, but the coupling effects can be substantially reduced by the introduction of a cupronickel jacket around each wire or by increasing the interwire resistance through certain metallurgical steps. For very fast $\dot{B}$ application ($\dot{B} > 0.01$ T/sec) an approach intermediate between complete organic wire insulation and metal filling may suffice. This approach utilizes a well-compacted cable or braid that contains bare wires in electrical contact but no additional filler matrix.

An area of some concern in future applications of superconducting magnets is the effect of nuclear irradiation on magnets operating in accelerator and, in the long run, controlled thermonuclear reactor environments. High irradiation dose rates in a superconducting magnet have two consequences: heat deposition that may be sufficient to drive the magnet normal, and degradation of both superconducting and normal material properties. Reliable data on performance degradation from radiation heating are only now becoming available, as various superconducting accelerator beam line magnets are brought into operation, and these data will require careful evaluation. Radiation damage to superconductors is treated in detail in the chapter by Sweedler, Snead, and Cox, this volume. Irradiation affects the electrical and mechanical properties of superconductors, stabilizers, and insulators. The residual resistivity of high-purity copper or aluminum at low temperature can increase by several orders of magnitude through radiation exposure. Such exposure can affect superconductor stability and can make these metals more prone to mechanical embrittlement. Copper–nickel appears unaffected by irradiation, as do the metallic stabilizing alloys such as indium–tin. The intermetallic diffusion bond layer between the superconductor and the normal metal matrix tends to increase in thickness, reducing the thermal diffusivity across the layer. Perhaps most serious is the effect of irradiation on coil organic insulation and structural materials. As a class, plastics have a very low radiation damage threshold, while glass fibers and glass fiber epoxies show little degradation.

F. Multifilamentary A15 Composites

At present the most widely used superconductor, and essentially the only superconductor available in multifilamentary form, is the ductile alloy niobium titanium. There are limits on the superconducting properties of this conductor, however. Approximately 3×10^9 A/m^2 appears to be the upper practical limit on the attainable critical current density at 4 T, and at 10 T one tenth of this value. The upper critical field is approximately 12 T, although in a practical device using this conductor the maximum operating field is in the region of 5–6 T. Perhaps the most serious limitation lies in the narrow operating temperature margin of Nb–Ti. Although the critical temperature is about 10 K, in a magnet operating in the temperature range 4.2–4.5 K fluctuations in temperature of a few tenths of a degree can be sufficient to cause a quench.

Substantial improvements in all respects are offered by the compound niobium tin [22], Nb_3Sn. It has an upper critical field as well as critical temperature almost twice that of Nb–Ti, a critical current density of approximately 2×10^{10} A/m^2 at 4 T, and at 10 T the critical current density is still about 5×10^9 A/m^2. In time it should be cheaper than Nb–Ti, at least on an ampere meter basis. The compound vanadium gallium (V_3Ga), another member in the A15 crystal structure family, has rather similar superconducting properties. A main difference between this compound and Nb_3Sn is a flatter J_c versus field curve with somewhat lower current density at low and intermediate fields but exceeding Nb_3Sn above 12 T. These materials, consequently, promise operation at higher temperature (with simplified refrigeration and larger temperature reserve), higher operating fields, and higher current densities for a given field.

Unfortunately, because of their extremely brittle nature it has not been possible to manufacture filamentary conductors from the A15 compounds until very recently. Instead, they have only been available in flat tape configurations consisting of a Nb_3Sn or V_3Ga superconductor layer plated with silver or high-purity copper for stabilization, and supported by a stainless-steel substrate for mechanical strength; the differential contraction between the superconductor and substrate during cooldown puts the brittle superconductor under compression. In spite of the dynamic stabilization, these tapes tend to be unstable in the presence of fields perpendicular to the flat surface of the tape. Greatly improved stability should be realized from niobium tin or vanadium gallium in filamentary form, and vigorous and promising development programs with this as a goal are currently being pursued at a number of laboratories in this country and abroad [23–25]. Basically, the low ductility of A15 compounds is circumvented by delaying the formation of the superconducting compound until all the extrusion and drawing steps have been completed. Billets of pure Nb or V rods embedded in a Cu–Sn or Cu–Ga bronze matrix are drawn down in much the same way as Nb–Ti in a copper billet. After drawing to final size, the wire is axially twisted and then heat treated, during which Sn or Ga from the bronze matrix diffuses into the core material to form filaments of Nb_3Sn or V_3Ga. A complication encountered in this "bronze" or "solid state diffusion" procedure is the rapid work hardening of these bronzes, necessitating frequent annealing during the drawing process. A new process, known as "external diffusion," overcomes this problem in the case of Nb_3Sn [26] (but has not yet been successfully accomplished with V_3Ga). In it, Nb is first drawn down in a pure copper matrix, which is much easier to work than bronze. Afterward, the composite is coated with a layer of Sn by a dipping technique, and is then heat treated in several steps to form the bronze matrix and the compound, respectively, by diffusion. These techniques for producing multifilamentary A15 compounds in bronze matrices are described in the chapter by Luhman, this volume.

Although the critical current densities of the filament layers in these composites are very high, the overall current density depends on the maximum achievable proportion of superconductor in the finished composite [20]. The onset of resistive behavior is about the same as in the fine-filament Nb–Ti composites. A potentially serious problem is the high resistivity of the bronze matrix compared to that of pure copper. Although good from the ac loss standpoint, it can result in unstable magnet performance. In this respect gallium bronze is better than tin bronze by perhaps an order of magnitude [10].

The most serious problem associated with A15 conductors is their inherent brittleness. Although the problem was circumvented in the production of the superconductor, it reappears when the composite wire is incorporated into a cable or braid and also when the magnet is wound. The brittleness requires that the composite wire not be subjected to bend radii less than 1–2 cm. This limitation may be acceptable in some magnet coil configurations, but it does rule out cabling or braiding reacted strands. Instead the macrocomposite must first be produced from unreacted wire, and then the cable or braid must be heat treated to form the superconductor by either method of diffusion. The external diffusion route is also compatible with the technique of braid stabilization utilizing tin-based filler, as described in Section II,E.

One may simply avoid the problem of brittleness altogether by winding the coil itself before it is reacted and heat treated as a whole to form the superconductor in situ. The drawback of this "wind and react" procedure is the high reaction temperature required to form the superconductor, about 700°C, which necessitates having adequately heat-resistant structural coil support forms and bakeable coil insulation. Even more problematic is the need to maintain mechanical tolerances during the large temperature excursion of the magnet from reaction temperature all the way down to operating temperature. Thus one could argue in favor of heat treating after the braiding operation but prior to actual coil winding in a "react and wind" approach for magnets where field homogeneity is important. In this case it is important that the conductor geometry be chosen to minimize damage from bending. A flexible flat cable or a braid geometry might better be used than rigid or monolithic conductors of similar current capacity. (For instance, the braided conductor discussed below in Section III,C, page 82.)

III. Applications

A. General-Purpose Magnets

The principal differences between conventional copper–iron or resistive magnets and superconducting ones are twofold: (a) the field and current are not independent variables in superconducting magnets; and, of more practical significance, (b) the current density of superconducting magnets is higher than that of conventional magnets. The task of the superconducting magnet designer is to optimize the coil configuration for a required field or field distribution and the available current density. The present discussion is not intended to be a comprehensive review of recent progress in superconducting magnet design, which is well covered in several extensive articles on the subject [27, 28]. The previous sections stressed certain limitations of the conductor that are a consequence of the intrinsic nature of superconducting materials and affect magnet performance and, conversely, various conductor requirements desirable from the point of view of practical magnet design. A brief survey is now given of particular applications selected to further emphasize these aspects and illustrate principles; these applications are listed roughly in order of increasing conductor sophistication.

First, it may be useful to reiterate the major advantages of superconducting magnets over conventional ones: their modest or negligible power requirements, and their high current density and field capability, which permit very compact devices to generate high fields over a large fraction of the volume enclosed by the winding. In spite of the high cost of present superconducting materials, which is a significant fraction of the magnet cost, the capital cost of a superconducting system may still be less than that of a conventional system of similar rating when the cost of the power supply of the latter system is accounted for, and the low operating cost favors the superconducting system (this advantage may be partially offset by the running cost of a helium refrigerator). However, even though a superconducting approach may not always have a clear-cut economic advantage over a room-temperature system it may nevertheless be desirable or even necessary in special applications not feasible or highly impractical with a conventional system. Moreover, with expanded use the cost of superconductors can be expected to decrease, which is in contrast to a likely increase in the cost of electric power.

The earliest, and still most widespread, application of superconductors was in compact high-field general-or special-purpose magnets for laboratory use. These magnets were predominantly seen in physics and especially solid-state physics laboratories, but more recently have also been used in chemical and biological laboratories and in a multitude of other settings. They are axially symmetric coils, typically simple solenoids wound with partially or intrinsically stabilized Nb–Ti conductors and designed to produce fields in the range of 5–10 T over a working bore a few centimeters in diameter.

Solenoid magnet design has been treated in considerable detail elsewhere [29]; here are noted several elementary but useful relationships that form the starting point for the design. The field at the center of a uniformly wound solenoid is given by

$$B_0 = \mu_0 J r_i \beta \ln \left(\frac{\alpha + \sqrt{\alpha^2 + \beta^2}}{1 + \sqrt{1 + \beta^2}} \right)$$

$$\equiv \mu_0 J r_i F(\alpha, \beta) \tag{8}$$

where B_0 is the magnetic field in teslas, $\mu_0 = 4\pi \times 10^{-7}$, J is the overall current density in the solenoid in amperes per square meter, r_i the inner radius in meters, α the ratio of outer radius r_o to inner radius, and β the ratio of coil length L to $2r_i$. In terms of the number of turns N and the current per turn I (amperes),

$$J = \frac{NI}{(r_o - r_i)L} \tag{9}$$

The function $F(\alpha, \beta)$ depends only on the dimensionless quantities α and β and can be obtained approximately from Fig. 7. Values α and β chosen along the diagonal line represent the coil configuration that achieves maximum field for a given volume of conductor. The maximum field in the solenoid is also the maximum field seen by the conductor; it occurs at the inner winding radius in the central plane of the solenoid, and can be obtained from Fig. 8, which shows B_{max}/B_0 as a function of β and for various values of α. In a superconducting solenoid, an important design consideration is that the peak field not be significantly higher than the central or useful field, so that the potential performance of the device is not limited.

A wide selection of superconducting solenoid magnets is now commercially available, and they tend to be quite reliable. A common practice is the use of solenoids that are constructed in several concentric sections, each separately optimized. [Nested coils may be treated with Eq. (8) by

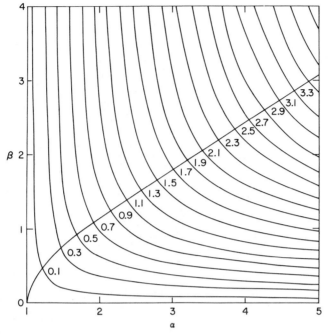

Fig. 7. The function $F(\alpha, \beta)$ plotted against the geometry terms α and β. The diagonal curve represents the minimum volume of superconductor for a given field value.

superposition.] In "hybrid" magnets of this type the outer sections may be wound from Nb–Ti and the inner sections from a higher-field material such as niobium tin tape. These magnets generate fields of 15 T or even higher. Coil subdivision by separately excited sections can have a conductor stabilizing effect, due to the fact that most flux jumps responsible for degradation occur at low fields, in addition to allowing current optimization of the conductor. In "field-stabilized" coils the outer sections are used to apply a biasing or stabilizing field to the inner section. This biasing results in a higher local field there, for the same transport current, with smaller flux jumps, which allows operation closer to the critical current. This procedure also assures that the current density of the outer sections is not limited by the short sample current of the high-field section, which may be less than the normal degraded current. Finally, coil subdivision permits separate operation of one or more sections, a factor that is sometimes useful.

Field homogeneity, commonly 1–0.1%, is often important and is realized by the use of a split coil configuration, which in addition provides radial access to the useful field region. [The field in the gap is simply calculated using Eq. (8) by subtracting a solenoid with the dimensions of the

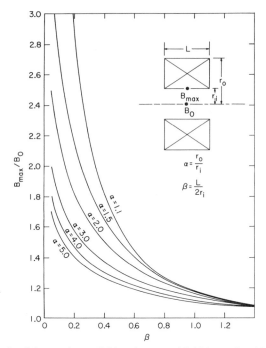

Fig. 8. The ratio of the maximum field to the central field in a solenoid as a function of β for various values of α.

gap from another solenoid of length equal to the combined length of both coils and the gap.] If very high field homogeneity (0.01% or better) is called for, further coil shaping or correcting with auxiliary compensating windings is necessary (again using superposition techniques, perhaps with the aid of tables or computer codes). Fine filaments are essential in this case in order to minimize the effects of magnetization currents occurring at low fields due to superconductor diamagnetism. Exceptional time stability is achieved by operation in the "persistent mode"; this requires a thermal switch across the current leads by which they can be short-circuited and the magnet isolated from the external current supply.

If the conductor is in the form of a low-current wire or cable (rated at, say, 100 A or less) the coil is usually layer wound with a pitch sufficient to assure turn-to-turn insulation. It may incorporate foils of anodized aluminum or Mylar–copper–Mylar between the layers for improved cooling as well as additional dynamic stabilization. (Copper shorting strips are also often placed across each layer to ensure that all turns are connected by a low-resistance path. This path lengthens the quenching time and thereby prevents excessive voltage buildup and arcing during the resistive transition). Conductors of higher current-carrying capacity, such as the wide cables or braids discussed in Section II,D, and the flat niobium tin or vanadium gallium tapes, require disk or pancake construction, where each disk is interconnected at the inside and outside to form a single current path. Since tape conductors are more prone to flux jumping instabilities in the presence of an external field perpendicular to their broad surface, the disks in the end of a solenoid, where large radial field components exist, may be spaced farther apart for better cooling and increasing the ratio of central to maximum radial field, or extra normal stabilizing material may be incorporated here.

Another stabilizing technique is the utilization of superfluid helium by pumping on the helium bath. The enhanced coil performance often realized by operating below the "lambda point" (2.18 K) depends on two factors: (a) the higher critical current and higher operating margin at the lower temperature, and (b) the increased thermal conductivity and low viscosity of helium in the superfluid state. However, operation below the lambda point does not invariably lead to enhanced performance; in fact, sometimes degraded performance is experienced, particularly in tape-wound magnets. In this case the improvement in heat transfer appears to be offset by increased flux jumping that accompanies the increased low-field current density. Stabilization by interleaving also has the serious drawback of compromising the overall current density vital in small coils. In any case, the need for these various coil stabilization methods is now lessened by the availability of intrinsically stable multifilamentary conductors.

Examples of magnets of this type are illustrated in Figs. 9–11. Figure 9 shows a 10-T split coil used for neutron scattering experiments. The neutron beam enters and leaves at right angles to the magnet axis, and the split-coil geometry provides 360° access to the target volume. The two sections are held apart by a bridge structure that must be strong enough to withstand the compressive force between the sections (of about 5 tons) without unduly attenuating the beam. The coils are wound from tape conductor consisting of niobium tin on a Hastelloy substrate and silver plated for stabilization.

Another excellent example of the use of superconducting magnets in applications where conventional magnets are precluded is furnished by a high-energy physics experiment carried out at Brookhaven National Laboratory. The experiment [30] was devised to measure the magnetic moment of the short-lived Ξ^- particle, which is created when protons from the Brookhaven synchrotron strike a hydrogen target. It called for a magnet capable of producing a product of field times distance along the flight path of the particle of 1 T m. Since the flight path of the particle be-

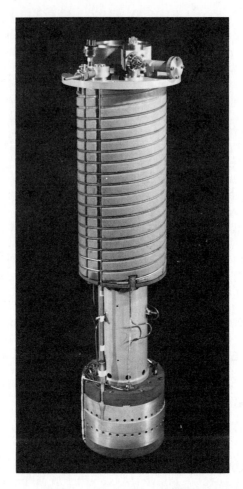

Fig. 9.   Helmholtz coil for neutron scattering experiments.

fore decaying was only 0.1 m, a field of at least 10 T was required. The superconducting magnet used was a 12.5-T solenoid consisting of four separately powered concentric sections, clearly seen in Fig. 10. Vapor-deposited niobium tin ribbon was again used for the conductor. This magnet was designed to operate in the persistent mode; the switches are housed in the insulating ring around the magnet.

The solenoid in Fig. 11 is one of the highest-field superconducting magnets constructed to date, producing 15.8 T at 4.2 K and 16.5 T at 3.0 K [31]. It utilizes copper-stabilized niobium tin tape conductor wound in modules or double pancake coils. The bore is 2.6 cm, outer diameter 23 cm, and the magnet is 27 cm long.

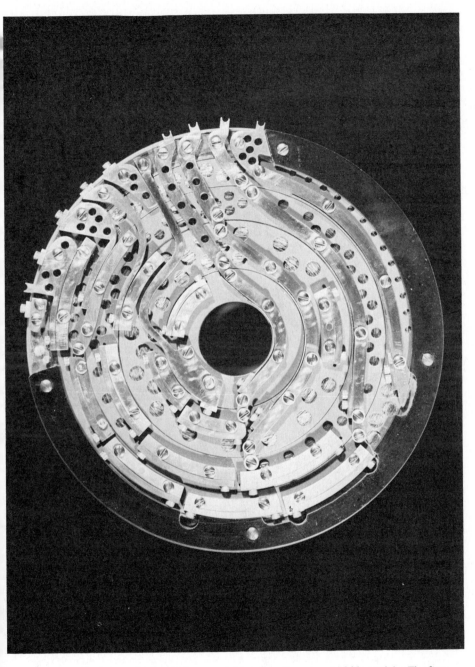

Fig. 10. End view of solenoid for measuring magnetic moment of Ξ^- particle. The four coil sections are clearly discernible.

Fig. 11. One of the highest-field superconducting magnets constructed to date. Nb$_3$Sn magnet, 158 kG at 4.2 K, 165 kG at 3.0 K, 26-mm bore, uniformity of 5×10^{-4} in 5 mm DSV at rated field. [From Fietz *et al.*, *IEEE Trans. Magn.* **MAG-10,** 246 (1974).]

B. Large Magnets

The magnets of Section III,B typify a class of compact high-field superconducting coils where full stabilization is ruled out because of overall current density considerations. In contrast are very large coils whose economics are relatively unaffected by overall current density and where fully stabilized conductors are imperative for magnet protection in case of a quench, due to the extremely high stored energy (10^8–10^9 J). Full stabilization also assures that the magnet performance can be reliably predicted from short sample tests on the conductor.

Examples of such large multimegajoule superconducting magnets are those in bubble chamber systems found in a number of high-energy physics installations in the United States and abroad. In these chambers, tracks of bubbles are formed in the wake of individual traversing charged particles that emerge from a proton synchrotron. A surrounding magnetic field forces the particles to follow a curved path whose direction and amount of curvature is a measure of the particle charge and momentum. Because conventional magnets, while technically feasible here, pose nearly prohibitive capital and running costs due to the large field volumes required, this was the first application of superconducting magnets on a large scale. Consequently, bubble chambers provided an important impetus for superconducting magnet technology. The pioneer was the 12-ft chamber at Argonne National Laboratory. This chamber utilizes an 80-MJ split-coil magnet designed for a relatively modest 2 T and has seen intermittent operation since late 1968. A larger and more recent one is the 3.7-m liquid hydrogen European bubble chamber (BEBC) [32], operating at the European Organization for Nuclear Research (CERN) in Switzerland (Fig. 12). Here the magnet is also of split-coil type. It was constructed from pancake coils measuring 4.7 m in inner diameter and weighs 1.66×10^5 tons assembled. It produces a central field of 3.5 T (the maximum field in the winding is then 5.1 T), corresponding to an operating current of about 6000 A, and has a stored energy of 800 MJ. A 2000-ton shield of low-carbon steel surrounds the magnet at some distance in order to reduce the stray magnetic field. (In some cases, notably in the Argonne magnet, the iron shield also serves to enhance the field produced by the coil alone.) A conventional magnet would have required 70 MW of power to produce the same field volume. In contrast, the CERN magnet needs only 360 kW installed power, primarily for the refrigerator, which has an output of 900 W.

Approximately 65 km of conductor was required to wind the BEBC magnet. This conductor is a rectangular copper-stabilized Nb–Ti composite, 6×0.3 cm in cross section, and it contains 200 superconducting

Fig. 12. Erection of the cryostat for the BEBC magnet at CERN, Switzerland. The split-coil magnet is one of the largest superconducting magnets ever built, storing 800 MJ at the peak field of 3.5 T. (Courtesy European Organization for Nuclear Research.)

strands. The copper-to-superconductor ratio is 26:1 and the overall conductor current density is 1.35×10^7 A/m². Each turn in the winding is sandwiched from five components: a milled cooling strip, the superconductor, a polyamide insulating strip, an aluminum heating strip (to quicken the transition to the normal state), and a stainless-steel reinforcing strip. The contract for the conductor was shared between a German and a Swiss firm.

The dominating concern in the design of magnets of this size is the problem of stresses, both electromagnetic and, to a lesser extent, thermal. At 3.5 T and 6000 A, there is a vertical attractive force of 9000 tons between the two coil halves of the BEBC magnet. The radial "hoop stress," resulting in longitudinal tensile force in the conductor, is about 10^8 N/m^2 and must be shared between the conductor and the prestressed (to avoid loss of tension during cooldown) steel reinforcing strip. (Sometimes it is more practical to provide extra copper or superconductor than to provide a secondary support structure.) Thermal stresses from differential contraction in such large coils, often comparable to the magnetic stresses, make slow and uniform cooldown advisable.

Another large magnet at CERN, the Omega spectrometer magnet [33], has a central field of 1.8 T and a bore diameter of 3.5 m. It uses forced flow of supercritical helium through a hollow Nb–Ti composite conductor, 1.8 × 1.8 cm^2 with a coolant passage 0.9 × 0.9 cm^2, for improved cooling and added stabilization, thereby taking advantage of the absence of two-phase flow and unstable critical heat flux in supercritical helium. This is an old idea first put into use on a large scale at CERN. A considerably smaller CERN bubble chamber magnet designed for a specific type of experiment, that of the HYBUC chamber [34], is a good example of reliable hybrid-type magnets generating much higher fields. It operates at 11 T, and consists of a niobium tin coil of 17.8-cm inner diameter producing 5 T alone, mounted inside a niobium–titanium coil of 28-cm inner diameter and designed to generate 7 T by itself. Here the Nb–Ti conductor is of the twisted filament type.

Even larger superconducting magnets are conceived both for inductive energy storage in connection with fusion reactors and for power network systems. In some of these applications, no significant further advance in superconducting magnet technology beyond the bubble chamber level appears necessary; rather it is a question of utilizing conventional materials technology and economic competition with proven and reliable systems already on hand. Because of their high energy density, magnets designed for energy storage may ultimately be competitive with other schemes, such as capacitive storage, flywheels, or pumped hydropower stations, although their advantage would probably be confined to very large central-station systems. They are especially of interest for network load-leveling or peaking power storage. Various coil geometries seem practical, including toroids, spherical coils, and simple solenoids, wound perhaps from hollow conductors similar to the CERN Omega magnet. One version, and the most ambitious proposal contemplated [35], is for a magnet system storing up to 30,000 MWh, or 10^8 MJ (comparable to the largest present U.S. pumped storage system). The coils would be 50 m in radius, and containment of the enormous magnetic forces would necessitate coil burial in granite bedrock. (Another problem is coping with the very considerable stray field.) More modest systems intended for transient response to load fluctuations (of typically 1-sec duration) might employ coils not significantly larger than the aforementioned bubble chamber magnets. However, they would have to be capable of faster rates of field change, a consideration that poses greater demands on the conductor (Section II,C), and even more so in the case of magnetic energy storage for fusion applications, where millisecond discharge times will be necessary.

C. Transverse-Field Magnets

The area in which superconducting magnets have their greatest immediate potential is in high-energy physics. Experiments in this field require large numbers of accelerator magnets, beam transport magnets, and other special-purpose magnets, all of which consume enormous quantities of electric power. Superconducting replacements for these magnets hold the promise of substantially relieving this heavy power burden. Moreover, the compactness and the higher fields of superconducting magnets have a number of important technical advantages in updating existing installations. Thus the extremely high-energy accelerator facilities contemplated for the future depend critically on the rapidly evolving state of the art of superconducting magnets.

Various examples of superconducting magnets used in experimental areas of high-energy physics laboratories have already been noted, including the bubble chambers of Section III,B. These magnets are of rather simple axially symmetric geometry, basically solenoids or split coils. More numerous, and of quite different configuration, are the magnets associated with the particle accelerator itself [36], usually a proton synchrotron. The essence of the synchrotron is a circular array of deflecting and focusing magnets, or dipoles and quadrupoles, that confine the primary beam to a circular path within a ring-shaped vacuum chamber as it is being accelerated. Because of the higher available field strength, the use of superconducting magnets here can reduce the system cost by reducing the ring diameter for a given maximum particle energy, and for a given ring size, higher maximum energy is possible. Since the particles are injected into the ring and accelerated to full energy in pulses, these are cycling magnets with a repetition period of typically 1 sec. In addition to the accelerator magnets, the channeling of protons ejected from the synchrotron to external targets as well as to experimental areas of secondary particles from targets within or external to the synchrotron ring requires a multitude of dipole and quadrupole dc "beam transport" magnets [36]. These are also major power consumers since they require power about equivalent to the pulsed magnets of the accelerator ring itself. Superconducting beam transport magnets not only can reduce this load, but their higher deflecting fields and field gradients offer the possibility of using shorter beam lines with consequent potential savings in capital and operating costs.

Superconducting accelerator and beam transport magnets are of the "transverse field" type, also known as "race track" magnets or "saddle coils." In this type of magnet, the field direction is transverse to the magnet axis (not axial as in a solenoid), or is perpendicular to the path of motion of the particles being deflected or focused [37]. The conductor is distributed in one or more layers of segmented coils closely wrapped around the periphery of a circular aperture in such a way as to produce a uniform field (dipole magnet) or a uniform field gradient (quadrupole) in the aperture. The magnets may be surrounded by a close-fitting cylindrical iron return yoke, a practice alluded to in connection with the bubble chamber magnets. The yoke serves several functions. It enhances the central field produced by the coil alone (by a factor depending on its proximity to the coil layers), reduces the total stored energy, and also shields the stray field from the magnet.

A number of conductor arrangements can, in principle, produce the desired dipole field (or more generally a field of any desired multipolarity) in the magnet aperture, although most configurations are approximations of one of two basic or ideal types: the "intersecting ellipse" or "constant current density" configuration, and the "$\cos \theta$" configuration. In the former type, the coil shape is derived from the superposition of two elliptical regions of equal and opposite current density, with centers displaced to form a central aperture of uniform field. In the second basic type, a uniform field results from surrounding the aperture with a circular winding of uniform thickness and a current density varying azimuthally as the cosine of the angle from the median plane. In practice, these current distributions must be approximated by an integral number of conductors of finite size, which are grouped in blocks or layers and position optimized with the aid of a computer program. If necessary, this position optimization also takes into account the near presence of an iron shield of finite permeability.

These accelerator magnets are more demanding than most of those discussed earlier, and present the designer with some conflicting requirements. The ac losses from pulsing are no longer the predominant problem in the frequency range of interest, ≤ 1 Hz, due to the recent advances in conductor technology outlined. Furthermore, the field rise time in future accelerators is likely to be lengthened considerably. However, the problems stem principally from two sources. First, there exists the difficulty of adequately supporting the windings in a complex coil configuration against electromagnetic forces after cooldown in order to avoid conductor movement and minimize training. This difficulty is compounded when the magnet is cycled rapidly. Second, a very high field uniformity is required. The attainable homogeneity is also sensitive to field rate of change.

With regard to the first difficulty, the force distribution in a simple axially symmetric coil such as a solenoid may be dealt with fairly easily, at least in principle, since it consists mainly of a symmetric radial hoop stress that can be supported in part or entirely by the conductor itself in a tightly wound coil. However, the forces in saddle coils are more complex and conductor restraint must depend on a more elaborate support structure. Great care must be taken to ensure a close match of thermal contraction coefficient between the support materials and the conductor (including superconductor, stabilizer, and insulation); and while the support structure should be basically mechanically rigid for strength, it must at the same time be sufficiently porous to allow efficient removal of the heat from a variety of sources inevitably dissipated within the coil. Movement of individual conductor wires in a cable or braid may be effectively arrested by the stabilizing technique of metal filling, as discussed in Section II,E. Avoiding movement of the coil as a whole is more problematic, especially in the ends of racetrack coils. (Conductor motion of a few microns can cause a temperature rise of several degrees that, in the case of Nb–Ti in a 4-T field, leads to a 50% reduction in critical current density.) Coil support is usually provided by vacuum or pressure impregnating the winding with suitable materials, such as epoxy–filler systems. These materials are kept as porous as possible. The coil is then wrapped circumferentially with stainless steel, aluminum, or epoxy-loaded fiberglass bands or clamps. Finally, the finished coil is mounted in a cylindrical iron shield (desirable as well for other reasons already mentioned) by a shrink-fitting technique that ensures that the coil remains under a predetermined compression during cooldown [38].

These precautions have resulted in improved performance, and while they have not eliminated training, it is not as pronounced as in earlier accelerator prototype magnets. Typical experience is that the first quench upon initial powering of the magnet occurs at 80–90% of the design current, which in turn may be 90% of the short-sample current and reached after perhaps a dozen additional quenches. Finally, the cause of training is most likely not only mechanical motion of the conductor or coil, but may also be release of stored strain energy in the impregnants during cooldown (for this reason a waxlike impregnant may do better since it tends to form microcracks rather than store energy) [39]. Training has also been attributed to relaxation of stresses within the conductor itself [40].

The demand on field homogeneity over a large fraction of the cylindrical bore of these magnets is very high, generally better than one part in 10^3 over perhaps 70% of the aperture. In the case of accelerator magnets, this demand must even be satisfied under pulsed conditions. Here again, minimizing conductor motion (from magnetic forces *or* thermal contraction) is crucial, although field aberrations may result from a variety of sources, in addition to deviations from the ideal coil configuration because of errors in construction (field requirements dictate tolerances in conductor location of typically 0.05 mm). These include persistent diamagnetic screening effects (at low fields), iron saturation (at high fields), and, during rapid rates of field change, induced field components proportional to $\dot{B}$. None of these effects can be fully controlled, and it is necessary to incorporate correction windings, which are normally placed around the aperture inside the main coil winding. Note also that the field distribution is in reality three dimensional, due to the finite length of the magnet, and the important quantity from the point of view of the beam-steering effectiveness of the magnet is the field integral along the magnet axis, $\int B \, dl$. The field integral can be manipulated or corrected by axially displacing the ends of the various sections that form the main coil winding with respect to each other. Moreover, in saddle coils, the peak field occurs in the portion of the end winding where the conductors have a rather small radius of curvature. The peak field enhancement over the useful field in the aperture can also be affected by shaping the coil ends.

High-energy accelerators utilizing superconducting main ring magnets have not yet been constructed, although several new facilities based on them are in the design stage. Meanwhile, a number of superconducting dc beam transport magnets are operational in the various laboratories [41] where, in addition to their primary function, they provide much needed experience in operating complete magnet–Dewar–refrigerator systems. An example is the superconducting quadrupole doublet OGA (*optique à grande acceptance*) [42] installed in a pion beam line at the *Saturne* synchrotron at Saclay, France. One member is shown in Fig. 13. The field length is 68 cm in each quadrupole, and the useful aperture 20 cm in the first magnet and 30 cm in the second. The design gradients are 35 and 23 T/m, respectively, and correspond to peak fields at the conductor of 5 and 4.5 T. The conductor is a rectangular stabilized Nb–Ti composite, and the coil current approximately 1200 A in both magnets.

Other operational secondary beam line magnets include a system of three dipoles in the beam line to the BEBC at CERN (operating at 4.7, 3.4, and 4.7 T, respectively), a quadrupole doublet and a 4-T bending magnet at the Berkeley Bevatron, and a 16-T/m quadrupole for the synchrotron at the National Laboratory for High Energy Physics (NLHEP) at Tsukuba, Japan. A unique superconducting beam line

Fig. 13.   One member of the quadrupole pair OGA installed in the experimental area of the proton synchrotron *Saturne*. (Courtesy Centre d'Etudes Nucléaires de Saclay.)

system, utilizing a solenoid rather than a saddle coil, is the magnet for the muon channel at the SIN cyclotron, Switzerland. The magnet is 8-m long with an aperture of 12 cm, operates at 5 T, and is cooled by forced flow of supercritical helium [43]. At the Brookhaven Alternating Gradient Synchrotron, two 4-T dipoles, each 2 m long, serve to deflect the primary proton beam to a bubble chamber (also utilizing a large superconducting magnet) [44]. The most ambitious superconducting beam transport system undertaken to date is now operating in a new unseparated beam line at Brookhaven [45]. It utilizes four 4-T dipoles, each 2.5 m long with a 25-cm coil aperture (20-cm warm bore), and a stored energy per magnet exceeding 1 MJ. The magnet design is very similar to the longer (and smaller cross section) dipoles for Isabelle, described later.

Pulsed versions of these dipoles and quadrupoles, or prototype synchrotron magnets, are presently in an advanced stage of development at Berkeley, Brookhaven, and the Fermi National Accelerator Laboratory (FNAL) in the United States [46], at NLHEP in Japan, and at the Karlsruhe, Rutherford, and Saclay laboratories in Europe [47]. Recently, the European efforts have been extended in other technological areas outside high-energy physics, particularly involving fusion power programs. The Berkeley pulsed magnet program culminated in the Experimental Superconducting Accelerator Ring (ESCAR) intended to furnish further experience with superconducting magnets in a true accelerator environment; this program has been discontinued, however. The FNAL program, known as the Energy Doubler/Saver Project, is on a very ambitious scale, aimed at doubling the peak energy of the present 500-GeV proton synchrotron ring there, and cutting yearly electricity costs (currently over $7 million). Prototype dipoles are already being fabricated; about 800 will be needed in all.

Illustrative of the state of the art are the full-scale prototype dipoles designed for a 400-GeV proton–proton colliding beam accelerator system Isabelle proposed at Brookhaven. These magnets [48], one of which is pictured in Fig. 14, are 4.75-m long with an inner coil diameter of 12 cm and they are designed for 5 T over a working (warm bore) aperture of 8 cm. The current corresponding to 5 T is 4000 A. The magnets have a peak

Fig. 14. Prototype dipole for the proton–proton colliding beam facility Isabelle. The magnet is shown inserted in its stainless-steel support tube, ready for mounting in the cryostat seen in the background.

stored energy of approximately 850 kJ and weigh 6900 kg including the laminated iron shield. The method of construction is indicated in Figs. 15 and 16. Several alternative dipole configurations pursued in other laboratories are also shown schematically in Fig. 15. The Brookhaven coils consist of a single layer of discrete blocks containing a wide conductor braided from superconducting wires. This block configuration is a step-function approximation to the ideal cos θ current density distribution. The conductor is distributed in equal angular sections around the aperture and the current graded appropriately in the various sections by interleaving insulating spacer turns. An important feature of the single-layer design is the high aspect ratio of the braided conductor. This permits a large number of turns in the winding cross section and thereby achieves a close approximation to the cos θ distribution. The conductor, 0.06-cm thick and 1.7-cm wide, is braided from approximately 100 twisted multifilamentary wires, each 0.3 mm in diameter and containing 500 Nb–Ti filaments in a copper matrix surrounded by a Cu–Ni jacket. It is filled with an Ag–Sn

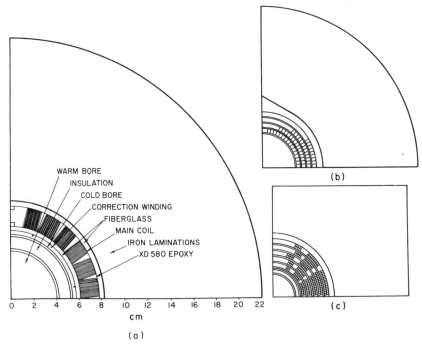

Fig. 15. Quadrant of dipole of the cos θ type based on a sector geometry with variable overall current density, (a) as proposed for Isabelle and, shown more schematically, two alternative dipole configurations of the constant current density type with several coil layers: (b) shell geometry and (c) modified sector geometry.

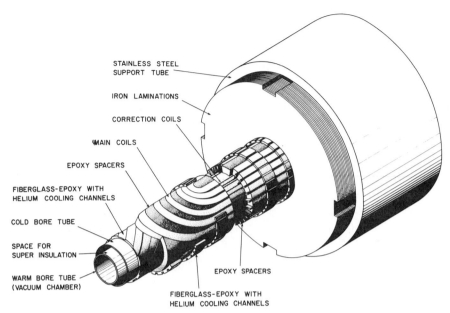

STAINLESS STEEL
SUPPORT TUBE

IRON LAMINATIONS

CORRECTION COILS

MAIN COILS

EPOXY SPACERS

FIBERGLASS-EPOXY WITH
HELIUM COOLING CHANNELS

COLD BORE TUBE

SPACE FOR
SUPER INSULATION

WARM BORE TUBE
(VACUUM CHAMBER)

EPOXY SPACERS

FIBERGLASS-EPOXY WITH
HELIUM COOLING CHANNELS

Fig. 16.   Isometric view of the Isabelle dipole magnet, showing the configuration of the coil ends. [From McInturff *et al., IEEE Trans. Magn.* **MAG-13,** 276 (1977).]

alloy for additional electrical and mechanical stability, and spiral wrapped with fiberglass-epoxy tape insulation. This tape also serves to bond all turns in a block together, yet the spiral pitch is such that the coil blocks are permeable to helium. Wedges molded from a silica-flour-filled epoxy separate the current blocks, and the two coil halves are held together on the bore tube (containing auxiliary field-shaping windings) with bands of fiberglass epoxy. The assembled coil is clamped firmly in a shield or "core" of low-carbon steel (also helium cooled), which contributes approximately 40% of the field strength and serves to maintain the coil under compression. In the most recent version of these magnets, the shield laminations, stamped in the form of unsplit annular disks, are contained within an accurate heavy-wall stainless-steel tube (see Fig. 14). The small clearance between laminations and the tube requires that the stack of laminations be inserted while the tube is maintained at elevated temperature; coil insertion into the shield thus assembled is accomplished with the coil precooled in liquid nitrogen and the shield at room temperature, ensuring an "interference fit" between core and coil at operating temperature.

This design, which has proven very successful from the magnet performance point of view (i.e., minimizing training), embodies most of the features that have been mentioned: a metal-filled conductor assuring a high degree of stability; coil support rigid yet having sufficient porosity for good cooling; avoidance of excessive amounts of epoxy impregnants, which tend to crack or store strain energy; and careful attention to proper matching of thermal contraction coefficients of all materials. In addition, the relatively simple design based on a single layer of high current capacity conductor with good thermal coupling between turns lends itself to accurate mechanical assembly and efficient heat transfer, and is electrically sound in terms of inductance, charging voltage, and quench tolerance. (The considerable magnetic stored energy can be dissipated in the coil without damage during a quench, since the quench front or normal zone propagates azimuthally at a rate sufficient to ensure that the energy is deposited in a large fraction of the coil winding.) Finally, the mechanical properties of the braided conductor render it potentially an excellent candidate for use with Nb_3Sn multifilamentary wires simply substituted for the present alloy composites when such wire becomes available in quantity.

The Isabelle magnets will be cooled by forced circulation of high-pressure supercritical gaseous helium [49], as in the case of the Omega and SIN magnets in Switzerland noted earlier, with the coolant circulating in spiral passages or slots in the fiberglass bands on the outside and inside surface of the coils; the stainless-steel support tube referred to earlier also forms the outer wall of the helium pressure vessel. For large magnets or magnet systems, refrigeration based on forced flow cooling is considerably more efficient, economical, and reliable than the "pool boiling" system in which magnets are simply immersed in a liquid bath—primarily because the temperature of magnets utilizing forced flow cooling is not sensitive to return line pressure drop (as in a pool boiling system where this pressure drop determines the temperature of the helium bath) and because the heat transfer to the coolant is not limited by the onset of film boiling. (This mode of refrigeration is also particularly well suited with the higher critical temperature Nb_3Sn-based conductors.)

D. Electrical Machines

The application of superconductors in rotating electrical machinery is an area of prime importance [50]. A number of model motors and generators have been constructed and demonstrated as feasible, although they have mainly been relatively small experimental units. It has not yet been shown that superconducting machines can compete favorably, from the point of view of capital and operating costs as well as that of performance, with conventional machinery that is based on a century of development. Favorable competition will only be possible when present models are scaled up to the larger systems for which their principal advantage is forecast. Moreover, these models have been limited to partially superconducting versions of conventional machines and to machine types dictated by the limited superconductor state of the art at the time they were initiated. Future designs will presumably exhibit more imagination and take greater advantage of both the sophisticated magnet technology evolved from the high-energy physics accelerator magnet program outlined above, and the newer high current density A15 composites now becoming available.

In this section, the present approach to superconducting machinery is summarized and less-conventional approaches are touched on briefly. With regard to the choice of ac versus dc, the use of superconducting excitation windings in ac machines implies a significant increase in field strength and consequently in machine output (within mechanical limits), whereas in conventional dc machines, the limitation is not determined by field strength, but is set by commutation requirements. Therefore, superconductors do not offer such clear-cut gains in dc applications. On the other hand, the main superconductor limitation, high-frequency alternating current losses, has restricted the application of superconducting windings to those configurations that are essentially dc in nature, or to dc homopolar machines and to the field windings of ac machines. Two applications appear most promising at present. One is in large central station power plants, where the main advantages are reduction of unit size and weight, increased unit rating, and higher operating efficiency. The other is in mobile applications, specifically in lightweight airborne ac generators and ship propulsion systems, and includes motors as well as generators.

The earliest significant machine development work was expended on homopolar machines [51]. Here the field winding is a solenoid (in disk-type machines) or split coil (in drum-type machines), producing a field parallel to the axis of rotation; it is not only a simple coil configuration, but is also a truly dc coil. This is not the case in heteropolar commutated dc machines, because there the field winding is subjected to a time-varying field produced by the armature current. While it appears practical to shield the superconductor from the ac field (some external field screening is also necessary in the homopolar machine and indeed in all machine types, ac or dc), the homopolar machine has the additional advantage, aside from the basic simplicity of avoiding commutation, of the field winding not being subject to a torque reaction. Compensation for the reaction forces in heteropolar machines is possible, but must be done at the cost of a substantial increase in refrigeration load due to the additional field winding support necessary. Finally, a stationary field winding (in all types of dc machines) eliminates many of the mechanical and cryogenic complications encountered in ac machines.

Two main drawbacks have traditionally been associated with homopolar machines. One, the difficulty of achieving reasonable terminal voltage without introducing multiple armature conductors, has now been largely overcome by the use of segmented slip rings. More intractable is the serious problem of achieving efficient current collection during current transfer between moving and stationary contacts. The various homopolar machine development programs differ largely in the current collection system adopted. Some utilize solid brushes, including the newer metal-plated carbon fiber materials, others use liquid metals; all schemes limit the brush current density and suffer from various practical complications. Another difference between various homopolar machine designs lies in the different methods of screening the field winding from external magnetic fields, either by (superconducting) screening coils or by an iron shield. Utilization of a heavy iron shield would appear somewhat redundant in a superconducting machine, but in applications where weight is not so important, it can also serve to enhance the field produced by the windings alone, as in other magnet applications already noted, and can be used to shape the field distribution within the rotating armature.

Noteworthy among the various experimental superconducting homopolar machine programs are the pioneering efforts of the International Research and Development Company (IRD) in England. A 50-hp demonstration disk-type motor, completed in 1966, was followed by a 3250-hp motor, tested under full load at Fawley power station in 1971 (Fig. 17), and more recently by a homopolar motor–generator set (including a 1-MW dc generator and a 1350-hp motor) that was designed for marine propulsion [51] (Fig. 18). Other development programs include one at Laboratoire de Génie Electrique de Paris and Laboratoire Central des Industries Electriques in France, where a 60-kW motor has been tested, and one at Toshiba Electric in Japan where a 10-kW generator has been succeeded by a 3000-kW disk-type generator intended for a copper electrolyzing plant. Another program is pursued by the United States Navy at the Naval Ships Research and Development Center in Annapolis, where a 1000-hp motor has been tested and both dc and ac–dc hybrid ship propulsion units in the 20,000–40,000-hp range are in the design stage, in cooperation with General Electric and Garrett Corp. Both disk- and drum-type machines are considered; the former are favored for low-speed motors, while drum-type designs are better suited for high-speed generators and motors. Finally, a 1-kW commutated dc machine with superconducting field windings and a warm iron yoke has been constructed at the Leningrad Polytechnic Institute in the Soviet Union.

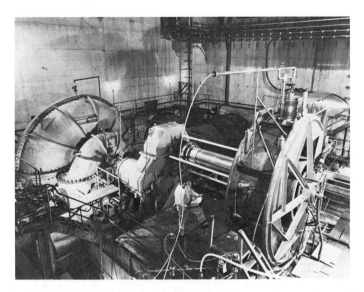

Fig. 17.   The Fawley homopolar motor (3250 hp at 200 rpm). (Courtesy International Research and Development Co., Ltd., United Kingdom.)

As noted, superconductors have their greatest potential in ac machines, primarily because maximum advantage can be taken of their high current density. However, a lack of adequately stabilized low-loss conductors has delayed their exploitation in this area [52]. Even with the appearance of multifilamentary conductors, the excessive ac losses at power frequencies have limited their use to the field winding. (It cannot be ruled out that new design concepts may eventually circumvent the ac loss problem in ac machines.) The field winding may be stationary, which is advantageous for cryogenic and mechanical reasons, or rotary, which is electrically preferable. Rotating field windings are imperative in central station applications due to the difficulty of removing high power from a rotating armature through sliding contacts. In ac machines the winding takes the form of a saddle-type dipole or multipole coil and produces a field transverse to the axis of rotation. The machine rating scales with the product of the field strength generated by the field winding and the attainable armature current, which because of the high reluctance of the superconducting air core magnet design can be increased in proportion to the excitation field, but only within limits.

Perhaps the major limitation is mechanical, due to the necessity of shielding the superconducting rotor from armature ac field components and transient flux changes. The rotating shield, normally conducting and bulky since it must also be strong enough to absorb fault forces, seriously reduces the armature–field magnetic coupling and degrades machine per-

Fig. 18. Marine propulsion system (1-MW dc generator, 1350-hp motor). (Courtesy International Research and Development Co., Ltd., United Kingdom.)

formance. A number of other problems associated with the superconducting field winding must be faced, of which a few are noted. Some of these problems have already been encountered in the dipole magnets considered earlier, but they are compounded here due to the high-speed rotation. The problem of coil or conductor motion is especially tricky because of the rotation and the limited space available for coil restraint since here the working field region is *outside* the winding and any external support structure limits the effective utilization of the generated flux. The need for such support and shielding means that the maximum field seen by the conductor is considerably higher than the effective field at the armature. The conductor must also tolerate substantial surge currents under certain operating conditions. Another difficulty associated with a rotating superconducting field winding is the pressure gradient set up from the accelerating forces and work done on the helium as it flows radially outward, causing an increase in helium pressure and consequently a temperature rise. Reduced coil stability and performance may also result from the effect of the centrifugal force on the convection and boiling heat transfer properties at the conductor–liquid interface. In addition, there are problems such as the limited lifetime of seals used for liquid helium transfer to the rotor, and developing current connection with low heat leak between the field winding and ambient-temperature slip rings.

The pioneer among superconducting alternators is the 8-kVA model that was constructed at AVCO in 1966 [53]. It featured a stationary field winding and normal rotating armature, as did a number of its successors. These included a 100-kVA machine at the Kurchatov Institute of Atomic Energy in Moscow, a 21-kVA machine at the Zentralinstitut für Festkörperphysik und Werkstofforschung in Dresden, East Germany, and one rated at 30 kVA at Fuji Electric, Japan. The first synchronous machine exhibiting a rotating superconducting field winding, rated at 45 kVA and rotating at 3600 rpm, was constructed at MIT in 1969. It has more recently operated as a synchronous condenser, (i.e., consuming and producing no power) connected to the laboratory power supply [52, p. 285]; its success led to the construction of a 2-MVA machine that has undergone tests there since 1973. Other machines of this type in the megavolt ampere range include a 1-MVA machine under construction at the Institute for Electrical Machines in Leningrad, and a 5-MVA machine built and tested by Westinghouse in 1973, attaining full voltage, current, and speed, but not all simultaneously. A representative design for a 1300-MW, 3000-rpm generator under development at IRD is shown in Fig. 19 [54]. Westinghouse has also constructed a 1-MVA 400-Hz machine for airborne application (the rotor operating at 12,000 rpm), and more recently load-tested a 5-MVA 400-Hz generator. The first machine operating under a real load and utility conditions will probably be a 20-MVA generator under construction at General Electric.

In the future, as already noted, superconducting machine designs will undoubtedly take fuller advantage of the rapidly developing state of the art of superconductors and superconducting magnet technology, and incorporate design features compatible with the various superconductor limitations. For example, while present superconducting alternator designs have been visualized mainly in terms of conventional, albeit superconducting, high-speed two- or four-pole field windings, future designs may employ excitation windings of high multipolarity, containing perhaps 30–60 poles. These would allow greatly reduced rotation speeds and simplify the severe cryogenic and mechanical complications hitherto encountered. Forced convection cooling will have to be carefully evaluated. The ac losses in commercial materials have been substantially reduced, and

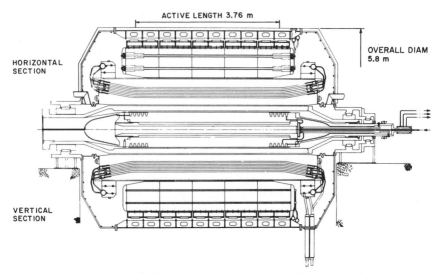

Fig. 19. Cross section of 1300-MW ac generator. (Courtesy International Research and Development Co., Ltd., United Kingdom.)

even Nb_3Sn now appears promising from this point of view, at least in certain applications, again demanding a reevaluation of machine design based on these materials. If necessary, the loss problem may be sidestepped by using low-frequency generation followed by static frequency multiplication [55]. However, the challenge is broader than simply making optimum use of the special properties of superconductors and matching available conductors with designs for a particular device, especially in power network applications. Not only must specific prototype designs and components be examined under realistic utility conditions, but many overlapping system considerations must receive close scrutiny, such as refrigeration and the interfacing of superconducting components with existing conventional systems.

E. Other Applications

Lack of space precludes more than passing reference to a host of other potentially important applications of superconducting magnets. Some of these, particularly in the industrial area, require little sophistication in conductor or magnet technology, and the first large-scale commercial impact may occur here. Other long-term applications, such as in magnetohydrodynamic and fusion power generation, pose requirements somewhat beyond the present superconductor state of the art. In any case, the technical feasibility is harder to assess in these areas, but they are also areas where superconductivity may ultimately be of the greatest benefit for mankind.

Among the industrial applications, several are of immediate interest and depend on high-field-gradient separation methods; these include cleaning processes (e.g., kaolin, iron ore, and coal), ore recovery, water filtration, and scrap separation. High-field magnets may also be important in some areas of industrial process control, such as in polymerization reactions and sterilization [56]. Electrical applications, besides power generation and energy storage, may include transformers, rectifiers, and above all, power transmission—an area with a significant ongoing world-wide superconducting development effort and treated in detail in the chapter by Bussière, this volume. Superconducting elements are obvious candidates for use in electron microscopes (see Pande, this volume), because of the high field gradients possible and the high current stability attainable in the persistent mode. The utilization of superconducting magnets in medicine looks very promising, especially for intravascular catheter propulsion and guidance (several systems are presently operational), cancer detection by nuclear magnetic resonance differentiation between malignant and normal tissue, and perhaps magnetic "starving" of tumors [57].

High-speed ground transportation based on magnetic suspension has received considerable attention [58], although development efforts have been limited except in Germany and Japan, where they are actively pursued as a matter of national policy. Its future will depend more on socioeconomic constraints than on strictly technical considerations. To compete with existing highly developed air and highway systems such a mode of transportation must feature high speed, competitive costs, and convenient service. Magnetically suspended rail systems appear superior to tracked air cushion trains from the standpoint of lift/drag ratio, payload capability, track clearance and tolerance, power collection, and noise. All superconducting schemes proposed employ basically an onboard superconducting magnet and a track or guideway of normal conducting loops or a metallic sheet. Magnetic fields generated by induced eddy currents in the track oppose changes in the primary applied field of the magnets in the moving vehicle. This results in a repulsive levitating force between magnets and the track. Such a repulsive system is inherently stable. The superconducting magnet and conductor requirements appear modest. Reliability is crucial; other considerations include a need for low weight and the need for shielding the conductor from external ac fields and possibly the passenger compartment from the primary dc field. Aluminum-stabilized conductors are especially attractive for weight reasons.

The role of magnetohydrodynamics (MHD) in future large-scale power generation is at present difficult to foresee; at any rate, it looks promising in special applications such as in airborne and space vehicles, and light-weight superconducting magnets are very attractive for this purpose [59]. Field strengths can be kept reasonably low, perhaps 5 T, though there are advantages in higher fields, and the magnet geometry is a fairly conventional saddle or axially symmetric type. Ambitious development programs on superconducting magnets for MHD have been underway for some time in Japan and in the Soviet Union. More recently, a strong cooperative program has been initiated between the United States and the Soviet Union.

Superconductors will be essential in controlled thermonuclear reactor (CTR) applications because of the need for very high fields (the reaction rate is proportional to a high power of the field strength) over large vol-

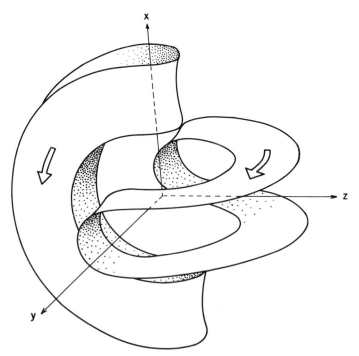

Fig. 20. Current path shown schematically in the mirror coil system (also known as a "Yin–Yang" coil).

umes (tens of cubic meters) [60, 61]. The field requirements depend strongly on the fuel cycle and reactor type assumed. With the deuterium–tritium cycle, conductors not very different from those presently available may suffice. Both high-current, high-aspect-ratio metal-filled cables or braids (or possibly flattened monolithic conductors) for large dc pancake windings and mixed-matrix ac conductors for pulsed application will be needed. High-purity aluminum-bonded Nb_3Sn is again an especially promising candidate [62], and cooling by forced flow of single-phase supercritical helium can be anticipated for maximum efficiency and reliability. Nevertheless, as in energy storage systems, conventional materials problems and structural considerations, will dominate the overall reactor cost rather than strictly magnetic ones.

In view of the long-term stakes involved, the present development effort on superconducting magnets for fusion is modest. This work is supported mainly in parallel with feasibility experiments utilizing conventional magnets, and is directed toward the next generation of large-scale demonstration devices. Basically, two types of systems for magnetically confining the plasma exist, depending on whether the field lines of force close within the reaction chamber or pass through the chamber walls.

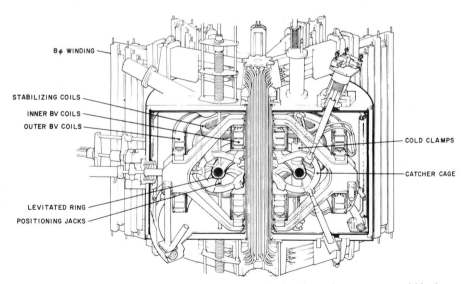

Fig. 21. The Culham superconducting levitron, showing the main components within the 1.7-m diameter vacuum chamber. These include the superconducting core conductor (the "ring"), vertical field windings (BV) (which are also superconducting), and toroidal field coils (Bϕ) of conventional copper design. (This figure is taken from the article entitled The Culham Superconducting Levitron written by S. Skellett and appeared in Vol. 15 of *Cryogenics,* published by IPC Science & Technology Press Ltd., Guildford, Survey, UK.)

"Open" cylindrical or spherical systems produce a stable magnetic well or minimum-B confinement field at the chamber center by the use of magnetic mirror end coils. "Closed" toroidal systems, which avoid the end losses associated with open-ended systems and promise long confinement times, consist of a basic confinement field produced by a toroidal winding plus either a helical stabilizing winding (in the case of the Stellarator), or one of several varieties of circulating currents established within the plasma, i.e., a ring current induced by transformer action (Tokamak), an unsupported superconducting ring (Levitron), or a sheath of relativistic electrons (Astron). A different appproach, also based on a toroidal configuration, is the theta-pinch confinement system, which produces plasma heating through several stages of rapid implosion and compression and utilizes an energy storage device. Some of these are steady state devices but require very high fields (> 10 T), such as the mirror systems, and complicated winding configurations, especially "baseball cusp" mirror systems and the Stellarator. Others are pulsed devices designed for

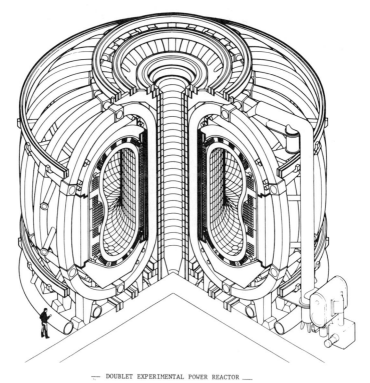

— DOUBLET EXPERIMENTAL POWER REACTOR —

Fig. 22. Overall view of a conceptual experimental power reactor (EPR). (From Baker *et al.*, IEEE Publ. No. 75CH1097-5-NPS.)

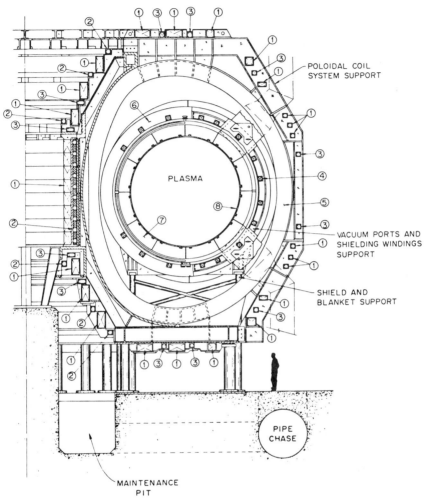

Fig. 23. Cross section of Oak Ridge experimental power reactor (EPR) reference design. The oval-shaped toroidal field coils have a vertical bore of approximately 10 m and are designed for a maximum field at the winding of 11 T (4.9 T on the plasma axis). The pulsing poloidal field coil system consists of four groups of coils: ohmic heating primary coils, shield-vertical field coils, decoupling coils, and trim-vertical field coils. All except the shield-vertical field coils are superconducting. The plasma current is parallel to the toroidal magnetic field and opposite to the ohmic heating coils in the burn phase: 1, air core primarily winding; 2, decoupling windings; 3, vertical field trim windings; 4, shielding vertical field windings; 5, toroidal field coils; 6, shield; 7, blanket; 8, vacuum containment. (From Oak Ridge National Laboratories Report No. ORNL-5275.)

very-high-$\dot{B}$ operation, notably the Tokamak, and the theta-pinch approach requires a high pulse rate as well as very efficient superconducting inductive energy storage coils and transfer equipment.

Several types of minimum-B coil systems using superconducting windings have been constructed. One is the baseball cusp magnet at the Lawrence Livermore Laboratory, where the current path resembles the seam in a baseball (Fig. 20). The spherical coil is 2 m in diameter, is wound with Nb–Ti, and produces 2 T at the center and 7.5 T at the winding. In the IMP experiment at Oak Ridge, a minimum-B field is produced by superposition of fields from a cylindrical mirror coil (wound with Nb–Ti) and a quadrupole coil (Nb_3Sn). Several model toroidal systems using superconducting windings have also been or are being constructed. These are mostly floating ring devices and include the Livermore and Culham (England) Levitrons (Fig. 21) [63]. A number of conceptual designs for prototype Tokamak devices based on superconductors have been carried out, and several laboratories, Oak Ridge in particular, have initiated a vigorous program for developing conductors and winding configurations suitable for Tokamaks. Figure 22 shows an overall view of an experimental power reactor (EPR) designed at the General Atomic Co., and Fig. 23 shows a cross section of the Oak Ridge EPR reference design. The first operating superconducting toroidal Tokamak device, at the Kurchatov Atomic Energy Institute, in Moscow, and known as T–7, was announced in late 1978 [64].

References

1. J. E. Kunzler, E. Bachler, F. S. L. Hsu, and J. E. Wernick, *Phys. Rev. Lett.* **6,** 89 (1961).
2. H. Brechna, "Superconducting Magnet Systems." Springer-Verlag, Berlin and New York, 1973.
3. C. P. Bean, *Rev. Mod. Phys.* **36,** 31 (1964).
4. P. S. Swartz and C. P. Bean, *J. Appl. Phys.* **39,** 4991 (1968).
5. A. R. Kantrowitz and Z. J. J. Stekly, *Appl. Phys. Lett.* **6,** 56 (1965).
6. C. Laverick, *Proc. Int. Symp. Magn. Technol., Stanford* p. 560 (1965).
7. C. N. Whetstone, *IEEE Trans. Magn.* **MAG-2,** 307 (1966).
8. P. F. Chester, *Proc. Int. Cryog. Eng. Conf., 1st, Tokyo* p. 147 (1967).
9. P. F. Smith, *Proc. Int. Conf. Magn. Technol., 2nd, Oxford* p. 543 (1967).
10. W. B. Sampson, *Proc. Int. Conf. Magn. Technol., 4th, Brookhaven National Laboratory* p. 487 (1972).
11. S. L. Wipf and M. S. Lubell, *Phys. Lett.* **16,** 103 (1965).
12. R. Hancox, *Phys. Lett.* **16,** 208 (1965).
13. P. F. Chester, *Proc. Int. Cryog. Eng. Conf., 1st, Tokyo* p. 147 (1967).
14. H. R. Hart, *Proc. Summer Study Supercond. Devices Accel., Brookhaven National Laboratory* p. 571 (1968).

15. M. N. Wilson, C. R. Walters, J. D. Lewin, and P. F. Smith, *J. Phys. D* **3**, 1517 (1970).
16. Y. B. Kim, C. F. Hempstead, and A. R. Strnad, *Phys. Rev.* **129**, 528 (1963).
17. G. H. Morgan, *J. Appl. Phys.* **41**, 3673 (1970).
18. D. B. Thomas and M. N. Wilson, *Proc. Int. Conf. Magn. Technol., 4th, Brookhaven National Laboratory* p. 493 (1972).
19. M. N. Wilson, *Proc. Int. Conf. Magn. Technol., 5th, Rome*, p. 615 (1975).
20. W. B. Sampson, P. F. Dahl, A. D. McInturff, and K. E. Robins, *IEEE Trans. Nucl. Sci.* **NS-18**, 660 (1971).
21. A. D. McInturff, P. F. Dahl, and W. B. Sampson, *J. Appl. Phys.* **43**, 3546 (1972).
22. D. Dew-Hughes, *Cryogenics* **15**, 435 (1975).
23. A. R. Kaufman and J. J. Pickett, *Bull. Am. Phys. Soc.* **15**, 838 (1970).
24. M. Suenaga and W. B. Sampson, *Appl. Phys. Lett.* **18**, 584 (1971).
25. M. Tachikawa, *Proc. Appl. Superconduct. Conf., Annapolis, Maryland* p. 371. (1972).
26. M. Suenaga and W. B. Sampson, *Appl. Phys. Lett.* **20**, 443 (1972).
27. W. A. Fietz and C. H. Rosner, *IEEE Trans. Magn.* **MAG-10**, 239 (1974).
28. *CERN Courier* **15**, 147 (1975).
29. D. B. Montgomery, "Solenoid Magnet Design." Wiley (Interscience), New York, 1969.
30. R. Cool *et al. Phys. Rev. D* **10**, 792 (1974).
31. W. A. Fietz, E. F. Mains, P. S. Swartz, E. G. Knopf, W. D. Markiewicz, and C. Y. Rosner, *IEEE Trans. Magn.* **MAG-11**, 559 (1975).
32. F. Wittgenstein, *Proc. Int. Conf. Magn. Technol., 4th, Brookhaven National Laboratory* p. 295 (1972).
33. G. Bogner, *Proc. Appl. Supercond. Conf., Annapolis, Maryland* p. 215 (1972).
34. *CERN Courier* **11**, 64 (1971).
35. R. W. Boom, G. E. McIntosh, H. A. Peterson, and W. C. Young, *Proc. Cryog. Eng. Conf., Atlanta, Georgia* p. 117 (1973).
36. W. Gilbert, *IEEE Trans. Nucl. Sci.* **NS-20**, 668 (1973).
37. W. B. Sampson, R. B. Britton, P. F. Dahl, A. D. McInturff, G. H. Morgan, and K. E. Robins, *Part. Accel.* **1**, 1973 (1970).
38. P. F. Dahl, *et al., IEEE Trans. Nucl. Sci.* **NS-20**, 688 (1973).
39. P. F. Smith and B. Colyer, *Cryogenics* **15**, 201 (1975).
40. F. R. Fickett, *Proc. Int. Conf. Magn. Technol., 5th, Rome* p. 659 (1975).
41. *CERN Courier* **15**, 150 (1975).
42. G. Bronca, J. Hamelin, J. Neel, J. Parain, and M. Renard, *Proc. Appl. Supercond. Conf., Annapolis, Maryland* p. 288 (1972).
43. G. Vecsey, J. Hovarth, and J. Zellweger, *Proc. Int. Conf. Magn. Technol., 5th, Rome*, p. 110 (1975).
44. J. Allinger *et al., Proc. Int. Conf. High Energy Accel., 9th, SLAC, Stanford*, p. 198 (1974).
45. K. E. Robins *et al., Proc. Particle Accel. Conf., Chicago, Illinois* p. 1320 (1977).
46. P. Reardon, *IEEE Trans. Magn.* **MAG-13**, 704 (1977).
47. D. B. Thomas, *Proc. Int. Conf. Magn. Technol., 5th, Rome* p. 499 (1975).
48. P. F. Dahl, *Proc. Int. Accel. Conf., 10th, Serpukhov, USSR* **II**, 192 (1977).
49. J. A. Bamberger, J. Aggus, D. P. Brown, D. A. Kassner, J. H. Sondericker, and T. R. Strobridge, *IEEE Trans. Magn.* **MAG-13**, 696 (1977).
50. J. L. Smith, J. L. Kirtley, and P. Thullen, *IEEE Trans. Magn.* **MAG-11**, 128 (1975).
51. A. D. Appleton, *in* "Superconducting Magnets and Devices" (S. Foner and B. Schwartz, eds.), Chapter 4. Academic Press, New York, 1974.
52. J. L. Smith and T. A. Keim, *in* "Superconducting Magnets and Devices" (S. Foner and B. Schwartz, eds.), Chapter 5. Academic Press, New York, 1974.

53. H. H. Woodson, A. J. J. Stekly, and E. Halas, *IEEE Trans. Power Appar. Syst.* **PAS-85,** 264 (1966).
54. A. D. Appleton, J. S. H. Ross, J. Bumby, and A. J. Mitcham, *IEEE Trans. Magn.* **MAG-13,** 770 (1977).
55. E. B. Forsyth, Private communication (1975).
56. J. Powell, *in* "Superconducting Magnets and Devices" (S. Foner and B. Schwartz, eds.), Chapter 1. Academic Press, New York, 1974.
57. S. J. St. Lorant, *Proc. Int. Conf. Magn. Technol., 5th, Rome* p. 393 (1975).
58. Special MAGLEV issue, *Cryogenics* **15** (1975).
59. P. Komarek, *Cryogenics* **16,** 131 (1976).
60. C. E. Taylor, *Proc. Appl. Supercond. Conf., Annapolis, Maryland* p. 239 (1972).
61. G. K. Hess, E. J. Ziurus, and D. S. Beard, (1975). *IEEE Trans. Magn.* **MAG-11,** 135 (1975).
62. C. D. Henning, R. L. Nelson, H. L. Leichter, and C. D. Ward, *Proc. Int. Conf. Magn. Technol., 4th, Brookhaven National Laboratory* p. 521 (1972).
63. S. Skellett, *Cryogenics* **15,** 563 (1975).
64. D. P. Ivanov, V. E. Keilin, B. A. Stavissky, and N. A. Chernoplekov, *IEEE Trans. Magn.* **MAG-15,** 550 (1979).

Metallurgy of Niobium–Titanium Conductors

A. D. McINTURFF

Isabelle Project Accelerator Department
Brookhaven National Laboratory
Upton, New York

I. Introduction 99
II. The Niobium–Titanium System. 100
III. Microstructure and Critical Current Density 102
IV. Theory of Flux Pinning in Niobium–Titanium Alloys 117
V. The Nb–Ti Conductors 120
   A. Conductor Design 120
   B. Conductor Fabrication 123
   C. Available Commercial Conductors in 1977 134
   References. 135

I. Introduction

For applications with magnetic fields below 8 T and temperatures below 5 K the ductile alloy superconductors based on niobium are almost exclusively used because of their ease of fabrication and excellent mechanical properties. The original commercial superconducting materials were single-filament Nb 25 wt. % Zr and Nb 33 wt. % Zr, and were marketed by SuperCon in 1962. These alloys were difficult to process since they were subject to severe galling and surface hardening during the rod-reduction and wire-drawing operations. Opertional fields for Nb–Zr were also limited to the 5–6-T range.

Two years later, Westinghouse promoted the first Nb–Ti alloy conductor. This material, despite its lower T_c (9.5 K compared to 11 K for Nb–Zr) was rapidly adopted by all of the superconductor manufacturers, not only because of its superior workability, but also because its upper critical field, 11.7 T at 4.2 K, was greater than that of Nb–Zr (9–10 T). The Nb–Ti solid solution alloys and some ternary alloys of Nb–Ti have almost completely replaced Nb–Zr, though the latter's superior current-carrying capacity at low fields has suggested its possible utilization in conduction of fault currents in superconducting power cables (see the chapter by Bussière, this volume) and switching circuits.

Increased ductility and ease of fabrication in superconductor design became of paramount importance with the advent to multifilamentary conductors, first fabricated by the MIT group [1] using a process developed by Levi [2] for obtaining Nb filaments in copper. (Later, its application was extended to Nb–Ti [3] filaments in copper.) The first commercial production of filamentary Nb–Ti in copper was that of Imperial Metals Industries, Ltd., United Kingdom. These composites and the subsequent development of aluminum-stabilized (Alcoa) Nb–Ti, with metallurgical barriers (Cu–Ni, or similar resistive elements) would not have been possible with the less ductile alloys of Nb–Zr. The high degree of development of superconducting devices today, based on reduction of interfilamentary magnetization current by twisting and the criteria of maximum filament size [4–8] to obtain (flux jump) stability, would have been difficult indeed without a ductile alloy. The predominant alloy in commercial production today is Nb 46 wt. % Ti. This discussion is concerned with the metallurgy of conductors based on alloys with Nb content from 35 to 55 wt. % in Ti.

This chapter has brief sections on the niobium–titanium system and its ternaries; the relationship between critical current density and microstructure; and theories of flux pinning in niobium–titanium alloys. An introductory section on conductor design leads into a thorough account of the fabrication of conductors, from the original production metallurgy of the Nb–Ti solid solution rods to the production of composite conductors of the ampere-to-multikiloampere class. Emphasis is placed on the practical aspects of the problems as they arise with regard to magnet applications.

II.   The Niobium–Titanium System

The phase diagram for the Nb–Ti binary system is given in Fig. 1. The very large area reductions obtained with extrusion and drawing are possible with the alloy in the β (bcc) phase. The two-phase $\alpha + \beta$ (α is hexagonal) region has the better superconducting current densities.

In 1962, Berlincourt and Hake [9] published data that gave the upper critical field H_{c2} as a function of Nb content at 1.2 K for the solid solution Nb–Ti alloys. Their data are presented in Fig. 2 and provided the basis for the work of Vetrano and Boom [10] on the effects of subsequent cold area reduction and precipitation heat treatments (PHT) in obtaining higher crit-

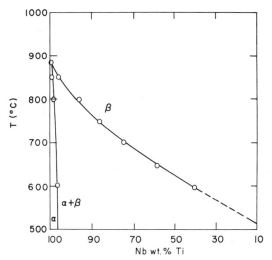

Fig. 1. The binary phase diagram for niobium and titanium.

ical current density from a given alloy. They obtained critical current densities in 3.0-T fields $J_c(3.0\ T) > 10^5$ A/cm^2 at 4.2 K with Nb 70 wt. % Ti after 99% cold area reduction and a 6-h 400°C PHT. The upper critical field H_{c2} was found to be 12.8 T at 1.2 K. One of the early (1964) commercial processes utilized in the production of single-strand wire was based

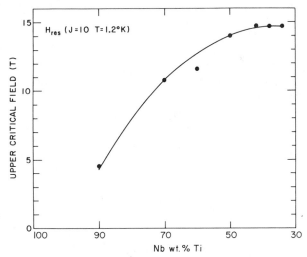

Fig. 2. The upper critical field as measured resistively at a current density of 10 A/cm^2 at 1.2 K as a function of the percent composition of niobium and titanium. (Data from Berlin court and Hake [9].)

on these results. A Nb 65 wt. % Ti alloy rod was β-phase solution annealed, copper jacketed, 99 + % cold area reduced, heat treated at 400°C for 3 h (PHT), and then 99% cold area reduced again. This process resulted in J_c(3 T) at 4.2 K of 1.24 × 10^5 A/cm² for 384 m of 0.635-mm-diameter single-core Nb 65 wt. % Ti wire. Subsequent Nb–Ti data have been obtained for different metallurgical histories by various workers on both Ti-rich [11–17] and Nb-rich [18–20] alloys. The bulk of the cold-work and heat-treatment data utilized in this chapter will be drawn from these papers.

Several general observations can be drawn from a study of the above-listed papers. Broad variations in metallurgical history for different alloys can lead to nearly the same $J_c(H)$ curve and certainly the same $J_c(H)$ at a given magnetic field. These same metallurgical histories can sometimes lead to markedly different physical properties of the alloy, such as increased hardness. The current densities at a given field of the PHT samples and titanium-rich cold-worked samples, provided area reductions > 99.9 + % took place after solution anneal, were size independent [21] after the self-field correction had been applied. The maximum critical current densities (4.2 K) are J_c(3 T) = 5 × 10^5 A/cm² and J_c(5 T) = 4 × 10^5 A/cm² [17] with the best commercially produced superconductors reaching ~0.7 of these values. The commercially produced Nb–Ti alloys have useful current densities (≥10^5 A/cm²) up to 7.5 T. Nb 50 wt. % Ti has a useful current density at fields slightly in excess of 8.0 T. At field intensities greater than 8.0 T, the operating temperature must be less than 4.2 K to be useful in applications requiring J_c > 10^5 A/cm². Critical temperatures are between 9.5 and 10.1 K, requiring that liquid or high-pressure helium near 4.2 K be the refrigerant.

The ternary additives such as Zr [22] and some of the interstitials [23] resulted in very fine-grain structures, therefore enhancing the critical current densities.

The ternary additives, such as Ta [24–26], have the added potential of greater high-field performance due to enhanced H_{c2}, but in 4.2-K tests hoped for gains have not been present so far.

III. Microstructure and Critical Current Density

It is a fact that the microstructures of these alloys do not correlate well with their superconducting properties with the one exception of cell size and critical current density. The reason for this general lack of correlation is that there is more than one competing mechanism for flux pinning. Candidates for pinning mechanisms include:

(1)   An interaction with the periodic stress field of the dislocation cell structure [27].

(2)   The free energy of the vortex at the dislocation cell wall is lowered because the coherence length is smaller (increase in κ) [19, 28–30].

(3)   The superconducting condensation energy of the cell wall is lower than the condensation energy of the matrix [29, 30–32].

(4)   The cell has a normal metal boundary or precipitates in it, resulting in a Josephson-type boundary [33].

Figure 3, a plot of Lorentz force versus cell size obtained by Neal *et al.* [20], shows an excellent correlation with cell size for Nb 44 wt. % Ti. They were unable, however, to correlate abrupt changes in the superconducting properties with 50°C changes in the heat treatment temperature. The lack of correlation can probably be accounted for in terms of competing mechanisms for flux pinning. Vetrano and Boom [10] obtained indirect evidence that pinning was at least partially due to α-Ti precipitates in Nb 66 wt. % Ti alloy. Later work [16] based on the use of Nb 65 wt. % Ti and Nb 60 wt. % Ti gives strong indirect evidence that α pinning is one of the main mechanisms in the 3- to 4-T range in these two alloys. Figure 4 shows the upper critical temperature as a function of precipitation heat

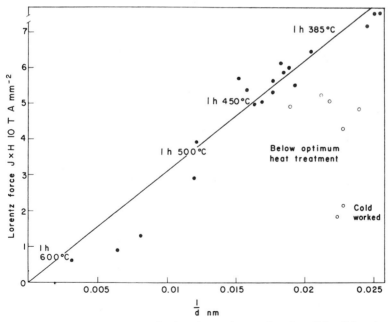

Fig. 3.   The Lorentz force **J** × **H** is plotted as an inverse function of the dislocation cell size for various samples of Nb 44 wt. % Ti with different metallurgical histories. (Data from Neal *et al.* [20].)

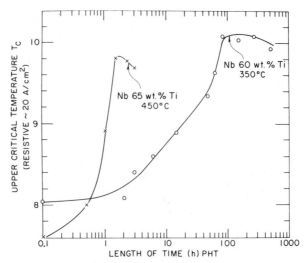

Fig. 4. The upper critical temperature as measured resistively at a current density of 20 A/cm² at near-zero magnetic field as a function of length of time of precipitation heat treatment for Nb 65 wt. % Ti at a temperature of 450°C and Nb 60 wt. % Ti at a temperature of 350°C.

treatment time for two different alloys and temperatures [16]. If the bulk superconductor determines T_c, then these alloys have changed Ti content during precipitation heat treatment; Ti has decreased according to the data of Suenaga and Ralls [24]. Figure 5 shows the upper critical temperature T_c as a function of alloy content for Ti, Nb, and Ta [24]. The critical temperature peaks roughly correspond to the critical current density peaks shown in Figs. 6 and 7 [16], which give the $J_c(3.0\ T)$ versus PHT time curves for various isotherms for Nb 65 wt. % Ti and Nb 60 wt. % Ti, respectively. X-ray diffraction data indicating the presence of α-Ti strengthen the aforementioned conclusions [18, 34]. Kramer and Rhodes have found the possibility of omega-phase precipitation to be involved in pinning [14].

The main conclusions, valid for most of the binary alloys, may be generalized as follows:

(a) $J_c(H)$ is enhanced by cold work (up to the point of mechanical damage). This is probably due to the refinement of the subcell dislocation structure and the presence of a high dislocation density. Figure 8 shows $J_c(H)$ versus the applied field H for Nb 60 wt. % Ti with curves of constant cold area reduction. $J_c(H)$ increases with increased cold area reduction until a peak at 91% area reduction is reached, after which the sample begins to break with further reduction and J_c consequently decreases.

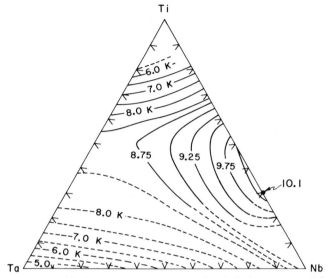

Fig. 5. The upper critical temperature as measured resistively at a current density of 30 A/cm² near zero magnetic field as a function of percent composition of the ternary alloy of niobium, titanium, and tantalum. (From Suenaga and Ralls [24].)

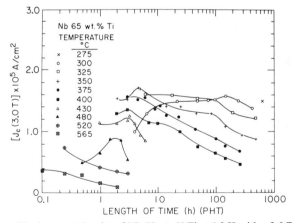

Fig. 6. The critical current density of Nb 65 wt. % Ti at 4.2 K with a 3.0 T magnetic field is mapped for various isothermal heat treatment curves as a function of time at temperature. (From McInturff and Chase [16].)

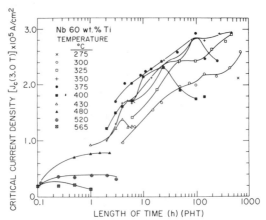

Fig. 7. The critical current density of Nb 60 wt. % Ti at 4.2 K with a 3.0 T magnetic field is mapped for various isothermal heat treatment curves as a function of time at temperature. (From McInturff and Chase [16].)

(b) $J_c(H)$ is enhanced by a precipitation heat treatment for some field range (in particular $\leq 0.8\ H_{c2}$). The highest J_c results from PHT in conjunction with the finest (smallest) dislocation cell structure.

(c) Once effective pinning centers in the form of α-phase precipitates have been formed, the pinning force is controlled by the cell size. The

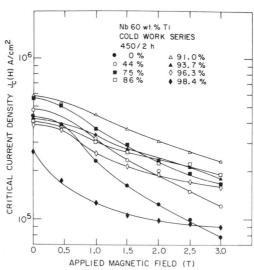

Fig. 8. The critical current density of Nb 60 wt. % Ti at 4.2 K is given as a function of applied magnetic field for various samples each with a different percentage of cold area reduction after a precipitation heat treatment of 450°C for 2 hr.

high-Nb content alloys, Nb 44 wt. % Ti, have been found to have α-titanium present in critical-current-optimized samples [18].

Highest J_c values are obtained when the cold area reduction is greater than 90%, thereby indicating a preference for α nucleation at the dislocation sites. Figure 9 is representative of a typical behavior of a Nb–Ti alloy (Nb 45 wt. % Ti) in which cold area reduction is initially followed by a PHT and then by further cold area reduction [18]. The monotonic increase in the critical current density with cold area reduction throughout this process would indicate that the cellular dislocation networks introduced by cold working are excellent pinning structures. With α phase being detected after heat treatment, the dislocation networks appear to act as nucleation sites for the α phase. It is of interest to note that the field dependence seems to be less pronounced in the cold-work-only samples than in those with a precipitation heat treatment.

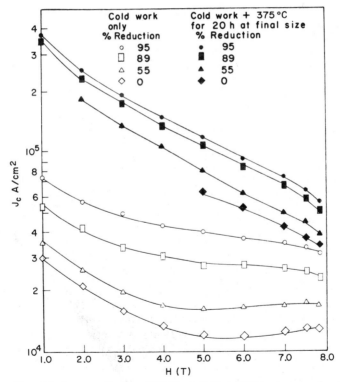

Fig. 9. The critical current density of Nb 45 wt. % Ti at 4.2 K is given as a function of applied magnetic field for various samples, each with a different percentage of cold area reduction both before and after a subsequent precipitation heat treatment of 375°C for 20 hr.

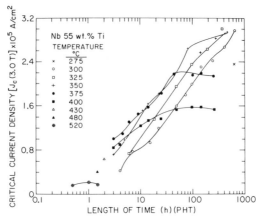

Fig. 10.   The critical current density of Nb 55 wt. % Ti at 4.2 K with a 3.0-T applied field is mapped for various isothermal heat treatment curves as a function of time at temperature.

The heat treatment of a severely cold-worked sample in the $\alpha + \beta$ region of the phase diagram (see Fig. 1) results in α-phase precipitation [18, 34]. Under the foregoing conditions, the α phase is fairly uniformly located throughout the entire bulk. Electron microscopy studies show α-phase precipitation at the dislocation sites.

The following series of figures are presentations of parametric sets of data obtained on various alloys of Nb–Ti, including those that are currently in production. Figures 6, 7, 10, and 11 are isothermal curves for $J_c(3.0 \text{ T})$ as a function of (PHT) heat treatment time for Nb 65 wt. % Ti, Nb 60 wt. % Ti, Nb 55 wt. % Ti, and Nb 50 wt. % Ti, respectively. These samples were drawn down to 0.38 mm from a 17-mm β annealed rod and

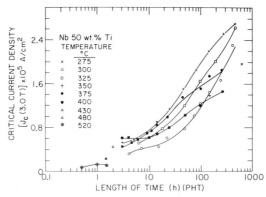

Fig. 11.   The critical current density of Nb 50 wt. % Ti at 4.2 K with a 3.0-T applied field is mapped for various isothermal heat treatment curves as a function of time at temperature.

then heat treated (PHT) for the prescribed temperature and time to give a data point in these figures. The general comments that can be made are

(1)   The optimum temperature for heat treatment seems to be from 325°C to 350°C. This is probably due to the competition between the α-phase precipitation rate and the annealing of the dislocation network.

(2)   The rate of α-phase nucleation is strongly dependent on the titanium percentage in the alloy.

The data on mechanical properties given in Table I can be combined with the critical current density data in Fig. 6 to obtain $J_c(3.0\ T)$ versus $\rho(11\ K)\ \Omega$ cm and hardness. If data of this nature are needed for other alloys of interest, reference may be made to a paper by Berger *et al.* [35] or to a data compilation by Grigsby [36].

The data presented in Figs. 6, 7, 10, and 11 are replotted in Figs. 12 and 13 for precipitation heat treatment temperatures of 300°C and 400°C, respectively, to present the behavior of J_c with respect to alloy content. J_c for the higher-percentage-titanium alloys seems to peak at shorter annealing times and at higher heat treatment temperatures than for alloys

TABLE I

PHYSICAL PROPERTIES OF Nb 65 Wt. % Ti

Temperature (°C)	Time (h)	$\rho_0\ (\mu\Omega$ cm) 300 K	Post-PHT $\rho_f\ (\mu\Omega$ cm) 300 K	11 K	Hardness KHN
As drawn	—	116	—	105	282
250	0.2	116	103	97.2	382
250	17	117	105	84.5	426
250	50	116	88	54.8	480
300	0.2	116	86	72.1	380
300	17	116	82	41	423
300	50	116	73	38	307
350	2	116	79	46	423
350	5	115	59	50	454
350	17	116	71	31.0	409
400	0.5	118	78	53	370
400	1.5	117	75	40	423
400	5	119	67	34	340
450	0.2	118	77	47	359
450	0.5	117	69	39	345
450	1.5	119	70	35	347
500	0.5	117	70	29	—
500	1.5	117	72	32	311
500	5	119	71	30.4	282

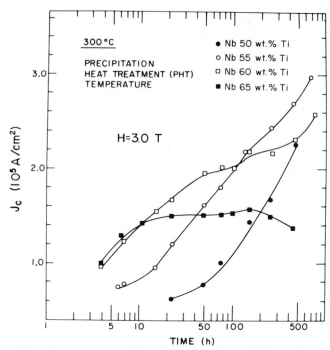

Fig. 12. The critical current density behavior of Nb–Ti as a function of alloy composition is illustrated with the critical current density at 4.2 K in a 3.0-T field obtained for various lengths of time at 300°C for the different compositions shown. Precipitates from the higher-Ti-content alloys dominate over the annealing out of the cold area reduction and the current density rises, but in the higher-Nb-content alloys this is not the case.

with higher niobium content. This can be explained by the rate of formation of α phase and the lowering of the transition temperature (T_c) once the maximum is reached. One problem with this interpretation is that the critical current density seems to increase for some time after the maximum transition temperature has been attained.

Intermediate heat treatments during cold work normally enhance the critical current density of the final wire. In the alloys with >50 wt. % Ti, care must be taken to avoid excessive hardening, which in turn limits ductility. The failure of an alloy to remain ductile after several heat treatment cycles during drawing is the limiting factor in the fabrication of these conductors. Figure 14 shows data for critical current density $J_c(3.0\ T)$ as a function of percent cold area reduction for different intermediate precipitation heat treatments for Nb 60 wt. % Ti and Nb 65 wt. % Ti. The samples were very easily broken after the peak current density was reached if further cold reduction was made.

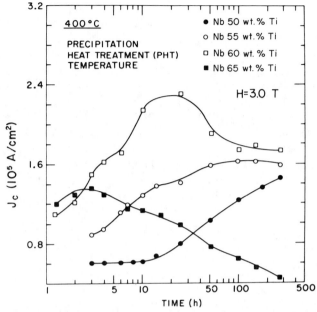

Fig. 13. The critical current density behavior of Nb–Ti as a function of alloy composition is illustrated with the critical current density at 4.2 K in a 3.0-T field obtained for various lengths of time at 400°C for the different compositions shown. The simple picture of the two modes of flux pinning in composition at 300°C seems to be more pronounced in the behavior of the high-titanium sample's superconducting current density actually peaking out.

It is of value to review, condense, and reinforce the points made in this section. If the data from Fig. 9 are considered, then the α phase is a very important flux pinner in the niobium-rich as well as the titanium-rich alloys. Probably the drawback in the Nb-rich alloys is the slow rate of α formation, allowing some annealing out of the dislocation structures and the removing of α-titanium from an already niobium-rich matrix which will decrease the transition temperature, as indicated by Figs. 4 and 5. Cold work has no effect on T_c, as is shown in Fig. 15 for Nb 65 wt. % Ti [16]. These data are consistent with the concept that the dislocation (cell) lines are flux-pinning structures and basically do not change the reversible superconducting properties of the alloy. As evidence for the existence of more than one pinning mechanism, the data from Figs. 16 and 17 may be considered. These figures plot J_c versus H curves for different metallurgical histories of Nb 50 wt. % Ti and Nb 60 wt. % Ti, respectively; in each case, the strongest low-field pinning alloy [greatest $J_c(H)$] was the poorest high-field pinning alloy. These observations and conclusions seem to support the explanations put forth by Ricketts [37], as empha-

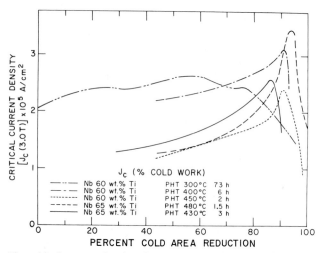

Fig. 14. The critical current density of Nb 60 wt. % Ti and Nb 65 wt. % Ti at 4.2 K in 3.0-T applied field as a function of percentage of cold area reduction for various heat treatments. Note that the lower-temperature heat treatments show very reduced dependence on the percentage of cold area reduction, whereas the higher-temperature treatments seem to start lower but ultimately to obtain higher values.

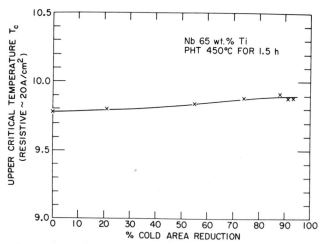

Fig. 15. The critical temperature as measured resistively at a current density of 20 A/cm² as a function of percentage of cold area reduction. This lack of dependence of the critical temperature on the cold work or dislocation structure would support the hypothesis that it is a property of the bulk superconductor only. This conclusion is strongly supported by the data shown in Fig. 4, which gave a strong dependence on the precipitation heat treatment (which means that as the α-titanium forms at the dislocation cell walls, the bulk becomes Nb rich and therefore has a higher critical temperature).

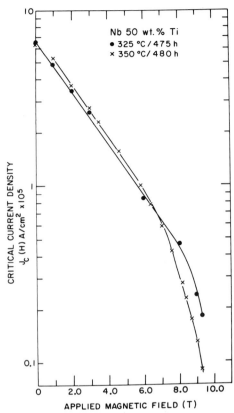

Fig. 16. The critical current density for Nb 50 wt. % Ti as a function of magnetic field is given for two different heat treatments. This clearly shows there are at least two flux pinning mechanisms in this alloy due to the switch in the relative positions of the curves for higher magnetic fields.

sized by plotting the Lorentz force $J_c H$ against H as shown in Fig. 18 [18], namely, α phase is an effective low-field (<5 T) pinner and that cold work yields a better high-field pinning structure. The data in Fig. 19, which gives $J_c(H)$ curves, vividly illustrate the conclusion above that cold area reduction is a stronger high-field pinning mechanism, and that the same alloy with a different metallurgical history can give similar current densities. Figure 19 contains some of the highest J_c reported for Nb–Ti alloys over 5 T applied magnetic field [16, 17]. If a large conductor is required for a specific application, large amounts of cold area reduction are not possible due to the limiting billet size. Thus, the higher-Ti-content alloys would offer a distinct advantage from a critical current performance criterion for a field less than 6.0 T. If the specific application requires fine

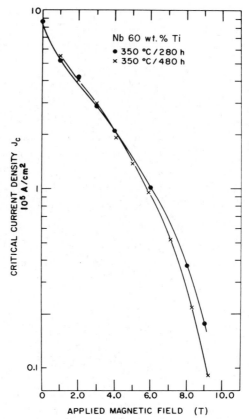

Fig. 17. The critical current density for Nb 60 wt. % Ti is given as a function of the applied magnetic field for two different lengths of time at the same PHT temperature. This change in relative position supports clearly the hypothesis that the precipitates are excellent low-field flux pinners but are not as effective at high field as the dislocation structure.

wire, then the ductility of the higher-Nb-content alloys seems the best answer within the constraints of the performance (J_c versus H) needed.

The highest $J_c(H)$ is produced in samples with metallurgical histories that result in the smallest dislocation cell structures. The generalization that a finer-structure alloy yields a higher current density, and that precipitation heat treatments will further increase the lower-field (<5 T) J_c, establishes the guidelines for future conductor improvement and development.

The addition of Zr to a Nb–Ti alloy will in general lead to a much finer structure [25] and thus to a higher $J_c(H)$. This statement is verified by data

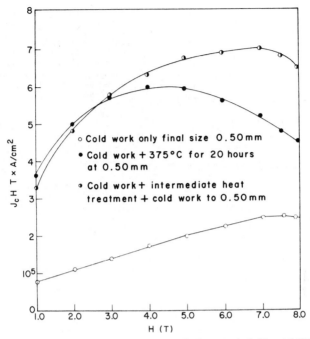

Fig. 18. The Lorentz force is plotted against applied magnetic field at 4.2 K for Nb 50 wt. % Ti, showing the enhancement achieved by a combination of low-field pinning by the precipitates and high-field pinning by the dislocation structure.

obtained for Nb 5 wt. % Zr 60 wt. % Ti by Doi *et al.* [22] where $J_c(5 \text{ T}) = 2.7 \times 10^5$ A/cm². This current density is about 15% higher than the best obtained in the similar composition binary alloy.

The following conclusions can be drawn with respect to the parameters of the ternary alloy of Nb–Ti–Ta and utilizing the data of Suenaga and Ralls [24] as shown in Figs. 5 and 20. A plateau of high upper critical fields (>12 T) occurs in the region of composition near Nb 10 wt. % Ta 40 wt. % Ti and is attributed by the authors to superconducting paramagnetism (Fig. 20). The critical temperature, as shown in Fig. 5, is lowered by the addition of tantalum. The current densities recently reported for the ternary alloy were, in general, lower than those of the binary counterpart [24, 33]. The author, utilizing a newer cold area reduction schedule and PHT map, has been able to obtain $J_c(3.0 \text{ T}) = 3 \times 10^5$ A/cm² and $J_c(5 \text{ T}) = 2.0 \times 10^5$ A/cm² at 4.2 K, which are similar to those of the binary counterpart, but the transition temperature was only 9.1 K. Data at higher fields will have to be obtained before the addition of Ta will prove

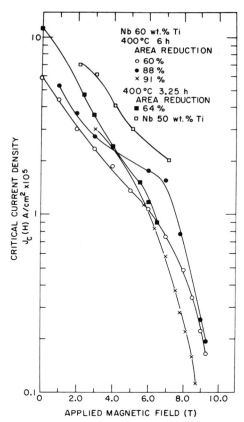

Fig. 19.   This critical current density is plotted versus applied magnetic field for some of the highest-performance critical current Nb–Ti alloys fabricated to date for fields in excess of 5.0 T.

to be an asset or not. It should be noted that one of the major impurities in most niobium ores is tantalum, and its intentional addition may have economic implications.

The interstitial addition of elements such as carbon, nitrogen, oxygen, normally reduces [23] the precipitation heat treatment time required to achieve a given J_c and rapidly harden the alloy. This rapid rate of hardening makes it very difficult to obtain the proper amount of cold area reduction to be useful in most applications. When fine filaments are required, the interstitials have the additional undesirable feature of causing an increase in the effective resistivity of the sample (see the chapter by Dahl, this volume) due to composition change or physical distortion of

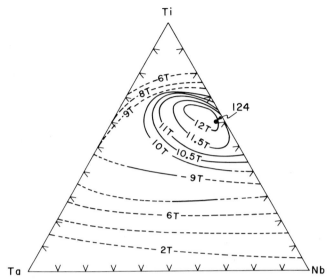

Fig. 20. This is an upper critical field map for the ternary alloy of Nb, Ti, and Ta at 4.2 K measured resistively with a current density of 30 A/cm². (From Suenaga and Ralls [24].)

the array because of increased hardness. This limits the smallest filament size attainable.

The quality of the alloy and its effects on the superconducting properties are primarily determined by the homogeneity and impurity content. If inclusions of the constituents are present (Nb or Ti), they have the effect of locally introducing premature field-dependent voltage well before the critical current is reached for the alloy. The cause of premature voltage can be mechanical in nature, and can include partial fractures, necking down of filaments, or any serious filament array distortion in a multifilament structure. Alternatively, local high Ti or Nb concentrations affecting the dislocation cell structure and/or precipitation rate (different T_c and H_{c2} locally) may change the basic superconducting parameters and give rise to premature voltage.

IV. Theory of Flux Pinning in Niobium–Titanium Alloys

In this section the theory of Hampshire and Taylor [19] will be used because this theory, particularly with the later low-field generalizations by Hampshire [28], is in agreement with a large fraction of the available data.

The number of adjustable parameters, however, is large enough that a reduction of their number or interpretation of their physical meaning as measurable parameters would be a desirable goal.

The theory assumes that flux-pinning centers are the walls of the dislocation cell structure, which due to their high dislocation density and therefore short normal electron mean free path, are regions whose Ginzburg–Landau κ is greater than that of the relatively dislocation-free cells. The cell size, which varies from several thousand down to a few hundred angstroms, is greater than the inter-flux-line spacing d ($d \approx 450$ Å at 1 T) with the exception of very low fields; therefore, several flux lines will intersect each cell. Each pinning center interacts with a volume of the flux-line lattice. Since the pinning arises from a change in κ, $\Delta\kappa$, the pinning force per unit length of a pinned flux line is given by Eq. (52) (see the introductory chapter by Dew Hughes, Section III,B).

$$f_p = \mu\phi_0(H_{c2} - H) \, \Delta\kappa/2.32\kappa^3 d \tag{52}$$

and the total pinning force by Eq. (1-56)

$$F_p = (f_p S_v/1.07)(B/\phi)^{1/2} = S_v H_{c2}^2 b(1 - b) \, \Delta\kappa/2.67\kappa^3 \tag{56}$$

where S_v is the surface area per unit volume of the cell structure correctly oriented for pinning. To a first approximation, S_v is proportional to $1/a$, where a is the cell diameter. This result of the theory is in agreement with the experimental results of Neal *et al.* [20]. Hampshire and Taylor were able to obtain a generally acceptable fit between this theory and their experimental results from a variety of the Nb–Ti samples over a range of applied field and temperatures.

Hampshire subsequently extended the theory to low fields by replacing the expression for the Gibbs function of a type II superconductor in the mixed state, Eq. (49), by the more general expression

$$G_m(H, T) = G_n(H, T) - \left[\frac{\mu_0(H_{c2} - H)^2}{2(2\kappa_2^2 - 1)\beta_a} + \frac{\overline{A}\mu_0}{2}\frac{H_{c2}^2}{\kappa_1^2}\exp\left(-\frac{\overline{B}H}{H_{c2}}\right)\right] \tag{1}$$

where β_a is the condensation energy associated with superfluid. This expression is only strictly valid close to T_c.) The second term on the right-hand side is the Gibbs function for the unit volume of the flux-line lattice. This quantity may be expressed, per unit length of flux line, as

$$g = -\left[\frac{\mu_0(H_{c2} - H)^2}{2(2\kappa_2^2 - 1)\beta_a} + \frac{\overline{A}\mu_0 H_{c2}}{2\kappa_1^2}\exp\left(-\frac{\overline{B}H}{H_{c2}}\right)\right]\frac{\phi_0}{B} \tag{2}$$

The change in g brought about as the flux line moves into a region of different κ, i.e., as κ changes by $\Delta\kappa$, is given by differentiating Eq. (2) with respect to κ. Assuming $\kappa_1 \approx \kappa_2 \approx \kappa$, we have

$$\Delta g = - \left[\frac{B}{2\mu_0} \frac{B_{c2} - B}{\kappa^3} + \frac{\overline{A}\overline{B}B}{B_{c2}\kappa} \exp\left(-\frac{\overline{B}B}{B_{c2}}\right) \mu_0 H_c^2 \right] \Delta\kappa \frac{\phi_0}{B} \qquad (3)$$

where $\overline{A} = 0.036$ and $\overline{B} = 7.35$.

The pinning force per unit length on an individual flux line $f_p = \Delta g/x$ where x is the distance over which the vortex energy changes.

At high fields, near to B_{c2}, $x = d/\pi$, while at low fields $x \approx 3\xi$, when ξ is the coherence length [38]. Utilizing Eq. (1-23), namely, $\xi = (\phi_0/2\pi B_{c2})^{1/2}$, we obtain, for the low-field pinning force per unit length of pinned flux line

$$f_p = \frac{\Delta g}{3\xi} = \frac{B_{c2}^2 \Delta\kappa \phi_0}{6\mu_0\kappa^3 B} \left(\frac{2\pi B_{c2}}{\phi_0}\right)^{1/2} [b(1 - b) + \overline{A}\overline{B}b \exp(-\overline{B}b)] \qquad (4)$$

The total pinning force per unit volume F_p for a material with a cell diameter a is

$$f_p = f_p \frac{W}{da} = \frac{(\pi\sqrt{3})^{1/2}}{6\mu_0} B_{c2}^2 b^{1/2}[(1 - b) + \overline{A}\overline{B} \exp(-\overline{B}b)] \frac{\Delta\kappa}{\kappa^3} \frac{W}{a} \qquad (5)$$

where W is the percentage of the cell wall correctly oriented for pinning.

All the foregoing quantities can be measured directly with the exception of $\Delta\kappa$ and W, and therefore a two-parameter fit can be used. The values obtained for W are plausible, but discrepancies do exist, up to a factor of 3, between high- and low-field values. $\Delta\kappa$ results from the difference in the normal-state resistivity between the relatively dislocation-free cells and the heavily dislocated cell walls [the relationship between κ and the normal-state resistivity has been given in Eq. (24)]. Deformation increases critical current density simultaneously by reducing cell size a and by raising the dislocation density in the cell walls, and thus increasing $\Delta\kappa$. Low-temperature recovery anneals, below 400°C, cause migration of dislocations within the cells to the cell walls, therefore increasing the differences between the cells and cell walls and thus increasing $\Delta\kappa$. Interstitial impurities may also segregate to the cell walls during heat treatment, therefore increasing $\Delta\kappa$. The presence of precipitates, such as α-Ti, carbides, nitrides, or oxides, in the cell walls will also increase $\Delta\kappa$. Precipitates formed prior to cold work will define the nodes of the cell structure. Precipitation as a result of heat treatment after cold work will nucleate on the cell structure. The theory described above can, therefore, explain qualitatively, and when sufficient microstructural information is available, quantitatively, the relationship between current density and composition, deformation and heat treatment described in Section III.

These results can be generalized to the bulk pinning force, given by

$$p_v = f(B/B_{c2}) B_{c2}^n \qquad (6)$$

In order to have a logical progression for various alloy preparations, a simple model for flux pinning is utilized. This model is based on the assumption that metallurgical irregularities in Nb–Ti-based alloy material cause various superconducting regions to exist. A pinning site would consist of separated volumes of superconductor resulting from the presence of such irregularities as grain boundaries, strain fields, different phases, dislocations, or impurities. The theory implies that the bulk superconductor should have as high a critical temperature and upper critical field as possible and that the metallurgical structure introduced to obtain flux pinning be as fine as possible. A maximum limit of this structure is set at a few thousand to a few hundred angstroms by the coherence of the superconductor, although the dislocation structure may act collectively on the flux lattice.

There have been several other models [33] that generally lead to these same conclusions, but the results computed from these models for various parameters do not agree with as wide a range of data.

V.   The Nb–Ti Conductors

A.   *Conductor Design*

Before a tentative production schedule is set up for an alloy, and prior to the conductor fabrication, the following conductor characteristics are established:

(1)   Current-carrying capacity for the superconductor matched to the cost and maximum field capabilities required for space limited magnets.

(2)   A reliable critical current performance for the designed use is usually obtained by setting the size of the filaments and by twisting the conductor.

(3)   Small-hysteresis cycle to minimize energy losses or remanent fields determines the maximum size of filaments and twist pitch of the conductor.

(4) T_c affects both the operating temperature and the winding structures. A high T_c allows for a more tightly closed and compact magnet due to the capability of withstanding higher temperature gradients. T_c is governed generally by the bulk alloy composition.

The conductor stability problem was discussed briefly in the introductory chapter by Dew-Hughes and reviewed in the chapter by Dahl. The theory is given in detail in the papers of Hart [34], Hancox [4], and others [5, 6,

25, 39]. A summary outline of their results as they are applicable here is as follows:

1. CRYOGENIC STABILIZATION

Normal metal is bonded to the superconductor to provide an alternate path for the current, i.e., to shunt the superconductor during the transition to the resistive normal state. The important parameter in this form of stability is the maximum ΔT reached during the shunting process.

2. ENTHALPY STABILIZATION

The heat capacity and/or energy stored in the individual superconductor and its surroundings are such that a normal front may not propagate and flux jumps therefore do not occur. The adiabatic criterion places an upper limit on filament size [5, 7]

$$d \leq \pi(3\bar{d}C_pT_0/2\mu_0)^{1/2}/J_c \tag{7}$$

for Nb–Ti, $d \approx 40~\mu$m, where $\bar{d} \equiv$ density, $T_0 = J_0/(\partial J_c/\partial T) \approx T_c/2$, and C_p is the specific heat per unit weight. The size of the multifilament composite strand is limited by the maximum stable field at the surface of the conductor, which is given by [34]

$$B(\text{surface}) \leq (2\bar{d}\mu_0 C_p T_0)^{1/2} \tag{8}$$

3. DYNAMIC STABILIZATION

The flux flow (catastrophic flux flow, the flux jump phenomenon, is discussed in Section II of the chapter by Dahl) is impeded by the presence of enough high- and intermediate-conductance material, which therefore slows down the growth of magnetic instabilities. The criterion for this form of stability for a ribbon-type conductor, i.e., multifilament strands braided into a ribbon with a metallic filler (see Fig. 21), is given by [6]

$$J^2W^2 \leq C_pT_0(\Delta/\Delta_n)^{1/2}2\pi$$

where W is the width of braid and Δ the diffusivity.

The hysteretic losses in a pulsed conductor are proportional to the size of the basic current-carrying element or to its effective size due to coupling. The coupling of multifilament strands in a composite braid, cable, or wire was first analyzed using the theoretical expressions developed at the Rutherford Laboratory [8]. These expressions have been pub-

BRAIDED ISABELLE CONDUCTOR

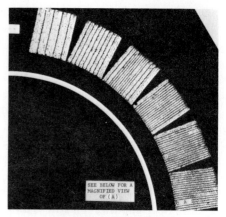

PHOTOGRAPH OF CROSS SECTION OF
QUADRANT OF
THE ISABELLE DIPOLE WINDING

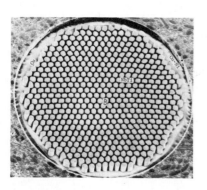

ELECTRON PHOTOMICROGRAPH OF A STRAND
FROM THE ISABELLE BRAIDED CONDUCTOR

PHOTOMICROGRAPH OF BRAIDED CONDUCTOR

Fig. 21. This is a composite photograph showing the ring magnet's braided conductor on the left and an actual photograph of a real magnet cross section (one quadrant). The photomicrographs below show finer degrees of detail with each higher magnification. (This is a proposed dipole for Isabelle [40].)

lished in a more detailed form by Morgan [41]. The critical parameter, that of maximum pitch length of the transposition, is

$$L_c^2 \leq 2\alpha^{1/2}[D/(d + D)J_c\rho d/\dot{B}] \qquad (9)$$

where α is the fraction of material occupied by the superconducting elements in the conductor, D the wire spacing (or filament spacing), d the effective wire diameter (or filament diameter), and ρ the resistivity of the matrix plus filler [e.g., Sn(Ag) or In Tl or Cu]; $\dot{B}$ is the rate of change of the magnetic field.

At this point, it would be of interest to consider the design of a few conductors for practical applications. A typical conductor is shown in Fig. 21 [42]. The early considerations are the ampere turn, physical size specifications, ramp rate $\dot{B}$, and maximum peak field in the winding as needed for the magnet performance. For this particular magnet the conductor specifications are

(1) ampere-turn = 4.3 kA at 5.5 T,
(2) ramp rate $\dot{B}$ = 30.0 mT/sec,
(3) maximum ramp (extraction) = 5 T/sec.

Stability against flux jumping required that the filament of Nb–Ti be ≤ 40 μm in diameter. The actual filament diameter is 8 μm, and the conductor is flux jump stable. The interfilament spacing is 11 μm. The fraction α of Nb–Ti is 0.4, and the maximum J_c is 10^6 A/cm². Therefore L_c, the critical pitch twist length calculated for an energy extraction rate of 5.0 T/sec, is 1.2 cm.

The twist pitch is 0.64 cm. The interstrand spacing is 0.304 mm and the effective strand diameter is 0.27 mm. The critical pitch for the braid strand (assuming 10% side-to-side contact) calculated for a ramp rate of 0.03 T/sec is 14 cm. The braid has a 12.5-cm pitch. (See Sampson *et al.* [43] for physical parameters such as effective resistivity.)

If a conductor is to be cryogenically stabilized using liquid helium as a coolant, a 24-strand cable with the parameters o.d. = 0.18 cm, $R(9\ \text{T})$ = 1.61 $\mu\Omega$/cm, and $K\ \Delta T$ = 0.3 in a 9-T field would have a stable current given by

$$I_{st} \leq K\ \Delta T\ S/R = (0.3 \times 1.04/1.61 \times 10^{-6})^{1/2} = 440\ \text{A} \qquad (10)$$

B. Conductor Fabrication

1. ALLOY PRODUCTION

The following would be a typical production schedule for fabricating Nb–Ti superconducting alloy material:

Step 1. Mix Nb_2O_5 with Al powder.
Step 2. Reduce this mix to Nb metal by thermal reaction.
Step 3. Remelt and electron beam melt to the purity required, for example, C ≪ 200 ppm, N < 100 ppm, O < 500 ppm.
Step 4. Roll to slabs and hydride.

Step 5.   Dehydride to obtain Nb powder.

Step 6.   Reduction of titanium tetrachloride with hot magnesium.

Step 7.   Powder the resulting titanium. "Sponge" to the proper grain size.

Step 8.   Mix Nb and Ti powder (and possibly add the ternary Ta or Zr if they are desired) and compact.

Step 9.   Consumable electrode arc melt at least twice. (If wire is to be fine, electron beam melt the alloy until desired uniformity is reached.)

Step 10.   Extrude Nb–Ti or ternary ingot into a rod.

Step 11.   Analyze and hot-swage the rod to near the size required for multifilament billet assembly. A typical analysis at this stage would be

Nominal:	*Ti*	*C*	*N*	*O*	*H*
	±1 wt. %	<300 ppm	<100 ppm	<400 ppm	<30 ppm

Step 12.   Vacuum anneal rod and water quench; this solution anneal is at 800°C for 4 h. It may be noted from the phase diagram given in Fig. 1 that this annealing heat treatment occurs in the single-phase β (bcc) region. It should be reemphasized that the alloys in this state are very ductile and easily worked. The annealed β phase, however, is very homogeneous, and has poor flux-pinning characteristics. The lower-temperature ($\alpha + \beta$)-phase mixtures are the best for high current densities [10] but tend to harden the alloy, and thereby make cold area reduction more difficult.

Step 13.   Centerless grind the rods.

Step 14.   Sandblast surfaces of rod to clean and cut it into the lengths desired for the billet.

The Nb–Ti has been arc melted, hot forged, sandblasted, pickled, ground, and β-phase annealed. This results in grain size, hardness, and interstitial content suitable to permit codrawing into a composite.

2.   COMPOSITE MULTIFILAMENTARY CONDUCTOR FABRICATION

The next step is to prepare an oxygen-free high-conductivity (OFHC) copper billet by drilling holes to hold the Nb–Ti rods or by taking an OFHC copper billet blank that has had the holes already cast into it. A common mode of billet fabrication is to use Nb–Ti rods assembled into copper tubes with a hexagon-shaped outer surface that are then placed into an outer can. This billet must be carefully cleaned prior to loading it with the Nb–Ti rods.

The billet is assembled into the desired geometry. Billet weights range from 81.6 kg (for a 20.32-cm o.d. billet) to a little less than 227 kg (for a 25.4-cm o.d. billet). After evacuation, a thick Cu nose cone (lid) is electron beam welded onto the billet. Cleanliness is essential during the assembly of the billet to prevent any hard constituents that might cause the Nb–Ti filament array to distort or fracture.

The billet is extruded after being heated to 600°C ($\sim$550°C for the high-titanium alloys to prevent titanium migration into the copper). An area reduction ratio as high as 50:1 is commonly used. The force that is required for the extrusion is given by

$$F = A_0 K \ln(A_0/A) \tag{11}$$

A_0 is the starting diameter and A the final diameter; K is a parameter dependent on billet temperature and constituents. For example, a Cu; Nb–Ti ratio of 3:1 requires a ram force of 1700 tons for a 150-mm-diam billet extruded to 22 mm in diameter ($K = 15$). For the fabrication of a large-cross-section conductor, it is necessary to extrude as large a billet as possible in order to get a fine dislocation cell structure for the flux-pinning optimization. This is especially true for the high-Nb-content alloys.

The extruded rod is now ready for cold area reduction using intermediate annealing (possibly PHT) heat treatments and a possible final PHT, as required by the alloy. In a few alloys, the intermediate annealing heat treatments result in some α-phase precipitation [16, 18]. The annealing and precipitation heat treatments will always be in the $\alpha + \beta$ two-phase region temperature range because the precipitation of nearly pure α hexagonal-close-packed titanium phase gives excellent pinning characteristics [10] (see Fig. 1).

3. HIGH-AMPERAGE CONDUCTOR FABRICATION

High-amperage conductors (HACS) can be divided into the following two categories: (a) a multiplicity of smaller strands of composites (this category is normally used when the magnetization of the overall conductor, i.e., coupling between filaments or strands, is a major consideration); (b) monolithic conductors.

It is in the application of these very-high-current conductors that the excellent mechanical strength of the ductile alloys is a major asset. Nb–Ti alloys that have been prepared for excellent superconducting properties routinely have yield points in excess of 1 GN/m².

It is often suggested by conductor fabricators that they can always supply suspension bridge supports if superconducting magnets or applications lag too far behind conductor production.

a. Stranded Composites. It is obvious from the dimensions of most Nb–Ti filament composites that a high-amperage conductor in the several-kiloampere range, for moderate-size pulsed magnetic field generation, will contain many strands. There are various methods for combining these strands. The two predominant forms utilized in pulsed magnets currently are compacted cables and braids. The equations that define parameters for stability and losses are analogous to those for the composite consisting of filaments in a matrix.

The insulation and separation of individual strands from one another are usually accomplished by one of two methods. One of the methods employed is the separate insulation of each strand with an organic coating such as Formvar with the possibility of a thermoplastic overcoat being added for mechanical rigidity. The HACs fabricated in this manner can be subjected to the highest rates of field change before magnetization coupling will occur. Magnetization coupling is defined as superconducting loop currents that cross from strand to strand, therefore effectively giving the combined width as the dimension in which the magnetization field is energized. These HACs are limited only by the $\dot{B}$ characteristics of the individual strands in that strand-to-strand currents do not exist. This method of insulation, however, gives rise to three problems:

(1)   The current cannot be shunted around a localized normal region, and therefore the weakest link sets the limit on the total HAC performance.

(2)   The heat transfer to the bath is slightly impaired in the case of a single coat and highly impaired in the case of the heavy thermoplastic overcoat.

(3)   The individual strand is not held as rigid mechanically as it would be by a metal filler. The thermoplastic coat helps, but if the forces are excessive it can become a source of heat when microfractures occur in the coating bond.

The second method is to insulate the strands with resistive metallic or intermetallic barriers such as those shown in the photomicrographs of strand cross sections in Fig. 22. The introduction of the metal filler has a major undesirable feature in the introduction of another limit for the maximum rate of change of magnetic field $\dot{B}$ for a given HAC geometry without magnetization coupling. This new limit is usually much lower than that of the individual strands. The advantages of bonding the strands (wires) together with a metallic filler are

(A)   Sensitivity to motion is greatly reduced.

(B)   The ultimate performance is the average or better than the average of that of the individual strands.

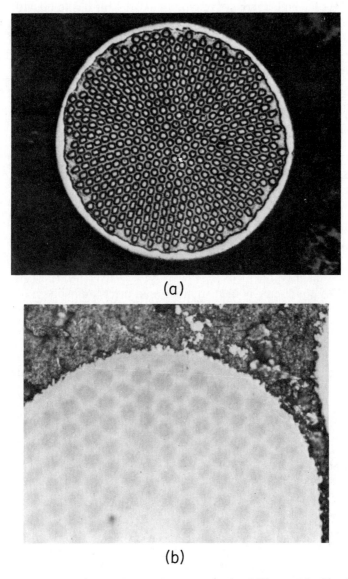

(a)

(b)

Fig. 22. These photomicrographs are of two strands of multifilament Nb–Ti and copper composite that have a metallic and an intermetallic resistance barrier to coupling currents that would normally cross from one strand to another in a multistrand magnet conductor when subjected to modest rates of field change. In (a), the insulation barrier is Cu 10 wt. % Ni. In (b) the insulation barrier is an intermetallic layer of Cu–Sn η phase plus Ag–In(Tl).

(C) A very high degree of reliability exists [this has to do with production of the only magnet coil of large size not to "train" (see the chapter by Dahl, Section II, this volume)].

(D) There is increased heat transfer, and the adiabatic and dynamic stability are enhanced by the presence of the metal filler.

The earlier metallic conductors, however, displayed an onset of magnetization coupling at very low rates of field change, thereby reducing their use to quasi-dc or slow rates $\dot{B} < 2.5$ mT/sec. With suitable metallurgical processing, it is now possible to have rates such as $\dot{B} > 600$ mT/sec [42, 44]. Figure 23 shows the loss per cycle versus magnetic field for various periods for the highest-rate conductor produced by the following process:

(1) Pretinned (Sn 4 wt. % Ag) copper and Nb 65 wt. % Ti composite strands are formed by cabling or braiding into a HAC with as short a transposition length as possible.

(2) The conductor is compacted to the final dimensions required for use in the magnet.

(3) The conductor is heat treated in an inert atmosphere (or micron-pressure vacuum) for 100–200 h at 325°C to form an η-phase tin–copper bronze.

(4) The conductor is filled with an In 10 wt. % Tl alloy at a bath temperature of 225–275°C.

(5) The conductor is resized to the final dimensions needed in the magnetic device.

If the HAC is to be subjected to rapidly changing magnetic fields, then an increase in the strand-to-strand resistance may be desirable. This in-

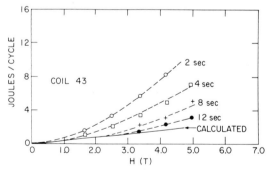

Fig. 23. The energy loss per cycle of a small solenoid is given as a function of the peak central field generated for various ramp periods. The strands were separated by an insulation barrier composed of an intermetallic layer formed from the reaction of the η-phase Cu–Sn bronze and the In(Ag)–Tl filler.

crease in resistance is achieved by heat treating the HAC at 300°C to form a reaction layer. The thickness of this layer, and therefore its resistance, is determined by the length of treatment time at this temperature (hours to ten of hours). The Cu–Sn η phase plus AgIn(Tl) compound forms around each strand in the HAC. The highest rates of change of field with the lowest losses obtained using the foregoing procedure are given in Fig. 23 for a metal insulated braid (660 A/turn at 5.0 T field). A Cu 10 wt. % Ni barrier is possible for moderate rates of field change $\dot{B}$, or rates up to 100 mT/sec. The Cu 10 wt. % Ni jacket is ductile, and thus allows a resizing operation as the final step. The following list summarizes the proposed, past, and currently utilized Nb–Ti multistranded HACs for pulsed magnet applications:

i. Cable (Multilayered, Stranded, Concentric, and Twisted). This form of HAC normally consists of pretinned twisted wires. The advantages of this configuration (when a metal insulator is employed) are the high packing density (ratio of strand volume to conductor HAC volume) and high rate of field change possible with a given strand-to-strand resistance. The high rate of field change is due to the low aspect ratio (width to height) and the very high rates of twisting possible. The drawbacks to this configuration are (a) its relative stiffness to winding around small radii, as required, for example, in the winding of a pole turn of a saddle coil, (b) the tendency of the cable to untwist when going around a turn, and (c) the modest self-field problem due to the fact that it is a two-dimensional transposition rather than a three-dimensional one, such as braid. This is a very good conductor configuration for multilayer structures such as solenoids. Figure 24 shows a photograph of one such conductor fabricated by Lawrence Berkeley Laboratory (3.0 kA/turn at 4.0 T) [45]. The rapid pulsing AC-III and AC-IV dipoles of Rutherford Laboratory's conductors are also shown in Fig. 24 [46].

ii. A Flattened Single Layer Cable (Rutherford Type). A stranded single-layer concentric flattened cable has the name "Rutherford type" [46] due to the extensive development of it by the Rutherford Laboratory superconducting group, Gallagher-Daggett in particular. A photograph of this type of multistranded HAC is shown in Fig. 24. The compacted flattened helical cable is two-dimensionally transposed and has a current capability range of hundreds to thousands of amperes per turn. The aspect ratio of cables produced to date seems to be limited to 12:1 or less, requiring, therefore, a multilayered structure for most dipole-type coil applications [47]. The twist pitch for cables of the multithousand ampere per turn class is limited to about 5 cm at present. The compaction or packing density of cables is limited to 90% and is the best in the multistrand HAC.

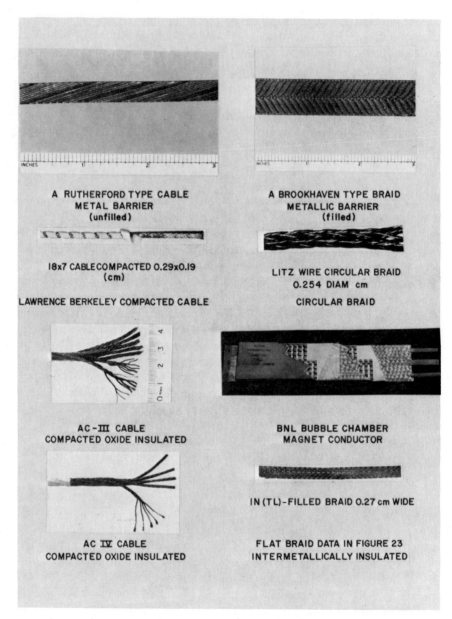

Fig. 24. A composite of photographs of various Nb–Ti-based conductors utilized in magnets that have been constructed in various high-energy physics laboratories.

The drawbacks are the relative stiffness to winding around small radii and the tendency to untwist while doing so. There is a modest self-field present due to the two-dimensional, rather than three-dimensional, transposition.

It has been necessary to fill with metal the pulsed-accelerator-type magnets constructed with the Rutherford-type cables to prevent them from training [47]. Most of the cables made to date exclude plastic or organic coatings, but do not normally fill the entire cable with metal after fabrication. These cables have the distinction of producing some of the highest ramp rate 5.0-T dipoles constructed to date [48]. Figure 25 is a schematic diagram of the sequence for production of "Rutherford" cabled conductor that was used to produce AC-V conductor [45].

iii. Circular Braid. In this configuration a multiplicity of strands is braided into a cylindrical geometric pattern. This configuration will accommodate a copper tube coaxially inside the braided strands, which then can be used to carry the helium refrigerant and thus eliminate the need for a Dewar. The stringent requirement that there may not be strand-to-strand shorts with Formvar (or organic) strand insulation was discovered by the diligent detective work of the Saclay [49] group. They measured unequal magnetization current patterns when strand-to-strand shorts were present. The very asymmetric magnetization currents lead to unstable behavior. This stringent insulation requirement, of course, is removed if the strands are soldered to the tube, but the maximum rate of

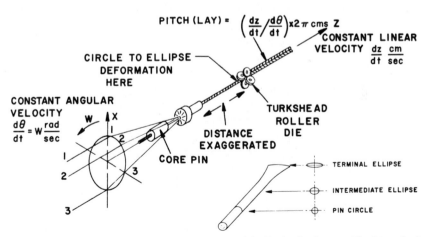

Fig. 25. A schematic drawing of the formation of the Rutherford-type cable shown in the upper right-hand side of Fig. 24. The multifilament composite strands are coming from the left, twisted about the core pin, and forced elliptical until they are rolled rectangular in the turkshead roller die.

field change at which magnetization coupling of the strands will not occur is greatly reduced. Figure 24 contains a photograph of a circular braided conductor.

iv. Flat Braid (BNL Braid). The flat three-dimensionally transposed compacted braided conductor has a current per turn that can range from the hundreds of amperes to multikiloamperes. The reason for this conductor being referred to as BNL type is that extensive development of this conductor has been carried on by the superconducting magnet group at Brookhaven National Laboratory [43]. Photographs of this type of conductor are shown in Figs. 21 and 24. The aspect ratio of the braids produced to date seems to be limited to 30:1; therefore, a single-layer dipole saddle coil that can produce bore fields in excess of 5.0 T is possible [40]. The twist pitch for multikiloampere per turn class conductors is limited to about 10 cm. The compaction of flat braids is limited to 80%. The drawbacks to flat braid include its lower packing fraction and, more important, the low rate of rise of field before magnetization coupling occurs (a consequence of the high aspect ratio) if the braid is metal insulated.

It has been necessary to fill the flat braids with a soft metal to limit or eliminate training of magnets wound with them. There has been at least one "perfect conductor" magnet (no strand-to-strand shorts and properly loaded coil structure) constructed of organically insulated strands with a plastic overcoat [43]. Figure 26 is a composite of photographs of the sequence of manufacture of the BNL braid [50].

b. Monolithic Conductors. The multikiloampere/turn monolithic conductors either with or without resistance barriers normally start the assembly stage with the largest size billet that can be extruded. The reasons are twofold. One reason is to limit the number of joints that must be made in the device and a second is to ensure a fine dislocation structure for the superconducting alloy of Nb–Ti. A typical 25.4-cm-o.d. billet will produce 143 m of 7.6 × 0.85 cm conductor or 430 m of a typical large solenoid conductor (7.5 × 0.32 cm) (see Fig. 24). Usually these monoliths are used in dc magnets (i.e., large solenoids) and thus have a high percentage of copper for they are cryogenically stabilized and therefore have small K (extrusion) values. The high copper percentage allows larger billets to be extruded for a given ram pressure. One of the major problems of the large monoliths is twisting; this is normally accomplished while the cross section is still round. In general, these are rolled to rather high aspect ratios before winding. Usually, a PHT is the last step and is the best procedure because it both anneals the copper and raises J_c at the same time. Figure 24 contains a photograph of a cryogenically stabilized conductor of the 7-ft BNL bubble chamber solenoid.

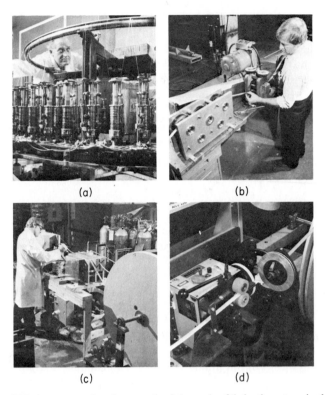

(a) (b)

(c) (d)

Fig. 26. This is a composite photograph of the major fabrication steps in the BNL flat braided conductor manufacture. (a) The strands are braided on a flat circular braiding machine. (b) The braid is cleaned and then rolled to size in the compaction step. (c) The conductor is made more rigid by soft metal filling. The braid is brushed during filling to get rid of excess metal. The conductor is then resized and cleaned before (d) it is helically insulated with glass tape.

If the monolithic conductor is to be used in rapidly changing magnetic fields, resistive barriers between filaments must be constructed to reduce losses. In practice, these magnetization cross-current barriers are made of Cu 10 wt. % Ni. Figure 2.4 is a photomicrograph of a pulsed conductor of this design. If a very complex barrier arrangement is desired, such as the conductor shown in the chapter by Dahl, then multiple extrusions are necessary. If the geometry involves a Nb–Ti filament with a Cu jacket and then a Cu–Ni barrier in a copper composite with a Cu–Ni jacket, the following process might be followed: The Nb–Ti solution-annealed (β-phase) rod is cleaned and placed in an OFHC copper tube that is then codrawn until the tube is in intimate contact with the Nb–Ti rod, or alternatively placed in a Cu tube whose outside shape is hexagonal. The combination of OFHC copper and Nb–Ti rod is placed inside a Cu–Ni sleeve or tube and this grouping is swaged into contact. The three-component (Cu–Ni, Cu, and Nb–Ti) structure is then codrawn to billet loading size and rolled into a hexagonal tube. One of the problems associated with winding this type of monolithic structure is that its relative stiffness makes it difficult, if not impossible, to wind around small radii and extremely difficult to twist. The K value for such a structure is large, limiting the initial billet size. The resulting short piece length necessitates a large number of splices. The Cu–Ni barriers require the joints to be fairly long in order to ensure low losses in the supercurrent transfer.

C. Available Commercial Conductors in 1977

The largest billets being extruded at present are 25.4 cm in diameter and weigh approximately 225 kg. There is the possibility of 30.4-cm-diam billets being available in the near future. The 25.4-cm upper diameter limit will produce 427 m of a large monolithic multikiloampere per turn conductor (7.6 × 0.32 cm) normally utilized in large dc magnets. This means that for a typical solenoid where 10–20 km are needed, a very large number of splices must be made. There are composites of copper and copper with Cu–Ni barriers surrounding Nb–Ti filaments from a few microns in diameter to millimeters available. The smaller-diameter filaments can number from a few to one or two thousand.

Both compacted cabled and braided conductors are commercially available from various firms. For the more modern refrigeration modes, such as forced flow, mixed phase, or high-pressure single-phase helium, there are hollow conductors that allow the helium to pass through an orifice in the middle of the conductor. The device then has the advantage of not needing a Dewar (enclosure for the refrigerant) but only a vacuum vessel.

The composite conductors can have cross sections on the order of square centimeters to square microns. The Nb–Ti filaments can be fabricated with dimensions ranging from centimeters to microns. Matrix ratios of copper to Nb–Ti can vary from 0.8:1 to 100:1.

The other parameter of interest is the twist pitch. In most practical cases, the pitch is limited to 5–10 per centimeter or a 45° angle, according to which is the most stringent condition. The twisting of the conductors is probably one of the most time-consuming steps of fabrication of fine wire.

The conductors whose intended use involves very high rates of change of magnetic fields, requiring therefore a complex Cu–Ni barrier structure, are normally limited to the 20.32-cm billet size because of the large ram force necessary to push the harder Cu–Ni. The 20-cm billet contains about one third the amount of conductor contained in 25.4-cm billets and piece lengths are limited accordingly.

The alloy most commonly used is Nb 44 wt. % Ti but there are some production runs with alloys containing as much as 56 wt. % Ti. Recently, aluminum-stabilized conductor has become available for use in applications where atomic number is important or weight is a consideration.

It is hoped that the foregoing discussions have shown how the remarkable versatility of these Nb–Ti alloys has allowed the magnet fabricator an almost unlimited latitude in design.

References

1. F. P. Levi, *J. Appl. Phys.* **31**, 1469 (1960).
2. R. M. Rose, H. E. Cline, B. P. Strauss, and J. Wulf, *J. Appl. Phys.* **37**, 5 (1966); *Trans. Am. Soc. Met.* **1**, 132 (1966).
3. M. E. Cline, R. M. Rose, and J. Wulf, MIT Rep. No. NASA CR-54103, NASA Contract No. NAS3-2590 (1964) (unpublished).
4. R. Hancox, *Phys. Lett.* **16**, 20 (1965); R. Hancox, Brookhaven National Laboratory Rep. No. 50155 (C-55) (1968) (unpublished).
5. P. F. Chester, *Rep. Prog. Phys.* **30**, 361 (1967).
6. Z. J. J. Steckley, Brookhaven National Laboratory Rep. No. 50155 (C-55), p. 748 (1968).
7. R. Hancox, *Proc. Int. Congr. Low-Temp. Phys., 10th, Moscow* **2**, 43 (1966).
8. J. D. Lewin, P. F. Smith, A. M. Spurway, C. R. Walters, and M. N. Wilson, *J. Phys. D* **3**, 1517 (1970); P. F. Smith, M. N. Wilson, C. R. Walters, and J. D. Lewin, Brookhaven National Laboratory Rep. No. 50155 (C-55), p. 913 (1968).
9. T. G. Berlincourt and R. R. Hake, *Phys. Rev. Lett.* **9**, 293 (1962); *Phys. Rev.* **131**, 140 (1963).
10. J. B. Vetrano and R. W. Boom, *J. Appl. Phys.* **36**, 1179 (1965).
11. Atomics International Investigations of Supercurrent Instabilities in Type II Superconductor, QPR No. 1, Rep. No. NASA-CR-76060 AI-66-87 (1966) (unpublished).
12. L. C. Salter, Jr., Atomics International Investigation of Current Degradation Phenomenon in Superconducting Solenoids, SR Contract No. NAS8-5356, NASA Rep. No. CR 74192-Al-6621 (1966) (unpublished).
13. C. N. Whetstone, G. G. Chase, J. W. Raymond, J. B. Vetrano, R. W. Boom, A. G. Prodell, and H. A. Worwetz, *IEEE Trans. Magn.* **2**, 307 (1966).
14. D. Kramer and C. G. Rhodes, *IEEE Trans. Magn.* **239**, 1612 (1967).
15. C. N. Whetstone and R. W. Boom, *Adv. Cryog. Eng., Proc. 1967 Conf., Stanford Univ., Stanford, California* (K. Timmerhaus, ed.), pp. 68–79. Plenum Press, New York, 1968.
16. A. D. McInturff and G. G. Chase, *J. Appl. Phys.* **44**, 2378 (1973).
17. J. Willbrand and R. Ebrling, *Metall* (*Berlin*) **7**, 677 (1975).
18. P. R. Critchlow, E. Gregory, and B. Zeitlin, *Cryogenics* **11**, 3 (1971).
19. R. G. Hampshire and M. T. Taylor, *J. Phys. F* **2**, 89 (1972).
20. D. F. Neal, A. C. Barber, A. Woocock, and J. S. F. Gridley, *Acta Metall.* **19**, 143 (1971).

21. A. D. McInturff, G. G. Chase, C. N. Whetstone, R. W. Room, H. Brechna, and W. Haldemann, *J. Appl. Phys.* **38**, 524 (1967).
22. Toshio Doi, Private communication, Hitachi Technical Bull. No. 2, Ref. #9950.
23. G. C. Rauch, T. H. Courtney, and J. Wulff, *Trans. Metall. Soc. AIME* **242**, 2263 (1968).
24. M. Suenaga and K. M. Ralls, *J. Appl. Phys.* **40**, 11, 4457 (1969).
25. K. Ishimara, Observation of Fiber Structure and Precipitates in Superconducting Ti–15Nb–5Ta Alloy by Transmission Electron Microscopy TMS Fall Meeting, Cleveland, Ohio (1970).
26. K. Ishimara, Mitsubishi Electric, private communication (1970).
27. C. S. Panda and M. Suenaga, *Appl. Phys. Lett.* **29**, 443 (1976).
28. R. G. Hampshire, *J. Phys. D* **7**, 1847 (1974).
29. A. V. Narlikar and D. Dew Hughes, *Phys. Status Solidi* **6**, 383 (1964).
30. D. Dew Hughes, *Mater. Sci. Eng.* **1**, 2 (1966).
31. D. Saint-James, G. Sarma, and E. J. Thomas, "Type II Superconductivity." Pergamon, Oxford, 1969.
32. A. V. Narlikar and D. Dew Hughes, *J. Mater. Sci.* **1**, 317 (1966).
33. A. D. McInturff and A. Paskin, *J. Appl. Phys.* **40**, 2431 (1969).
34. H. R. Hart, *Proc. Summer Study Supercond. Devices Accel. Brookhaven National Laboratory, Upton, New York,* BNL Rep. No. 50155 (C-55), p. 571 (1968) (unpublished); P. S. Swartz, H. R. Hart, and R. L. Fleischer, *Appl. Phys. Lett.* **4**, 71 (1964).
35. L. W. Berger, D. N. Williams, and R. J. Jaffee, *Trans. Am. Soc. Met.* **50**, 384 (1958).
36. D. Grigsby, Electronic Properties Information Center Rep. No. DS-148(S-1) USAF Materials Laboratory, 1968 (unpublished).
37. R. L. Ricketts, Ph.D. Thesis, MIT, Cambridge Massachusetts (1969) (unpublished).
38. E. H. Brandl, *Phys. Status Solidi* (b) **51**, 345 (1972).
39. S. L. Wipf and M. S. Lubell, *Phys. Lett.* **16**, 103 (1965).
40. 400 × 400 GeV Isabelle Proposal (1977).
41. G. H. Morgan, *J. Appl. Phys.* **41**, 3673 (1970).
42. A. D. McInturff, *Appl. Supercond. Proc.* p. 395, IEEE Publ. No. 72CHO 682-5-TABSC (1972).
43. W. B. Sampson, R. B. Britton, P. F. Dahl, A. D. McInturff, G. H. Morgan, and K. E. Robins, *Part. Accel.* **1**, 173 (1970).
44. A. D. McInturff, P. F. Dahl, and W. B. Sampson, *J. Appl. Phys.* **43**(8), 3546 (1972).
45. W. Gilbert, LBL, private communication (1972).
46. G. E. Gallagher-Daggitt, Superconductor Cables for Pulsed dipole magnets, Rutherford Lab Memorandum No. RHEL/M/A25 (1975) (unpublished).
47. B. Strauss, R. Remsbottom, and A. D. McInturff, *Proc. Isabelle Summer Study* (1975).
48. J. H. Coupland, *Proc. IEEE* **121**(7), 771 (1974).
49. A. Berryger *et al.*, The superconducting pulsed dipole moby, *Int. Conf. Magn. Technol., 4th Brookhaven National Laboratory, Upton, New York* (1972); J. Perot, Group for European Superconducting Synchrotron Studies (GESSS) Rep., Meeting Karlsruhe (1973).
50. A. D. McInturff, Conductor fabrication for Isabelle dipole magnets, *Proc. Tech. Appl. Supercond. Conf., Alushta, USSR* (1975).

Physical Metallurgy of A15 Compounds

DAVID DEW-HUGHES

Department of Energy and Environment
Brookhaven National Laboratory
Upton, New York

I. Introduction 137
II. Superconducting Critical Temperature T_c 142
   A. Experimental Results 142
   B. Phase Diagrams and T_c 144
   C. Impurity Effects and Pseudobinaries 153
   D. Theory of T_c in A15 Compounds 156
III. Upper Critical Field H_{c2} 158
IV. Critical Current Density J_c 162
   A. Microstrucutre and Critical Currents 162
   B. Theories of Flux Pinning in A15 Compounds 164
   References. 167

I. Introduction

Compounds with the A15 structure were first discovered to be super-conducting when Hardy and Hulm [1] found that V_3Si had a superconducting transition temperature (T_c) of 17.1 K. Nb_3Sn, with a T_c of 18.05 K, was discovered by Matthias and his co-workers in the following year [2]; its T_c was subsequently raised to 18.3 K [3]. Matthias's group then showed that the ternary compound $Nb_3(Al_{0.8}Ge_{0.2})$ was superconducting at just over 20 K [4] after suitable heat treatment. Subsequently, the T_c of binary compounds was raised to 18.9 K for Nb_3Al [5], 20.3 K for Nb_3Ga [6], and finally to 23 K for Nb_3Ge [7, 8]. The last figure currently represents the highest T_c known. At least 46 of the 76 known A15 compounds are superconductors. The occurrence of the structure is shown in Fig. 1. Originally, but erroneously, referred to as the "β tungsten" structure, crystallographers now classify it as the Cr_3Si structure. It generally

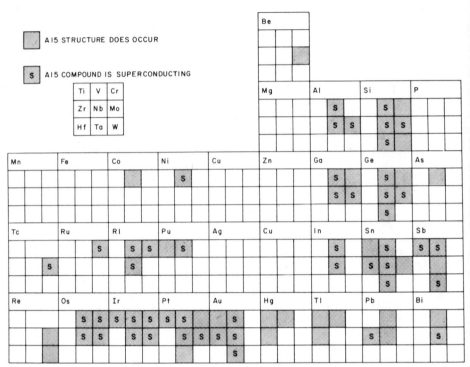

Fig. 1. Occurrence of the A15 (Cr₃Si) crystal structure.

occurs, with few exceptions, close to the A_3B stoichiometric ratio. The A atoms are the group IVA, VA, and VIA transition metals Ti, Zr, V, Nb, Ta, Cr, Mo, and W; the B atoms come mainly from periodic groups IIIB and IVB, and the precious metals Os, Ir, Pt, and Au.

The stability of the A15 structure is governed more by size considerations than by other atomic or electronic parameters. In this regard it is similar to other topologically close-packed structures [9]. It has a primitive cubic cell of eight atoms, and belongs to the space group $O_h{}^3$–$PM3n$. The cubic unit cell (Fig. 2) has six A atoms at $\frac{1}{4}0\frac{1}{2}$, $\frac{1}{2}\frac{1}{4}0$, $0\frac{1}{2}\frac{1}{4}$, $\frac{3}{4}0\frac{1}{2}$, $\frac{1}{2}\frac{3}{4}0$, $0\frac{1}{2}\frac{3}{4}$; and two B atoms at 000, $\frac{1}{2}\frac{1}{2}\frac{1}{2}$. The A atoms have a coordination number of 14; the CN14 polyhedron around each A atom contains two A atoms at a distance $\frac{1}{2}a$, four B atoms at a distance $\frac{5}{4}a$, and eight A atoms at a distance $2r_A = \frac{6}{4}a$. The B atoms have 12 nearest neighbors (CN12) at a distance $r_A + r_B = \frac{5}{4}a$. (a is the lattice parameter and r_A and r_B are the atomic radii in this structure of the A and B atoms, respectively.)

A conspicuous feature of the A15 structure is the chains of A atoms parallel to the three $\langle 100 \rangle$ directions. The interatomic spacing along these chains is 10–15% less than the distance of closest approach in a pure A crystal. These chains are indicated in Fig. 2.

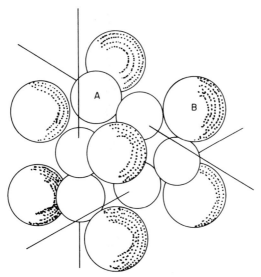

Fig. 2. The structure cell of the A15 crystal structure. The lines indicate the orthogonal A-atom chains.

The A15 structure requires that the radius ratio, based on the Goldsch-midt CN12 radii of the constituent atoms, must be close to unity [10]. Geller [11] derived a set of effective A15 CN12 radii, based on A–B contact distances, that can be used to predict lattice parameters of A15 compounds to within 0.03 Å. The Geller scheme has been revised by Johnson and Douglass [12], and more recently by Tarutani and Kudo [13]. Johnson and Douglass proposed that the A-element radius changes according as the B element is a transition or nontransition metal. Tarutani and Kudo propose that r_A be fixed for a given A element, that r_B, where B is a nontransition element, depends on whether the A element is from group IVA, group VA, or group VIA, and that when B is an fcc transition element, r_B is equal to the elemental metallic radius. Their values of r_A and r_B are given in Table I. Tarutani and Kudo claim that their scheme produces a unique set of lattice parameters. Values of calculated lattice parameters are compared with experimentally determined values in Table II.

Hartsough [14] has examined the stability of A15 phases as a function of various parameters and has summarized criteria for the formation of this phase. He concludes that the fact that A elements come only from transition element groups IVA, VA, and VIA, and B elements come from groups to the right of these in the periodic table up to group VB, reflects the operation of the electrochemical factor that results from the tendency of the A elements to approach a d^5 configuration and the necessity of the B element to supply more than one s,p electron. The tendency of A ele-

TABLE I

ATOMIC RADII[a] FOR ELEMENTS IN A15 COMPOUNDS A_3B
(AFTER TARUTANI AND KUDO [13])

Transition elements				Nontransition elements			
						r_B	
	r_A		r_B		$A(IVA)_3B$	$A(VA)_3B$	$A(VIA)_3B$
Ti	1.416	Re	1.38	Cd		1.503	
Zr	1.560	Ru	1.35	Hg	1.533	1.500	
V	1.331	Os	1.355	Al		1.421	1.330
Nb	1.459	Co	1.253	Ga		1.388	1.323
Ta	1.463	Rh	1.345	In		1.564	
Cr	1.290	Ir	1.357	Si		1.294	1.240
Mo	1.397	Ni	1.246	Ge		1.349	1.304
W	1.409	Pd	1.375	Sn	1.561	1.547	1.532
		Pt	1.387	Pb		1.510	
		Au	1.442	As		1.311	1.295
				Sb	1.563	1.506	
				Bi		1.585	

[a] All radii given in angstrom units.

ments to form the A15 phase is found to decrease in the order V, Nb, Cr, Ti, Mo, Zr, Ta, W, Hf, The radius ratio r_A/r_B, expressed in terms of the Goldschmidt CN12 radii, lies between 0.84 and 1.12, with a most probable value of 0.94. The electron-to-atom ratio (e/a) influences the composition at which the A15 phase is most likely to occur, the most probable values of the ratio being 4.75 or 7.0. These values are, of course, those of the Matthias rules [15] (see the introductory chapter by Dew-Hughes).

The A15 phases usually exist over a range of composition, which may or may not include the A_3B composition. The homogeneity range often includes the stoichiometric composition at high temperature, but falls to the A-rich side at lower temperatures. Several compounds do not form anywhere near to the ideal composition.

Many A15 compounds undergo a martensitic type shear transformation at a temperature above T_c. The new structure is tetragonal with an axial ratio (c/a) very close to 1. The structural instability is heralded by a lattice softening as the temperature is lowered. The elastic shear modulus $\frac{1}{2}(C_{11} - C_{12})$ becomes very small as the transition temperature is approached. The transformation is not seen in all samples of a given material. If the superconducting critical temperature is reached before transformation, $\frac{1}{2}(C_{11} - C_{12})$ remains constant at the value it reached at T_c, and on further cooling no structural transformation occurs. In samples

TABLE II

Comparison of Calculated versus Measured Lattice Parameters[a] of A15 Compounds (Data from Tarutani and Kudo [13])

B element	Ti Calc.	Ti Meas.	Zr Calc.	Zr Meas.	V Calc.	V Meas.	Nb Calc.	Nb Meas.	Ta Calc.	Ta Meas.	Cr Calc.	Cr Meas.	Mo Calc.	Mo Meas.
										A element				
Re													4.986	4.980[b]
Ru														
Os					4.780	4.808[b]	5.131	5.135			4.673	4.683	4.968	4.969
Co					4.681	4.681					4.678	4.678		
Rh					4.777	4.785	5.122	5.120			4.668	4.673		
Ir	5.003	5.000			4.790	4.788	5.133	5.133			4.680	4.678	4.970	4.968
Ni					4.674	4.708								
Pd	5.002	5.055			4.809	4.816								
Pt	5.035	5.033			4.822	4.817	5.157	5.155						
Au	5.092	5.094	5.482	5.482	4.879	4.876	5.202	5.202	5.224	5.224	4.709	4.711	4.991	4.989[b]
Cd					4.943	4.943								
Hg	5.188	5.188	5.559	5.559	4.940	4.94								
Al					4.857	4.836[b]	5.185	5.186					4.950	4.950
Ga					4.823	4.916	5.158	5.165			4.647	4.645	4.945	4.945
In					5.007	5.218[b]	5.303	5.303						
Si					4.724	4.725	5.080	5.16[b]			4.567	4.564	4.886	4.888
Ge					4.782	4.783	5.126	5.139[b]			4.629	4.632	4.932	4.933
Sn	5.217	5.217	5.583	5.65[b]	4.989	4.984	5.289	5.289	5.277	5.277			5.094	5.094
Pb					4.950	4.937	5.258	5.270						
As					4.742	4.743								
Sb	5.220	5.220	5.584	5.634[b]	4.946	4.945	5.255	5.262	5.257	5.259	4.620	4.620		
Bi					5.029	4.72[b]	5.320	5.320						

[a] All values given in Angstrom units.

[b] Measurement made on nonstoichiometric samples.

that do transform, at a temperature $T_m > T_c$, the value of $\frac{1}{2}(C_{11} - C_{12})$ undergoes an arrest at T_m. At lower temperatures it then increases (Nb_3Sn [16]) or continues to decrease at a lesser rate (V_3Si [17]). If the transformation occurs, the c/a ratio gradually changes from 1 as the temperature is lowered below T_m. At T_c, the change ceases and c/a remains fixed. For Nb_3Sn, $T_m \approx 43$ K and c/a ($T < T_c$) ≈ 0.9938–0.9964 [16]. For V_3Si, $T_m \approx 18$–25 K and $c/a \approx 1.0024$ [18]. Martensitic transformations have also been observed in V_3Ga [19], V_3Ge [20], V_3Sn [20], and $Nb_3Al_{0.75}Ge_{0.25}$ [21]. The lattice and electronic properties of A15 compounds have been reviewed by Weger and Goldberg [22], the elastic properties and the low-temperature transformation by Testardi [23].

II. Superconducting Critical Temperature T_c

A. Experimental Results

The critical temperatures of all the known superconducting A15 compounds are listed in Table III. In each case the best-substantiated value is recorded. The T_c of most A15 compounds is strongly influenced by degree of long-range order, composition, and the addition of third elements.

Hanak et al. [25] were the first to suggest that T_c could be affected by long-range crystallographic order (LRO). They were studying T_c as a function of Nb-to-Sn ratio in both vapor-deposited and sintered Nb_3Sn. T_c decreased as the niobium content was increased beyond the 3-to-1 ratio, but fell off more rapidly in vapor-deposited than in sintered samples. It was assumed that there was greater disorder in the vapor-deposited material, with Nb atoms located on the Sn sites and vice versa [26]. Subsequent work [27] suggested that it was most deleterious to have Sn atoms in the Nb sites and that it was important not to disrupt the continuity of chains of Nb atoms along the ⟨100⟩ direction. Further work [28, 29] showed that the reduction in T_c could be accounted for by Sn losses due to volatilization during high-temperature anneals.

An extensive determination of the effect of LRO on T_c of 26 binary A15 compounds was carried out by Blaugher et al. [30]. They showed that for compounds in which the B atom is not a transition element, T_c is maximized when all the A atoms are on A sites, all the B atoms are on B sites, and the LRO parameter S approaches 1. The largest effect is seen in V_3Au, in which increasing S from 0.80 to 0.99 increases T_c by 450% from 0.7 to 3.2 K [31]. When the B atom is a transition element, the compounds do not have the same sensitivity to ordering.

A15 compounds may also be disordered by fast-particle irradiation. These effects are discussed in the chapter by Sweedler, Snead, and Cox in

TABLE III

CRITICAL TEMPERATURE OF ALL A15 COMPOUNDS KNOWN TO BE
SUPERCONDUCTING (DATA FROM DEW-HUGHES [24])

Compound	$T_c(K)$	Compound	$T_c(K)$
Ti_3Ir	4.6	Nb_3Os	0.94
Ti_3Pt	0.49	Nb_3Rh	2.5
Ti_3Sb	5.8	Nb_3Ir	1.76
Zr_3Au	0.92	Nb_3Pt	10
Zr_4Sn	0.92	Nb_3Au	11
Zr_3Pb	0.76	Nb_3Al	18.9
V_3Os	5.15	Nb_3Ga	20.3
V_3Rh	0.38	Nb_3In	8
V_3Ir	1.39	Nb_3Ge	23
V_3Ni	0.57	Nb_3Sn	18.3
V_3Pd	0.08	Nb_3Bi	2.25
V_3Pb	3.7	$Ta_{4.3}Au$	0.58
V_3Au	3.2	Ta_3Ge	8
V_3Al	9.6	Ta_3Sn	6.4
V_3Ga	15.4	Ta_3Sb	0.72
V_3In	13.9	Cr_3Ru	3.43
V_3Si	17.1	Cr_3Os	4.03
V_3Ge	7	Cr_3Rh	0.07
V_3Sn	4.3	Cr_3Ir	0.17
V_3Sb	0.8	Mo_3Os	11.68
		Mo_3Ir	8.1
		Mo_3Pt	4.56
		Mo_3Al	0.58
		Mo_3Ga	0.76
		Mo_3Si	1.3
		Mo_3Ge	1.4
		Mo_2Tc_3	13.5

this volume. They are in entire agreement with the concept that any reduction in LRO, whether by radiation damage, thermal disordering, or composition effects, causes a reduction in T_c. Again the effect is strongest when the B atom is a nontransition element.

The effects of composition upon T_c cannot be unambiguously separated from those of LRO, as it is not possible to have $S = 1$ unless the composition of the compound corresponds to the ideal 3 : 1 ratio. S decreases automatically with any departure from stoichiometry.

Those compounds with a transition element B atom whose T_c is insensitive to the degree of LRO are also less sensitive to composition [30]. These compounds are called, by Muller *et al.* [32], "atypical" since they differ from the typical behavior of nontransition metal compounds. The equivalence of order and composition effects is again apparent. The atypical compounds, which probably include all the compounds based on Cr and Mo, as well as those in which the B element is Os or Ir, in general, have rather low values of T_c. It would appear that T_c is already degraded to such an extent by the B transition element that disorder or compositional deviations can have little further effect.

In view of the strong effect of composition upon T_c, the phase diagrams relevant to the A15 compounds assume great importance. A discussion is now given of the thermodynamic basis of the relevant phase diagrams, followed by the actual diagrams for the six high-T_c compounds, together with experimental results on compositional effects.

B. *Phase Diagrams and* $\mathbf{T}_c$

1. THERMODYNAMICS AND PHASE STABILITY

Only 6 of the 45 superconducting A15 compounds are, on the basis of their high critical temperatures, upper critical fields, and critical current densities, of actual or potential interest as conductors. Of these, V_3Si, V_3Ga, and Nb_3Sn all have a range of homogeneity that includes the A_3B composition. Maximum T_c is readily obtainable in bulk samples of these compounds. Nb_3Al and Nb_3Ga include the ideal composition only at temperatures so high that thermal disorder is excessive. Nb_3Ge does not exist in equilibrium at the stoichiometric ratio. These last three compounds, all with potentially higher values of T_c and H_{c2} than the first three, normally contain excess Nb and their T_c's are, as a consequence, degraded from the optimum values.

The metallurgical equilibrium diagrams for the systems containing the foregoing six phases are given in the following subsections. In the case of V_3Ga and Nb_3Sn, the A15 compound is followed, at increasing B-element concentration, by a series of compounds of much lower melting point. Nb_3Al, Nb_3Ge, and Nb_3Ga are all succeeded by a phase, σ, 5:3, or 3:2, that is more stable than the A15 phase. It is the presence of this stable phase that prevents the A15 phase from forming at the A_3B composition. This can be understood by reference to Fig. 3, in which, at a particular temperature, the Gibbs function ΔG for the Nb solid solution, the A15 phase, and the 5:3 phase is plotted versus composition for a hypothetical Nb–X binary system. At any composition, equilibrium, which corresponds to minimizing ΔG, is established by drawing common tangents to the curves. The points of intersection determine the phase boundaries. If the 5:3 phase is less stable (has a higher ΔG) than the A15 phase, the range of homogeneity of the latter includes the A_3B composition. If the stability of the 5:3 phase is increased (ΔG is lowered) with respect to the A15 phase, the A15-5:3 boundary is pushed to the left, and the A15 range of homo-

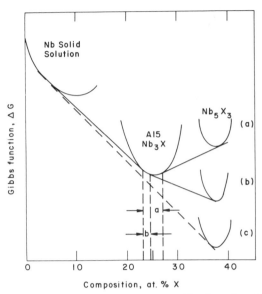

Fig. 3. Variation of Gibbs function ΔG with composition for a hypothetical Nb–X system containing A15 and 5:3 phases, with differing relative stability. (a) A15 more stable then 5:3, range of homogeneity of A15 includes 3:1 ratio. (b) A15 less stable than 5:3, range of homogeneity restricted to high Nb content. (c) A15 so much less stable than 5:3 that its formation is completely suppressed.

geneity is now restricted to Nb-rich compositions. If the 5:3 phase is much more stable than the A15 phase, the latter may disappear altogether.

Studies of the Nb–Al–Ge [33], Nb–Al–Si [34], and Nb–Al–Ga [35] ternary systems show that in each case there is a miscibility gap between the Nb₂Al σ phase and the 5:3 phases. The effect of this miscibility gap in the first two systems is to allow the A15 phase boundary to advance to the stoichiometric composition at the appropriate Al-to-Ge and Al-to-Si ratios. Were the miscibility gap to occur at higher concentrations of Ge or Si, a stable A15 phase with T_c closer to 23 K for sputtered Nb₃Ge (or to whichever value is appropriate for hypothetical stoichiometric Nb₃Si) should be formed. Neither alloying additions nor preparation conditions have yet been found that, in the bulk, allow the achievement of this desirable goal. As yet, the factors that influence phase stability are insufficiently understood allow any progress here. This problem has been previously discussed by the author [24, 36, 37].

2. Nb₃Sn

The niobium tin phase diagram is shown in Fig. 4 [38]. The A15 phase forms from a peritetic at 2130°C, with only about 18 at. % Sn. Below

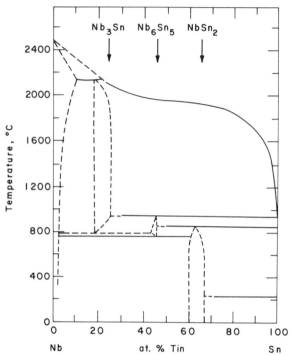

Fig. 4. Niobium–tin phase diagram. (After "Constitution of Binary Alloys" by Shunk [38]. Copyright © 1969, McGraw–Hill Book Company. Used with permission of McGraw-Hill Book Company.)

1800°C the range of homogeneity extends from ~18 to 25.1 at. % Sn, and thus just includes the stoichiometric composition. According to the phase diagram, the A15 phase is unstable below ~775°C. However, as described in the chapter by Luhman, this volume, Nb_3Sn can be formed by solid state reaction from Nb/Cu–Sn composites at temperatures of 700°C and below. Either the phase diagram is in error here, or the presence of copper in the bronze extends the stability range of the A15 phase to lower temperatures, perhaps by lowering the melting points of the Nb_6Sn_5 and $NbSn_2$ phases.

The early work on the effect of composition on T_c has been mentioned in Section A. The effect of excess tin on the superconducting properties cannot be studied since the range of homogeneity of Nb_3Sn does not extend to the tin-rich side of homogeneity.

3. V_3Si

Figure 5 shows the phase diagram for vanadium silicon [39]. The A15 phase forms congruently from the liquid at 1735°C. V_3Si is one of the few examples of an A15 compound that does form congruently from the melt, and is the only A15 material from which large single crystals have been prepared. The range of homogeneity of the phase has a maximum, from 19–25% Si, at 1800°C. Below 1200°C the phase is restricted to 24–25% Si. There is no reported study of the effects of composition on superconducting properties, presumably because of this limited composition range.

4. V_3Ga

The vanadium gallium phase diagram is shown in Fig. 6. This is the diagram given in Shunk [40], with the composition limits of the A15 phase modified by Das et al. [41]. The V_3Ga phase is formed by solid state reaction directly from the bcc solid solution at 1300°C and at a composition corresponding to the stoichiometric formula V_3Ga. The range of homogeneity is a maximum, from 21–31.5 at. % Ga, at 1010°C, and falls to 21–29 at. % Ga at 600°C. This wide range of homogeneity makes V_3Ga an ideal compound in which to study the effects of composition upon T_c. This was first done by Van Vucht et al. [42], who found T_c to have a maximum value at 25. % Ga and to fall off at either side of this composition by ~ 1 K per atomic percent change in composition.

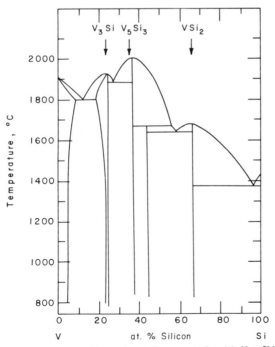

Fig. 5.   Vanadium silicon phase diagram. (After Moffatt [39]).

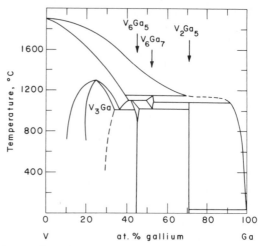

Fig. 6. Vanadium gallium phase diagram, after Shunk [40], with modifications to the A15 phase boundaries. (From Das *et al.* [41].)

The results of the more recent work of Das *et al.* [41] are shown in Fig. 7, where T_c is plotted versus composition. The maximum T_c, 15.3 K, occurs close to 25 at. % Ga. T_c falls off rather more steeply with excess gallium than with excess vanadium. Their variation of lattice parameter a_0 with composition is shown in Fig. 8; a_0 increases with increasing gallium content, the rate of increase being larger as 25 at. % Ga is exceeded.

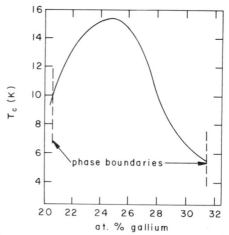

Fig. 7. Superconducting transition temperature T_c versus composition of the A15 compound in the V Ga system (After Das *et al.* [41]. Copyright by American Society for Metals and the Metallurgical Society of AIME 1977.)

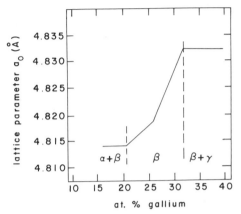

Fig. 8.   Lattice parameter a_0 versus composition of the A15 compound in the V Ga system. (After Das *et al.* [41].)

5.   Nb$_3$Al

The portion of the Nb Al phase diagram relevant to the phase relationships with the A15 phase is shown in Fig. 9, after Sveshnikov *et al.* [43]. Nb$_3$Al is formed by a peritectoid reaction from the bcc solid solution and σ(Nb$_2$Al) phase at 1730°C, and contains 26 at. % Al. The aluminum content of the A15 phase decreases with decreasing temperature; the homogeneity range at 1000°C is 19–22 at. % Al. The work of Moehlecke [44] has largely confirmed this diagram, though he was able to obtain the A15 phase with a maximum of only 24.5 at. % Al at 1730°C.

Moehlecke measured the superconducting critical temperature T_c and the lattice parameter a_0 in the Nb$_3$Al phase as a function of aluminum content. T_c varied from 10.2 K at 20.2 at. % Al to 18.6 K at 24.5 at. % Al, as shown in Fig. 10. Also shown in this figure is the decrease in a_0 from 5.196 to 5.184 Å over the same range of aluminum content.

6.   Nb$_3$Ga

Figure 11, from the work of Jorda *et al.* [45], shows most of the niobium gallium phase diagram. This is the most recent, and most careful, study of this system. Their A15 phase field is more restricted than those of other workers [46]. The diagram is also in conflict with that of Feschotte and Spitz [47], who describe the second intermetallic phase to form as Nb$_3$Ga$_2$, rather than Nb$_5$Ga$_3$ (as will be seen in Section II,B,7, both 5:3 and 3:2 compounds occur in the Nb–Ge system). The A15 phase forms,

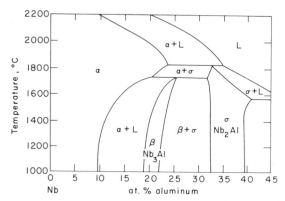

Fig. 9. Portion of the niobium aluminum phase diagram. (After Sveshvikov *et al.* [43].)

with 21 at. % Ga, by a peritectic reaction at 1860°C. The stoichiometric composition, 25 at. % Ga, is just attained at 1740°C. Below this temperature the range of homogeneity narrows rapidly and below 1000°C extends from ~19.7 to 20.6 at. % Ga.

An expanded version of the A15 phase field is shown in Fig. 12a. The critical temperatures of samples corresponding to compositions along the A15–Nb_5Ga_3 phase boundary are shown in Fig. 12b. These vary from 9 K for a specimen corresponding to 20.8 at. % Ga to 18 K for 24.3 at. % Ga. T_c for the latter specimen, quenched from 1740°C and subsequently annealed at <700°C, which allows an increase in LRO without precipitation of Nb_5Ga_3, was 20.7 K. The lattice parameter is plotted as a function of composition in Fig. 13.

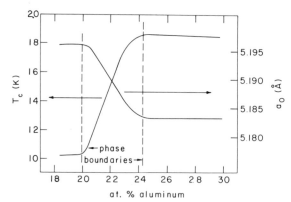

Fig. 10. Superconducting transition temperature T_c and lattice parameter a_0 versus composition of the A15 phase in the Nb–Al system. (After Moehleike [44].)

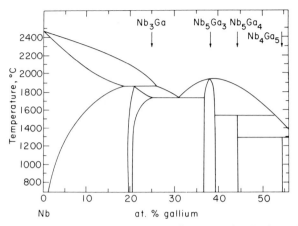

Fig. 11. Part of the niobium gallium phase diagram. (After Jorda *et al.* [45].)

7. Nb₃Ge

The niobium germanium phase diagram is shown in Fig. 14 [48], and the region corresponding to the formation of the A15 phase in greater detail in Fig. 15 [46]. The A15 phase forms, with a composition of 18 at. % Ge, by a peritectic reaction at 1900°C. The maximum germanium content is 22 at. %.at 1865°C. The range of homogeneity below ∼ 1500°C is from 15–18 at. % germanium.

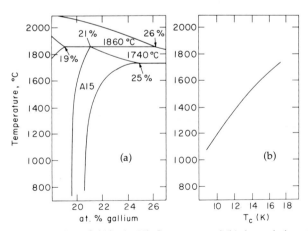

Fig. 12. (a) The A15 phase field in the Nb Ga system and (b) the variation of critical temperature T_c with composition along the Ga-rich phase boundary. (After Flukiger and Jorda [46].)

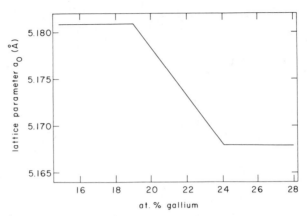

Fig. 13. Lattice parameter a_0 versus composition of the A15 phase in the Nb Ga system. (After Flukiger and Jorda [46]).

Slow solidification of a 25 at. % Ge melt results in a two-phase mixture with $T_c \approx 6$ K [49]. This T_c is presumably that of the A15 phase with 18 at. % Ge. Rapid quenching of the same melt produces a material with a broad superconducting transition extending from 6 to 17 K [50]. The quench retains some A15 phase containing up to 22 at. % Ge, giving rise to the higher T_c. The quenched structure is metastable up to 1000°C; annealing at 1100°C restores the equilibrium, 18 at. % Ge A15 phase with the low T_c.

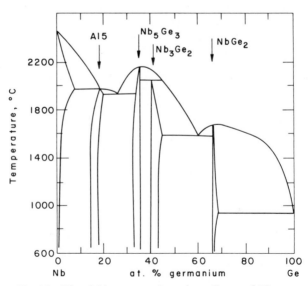

Fig. 14. The niobium germanium phase diagram [48].

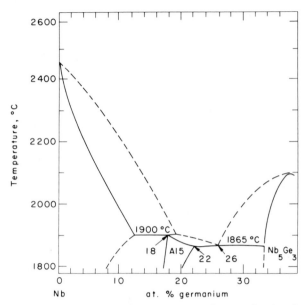

Fig. 15. A portion of the niobium germanium phase diagram showing the formation of the A15 phase by a peritectic reaction. (After Flukiger and Jorda [46].)

It has not been possible to raise T_c above 17–18 K in bulk samples either by quenching or by other means. Metastable stoichiometric, or near stoichiometric, Nb_3Ge can be prepared as thinfilms by sputtering [7, 8], chemical vapor deposition [51, 52], and electron beam deposition [53, 54]. These materials have critical temperatures in the range 21–23 K. Vapor deposition must be equivalent to a quenching rate that is sufficiently rapid to retain the stoichiometric Nb_3Ge phase. Figure 16 shows the relation between T_c and composition for sputtered films [55]. The T_c values range from 9 K for 15 at. % Ge to a projected 23.4 K for stoichiometric material. For a given composition, T_c is higher in thin films than in quenched bulk material. This is because thin films are deposited at substrate temperatures considerably below 1000°C, and the degree of LRO at these lower temperatures more nearly approaches unity.

C. Impurity Effects and Pseudobinaries

A third element added to the A15 structure may (a) substitute for the A element, (b) substitute for the B element, or (c) form a pseudobinary (a mixture of A_3B and $A_3'B'$) with either $A = A'$ or $B = B'$. In the typical A15 compounds, the substitution of an A atom by a third element, even

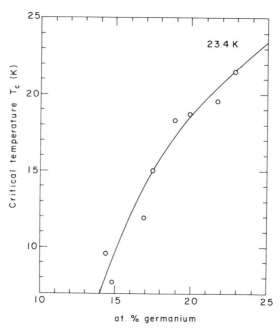

Fig. 16.   The critical temperature T_C versus composition for the sputtered films of Nb_3Ge. (After Rogowski and Roy [55].)

one in the same periodic group, invariably results in a decrease in T_c. Thus, once again, the importance of A-chain integrity in conferring a high T_c on the material is highlighted.

The substitution of some of the B atoms by a third element is usually not so deleterious, and may even result in a slight increase in T_c. Table IV lists examples of an increase in T_c by the substitution of a few percent of the B atoms. The increases in T_c are only a few tenths of a kelvin for a few atomic percent of the third element. The degree of order in the specimens is not reported; the addition of the third element may merely enable a closer approach to perfect stoichiometry or perfect LRO.

Results on pseudobinary systems are of great interest. If the pseudobinary is of the type $A_3B-A_3'B$, then T_c falls from that of either pure binary. When the A atom is the same and the B atoms are of different species, T_c either goes through a minimum or is a linear interpolation of the two values for the pure phases. This behavior is illustrated in Fig. 17 for several V-based A15 pseudobinary systems. V_3Si-V_3Ga has a minimum T_c value, as does V_3Si-V_3Al. V_3As-V_3Al and V_3Sb-V_3Al can both be extrapolated to a T_c of 16.5 K for hypothetical V_3Al, which has not yet been

TABLE IV

<small>SUMMARY OF SUCCESSFUL ADDITIONS OF THIRD ELEMENT TO AN A15 COMPOUND
($A_3B_{1-x}B'_x$) TO RAISE T_c (DATA FROM DEW-HUGHES [24])</small>

Compound	Initial T_c (K)	B' substitution	Amount (x)	New T_c (K)
V_3Ga	14.5	Al	0.1	15.0
V_3As	—[a]	Al	0.7	10
V_3Sb	0.8	Al	0.4	7
Nb_3Al		Si	0.13	19.2
Nb_3Al		As	0.4	19.2
Nb_3Al		Ga	0.2	19.4
Nb_3Al	18.0	Be	0.05	19.6
Nb_3Al	18.0	Cu	0.1	19.0
Nb_3Al	18.0	B	0.1	19.1
Nb_3Al	18.0	Ga	0.2	19.5
Nb_3Sn		Al	0.1	18.6
Nb_3Sn	18.0	Ga	0.05	18.35
Nb_3Sn	18.0	In	0.15	18.3
Nb_3Sn	18.0	Tl	0.1	18.25
Nb_3Sn	18.0	Pb	0.15	18.25
Nb_3Sn	18.0	As	0.05	18.2
Nb_3Sn	18.0	Bi	0.15	18.25

[a] Not superconducting.

produced as a pure A15 phase. V_3Ga–V_3Ge extrapolates to pure V_3Ge at 10 K compared with the experimental value of 6 K for nonstoichiometric $V_{77}Ge_{23}$ [56]. Films of V_3Ge with onset temperatures as high as 11.2 K have recently been prepared by getter sputtering [57]. The central linear portion of the V_3Ge–V_3Al curve extrapolates to stoichiometric V_3Ge at 10 K and to hypothetical V_3Al at 16 K. Thus, pseudobinary studies can be used to determine T_c for phases that either do not normally exist or that do not form at the ideal composition. T_c for ideal Nb_3Au has been estimated in this way [32].

The most interesting ternary system is Nb–Al–Ge since the 18.3-K record held by Nb_3Sn for 13 years was finally exceeded by an alloy in this system. Neither Nb_3Ge nor Nb_3Al can be formed at the stoichiometric composition in the bulk. However, at compositions close to $Nb_3(Al_{0.8}Ge_{0.2})$ stoichiometry can be just achieved at very high temperatures. Subsequent heat treatment produces specimens with critical temperatures close to 21 K [4]. This is the highest T_c yet achieved in a ternary compound.

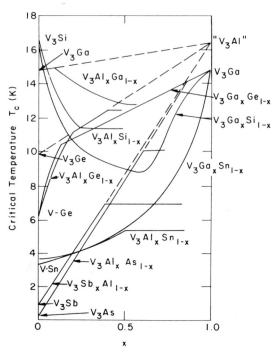

Fig. 17. Critical temperature T_c versus composition for various vanadium-based A15 pseudobinaries [24].

D.   Theory of T_c in A15 Compounds

The Bardeen, Cooper, and Schrieffer (BCS) microscopic theory of superconductivity relates critical temperature to the electron and phonon properties of the superconductor (see the introductory chapter by Dew-Hughes, this volume). Any factor that changes either the electron density of states or the phonon distribution may also be expected to affect the critical temperature. Conversely, any change in the critical temperature presumably results from a change in the electron density of states or the phonon distribution, or both.

Muller *et al.* [32], in their studies on various A15 compounds, investigated the fundamental question whether factors responsible for T_c variations in response to changes in LRO have their origin mainly in the electronic structure or in the phonon spectrum. Specific heats were measured; the lattice specific heats were found to deviate from a Debye function; they were analyzed by superimposing a delta function at an Einstein frequency ω_e on a regular Debye spectrum. Not more than 5% of the change in T_c with LRO could be attributed to changes in this phonon spectrum. On the other hand, there was an excellent correlation between T_c and the temperature coefficient of electron specific heat [58]. It was concluded that the change in T_c as the degree of LRO changes is an electron rather than a phonon effect.

The effects of order and composition, and the difference between transition and nontransition B elements, can be explained by considering the electronic structure of the A15 compounds. Weger was the first to suggest that the A-atom chains would lead to an essentially one-dimensional density of states distribution with a narrow peak in the d band at the Fermi level [59]. Clogston and Jaccarino had previously shown that many of the properties of these compounds were explainable by such a peak [60]. Labbé and Friedel extended these ideas and suggested the density of states curve that is shown schematically in Fig. 18a [61]. The high values of T_c that occur so frequently in compounds with this structure are presumed to result from the very high density of states at the Fermi level.

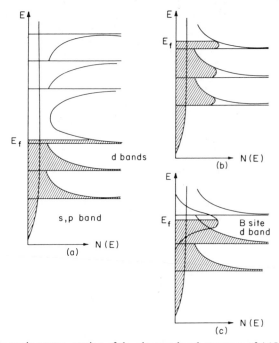

Fig. 18. Schematic representation of the electron band structure of A15 compounds. (a) Labbé–Friedel [61] structure for perfect A15 lattice showing the high density of states peaks in the d band. (b) Rounding off of these peaks due to disruption of A-atom chains (Labbé and Van Reuth [62]). (c) Competing d band introduced by transition metal atoms on B sites (Muller).

The high density of states arises from the A-atom chains and will be seriously affected by any interruption in their continuity. Classical A–B interchange-type crystallographic disorder, in which A and B atoms exchange positions, will cause chain disruption. Labbé and Van Reuth have shown how this leads to a rounding off of the d-band peak with a resultant decrease in $N(0)$ (Fig. 18b) [62]. Departures from the stoichiometric composition in the direction of excess B lead to B atoms taking up positions in the A chains with a similar result. Deviations in the direction of excess A are more commonly found in these compounds. The excess A atoms, positioned on the bcc B sublattice, produce a broader, competing d band, which removes electrons from the narrow A-chain band, and reduces $N(0)$, as shown in Fig. 18c. This model can also be used to explain why compounds in which the B element is a transition metal have generally low values of T_c and are relatively insensitive to the degree of LRO and to composition. The transition metal B atom will similarly form a d band, thereby removing electrons from the A-chain band [32, 56].

III.   Upper Critical Field H_{c2}

The upper critical fields of the high-T_c A15 compounds are some tens of teslas. The variation of H_{c2} with temperature for several A15 superconductors is shown in Fig. 19. Because of the high fields involved, direct measurement of H_{c2} at temperatures well below T_c is difficult, and $H_{c2}(0)$ is often calculated from other data. These compounds have been shown to be "dirty" type II superconductors, in that their BCS coherence length ξ_0 is very much greater than their normal mean free path l [63]. The theories of the upper critical field briefly reviewed in Section II,C of the introductory chapter by Dew-Hughes are applicable here.

The upper critical field at absolute zero $H_{c2}(0)$ is related to the experimentally determinable quantities T_c, γ, and normal state resistivity ρ_n by

$$H_{c2}(0) = 3.11 \times 10^3 \gamma \rho_n T_c \tag{1}$$

T_c is easily measured and γ is obtained from low-temperature specific heat data, but accurate values of ρ_n are not easy to obtain. The actual measurement is quite simple, provided a homogeneous specimen of uniform cross section is available. It is this requirement that is difficult to fulfill; A15 specimens are rarely homogeneous and usually contain traces of a second phase. Due to their brittle nature, it is also difficult to prepare samples of uniform cross section. Measured values of ρ_n are usually an overestimate, and calculated upper critical fields are too large.

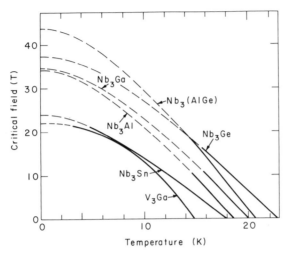

Fig. 19. Variation of upper critical field H_{c2} with temperature for several high-T_c A15 compounds [24].

Low-field, high-temperature (near to T_c) data can be extrapolated to give $H_{c2}(0)$, as follows

$$H_{c2}(0) = -0.693T_c \left[\frac{dH_{c2}(T)}{dT} \right]_{T \approx T_c} \tag{2}$$

$H_{c2}(0)$ values obtained in this way are given in Table V.

Using either Eq. (1) or Eq. (2) ignores the possibility of paramagnetic limitation of the upper critical field. If a compound is strongly paramagnetic in the normal state, then superconductivity is unlikely to exist beyond a field $H_p(0)$ given by

$$H_p(0) = 1.84T_c \tag{3}$$

$H_p(0)$ values are also listed in Table V. For the two vanadium-based compounds V_3Ga and V_3Si, and for the ternary, $Nb_3(Al_{0.7}Ge_{0.3})$, $H_p(0)$ is less than $H_{c2}(0)$, and these compounds are expected to be paramagnetically limited. For the other compounds, $H_p(0)$ is greater than $H_{c2}(0)$ by 2–4 T. In this case, the actual upper critical field should be given by

$$H_{c2}(0)^* = H_{c2}(0)(1 + \alpha^2)^{-1/2} \tag{4}$$

where $\alpha = \sqrt{2}H_{c2}(0)/H_p(0)$, and is also derivable from the temperature variation of H_{c2} near to T_c, since

$$\alpha = -0.533 \left[\frac{dH_{c2}(T)}{dT} \right]_{T \approx T_c} \tag{5}$$

TABLE V

UPPER CRITICAL FIELDS FOR HIGH-T_c A15 COMPOUNDS (DATA TAKEN FROM DEW-HUGHES [24])

Compound	T_c (K)	$-\left(\dfrac{dH_{c2}}{dT}\right)_{T=T_c}$ (T/K)	$H_{c2}(0)$ (T)	$H_p(0)$ (T)	α	$H_{c2}(0)^*$ (T)	$H_{c2}(0)^*_{max}$ (T)	$H_{c2}(4.2)_{obs}$ (T)	$H_{c2}(0)_{expt}$ (T)
V$_3$Ga	14.8	3.4	34.9	27.2	1.82	16.8	19.5	23.6	25.0
V$_3$Si	16.9	2.9	34.0	31.1	1.55	18.4	22.0	23.0	24.0
Nb$_3$Sn	18.0	2.4	29.6	33.1	1.28	18.0	23.4	26.0	28.0
Nb$_3$Al	18.7	2.52	32.7	34.7	1.34	19.6	24.3	29.5	33.0
Nb$_3$Ga	20.2	2.43	34.1	27.2	1.30	20.8	26.2	33.0	34.0
Nb$_3$(Al$_{0.5}$Ga$_{0.5}$)	19.0	2.4	31.6	35.0	1.28	19.3	24.7	31.0	32.5
Nb$_3$Ge	22.5	2.38	37.1	41.3	1.27	23.0	29.3	37.0	38.0
Nb$_3$(Al$_{0.7}$Ge$_{0.3}$)	20.7	3.1	44.5	38.0	1.65	23.1	26.9	41.0	43.5

Substituting this value for α into Eq. (2) yields

$$H_{c2}(0) = 1.3\alpha T_c$$

which, when further substituted into Eq. (4), gives

$$H_{c2}(0)^* = 1.3\alpha(1 + \alpha^2)^{-1/2}T_c \qquad (6)$$

Table V also lists values of α and $H_{c2}(0)^*$. It can be seen that the latter quantity is everywhere too small. It has been pointed out that the factor $\alpha(1 + \alpha^2)^{-1/2}$ can have a maximum value of 1 when α is increased to infinity [64]. Thus, it is possible to define a maximum value of $H_{c2}(0)^*$

$$H_{c2}(0)^*_{max} = 1.3T_c$$

This quantity is also listed in Table V, and though larger than $H_{c2}(0)^*$ is still seen to be too small. The theory of Maki [65] that produced Eq. (4) must be presumed not to hold for the high-T_c A15 compounds.

These calculated values for the various critical fields are compared with experimentally observed values at 4.2 K in Table V, and estimates of $H_{c2}(0)$ are extrapolated from experimental data. The vanadium-based compounds are clearly paramagnetically limited. However, the observed upper critical fields are greater than predicted from theory; the limit is not as severe as expected.

The niobium-based compounds all have observed upper critical fields that agree closely with values of $H_{c2}(0)$ calculated from the temperature variation of the critical field near to T_c [Eq. (2)]. This is reasonable, in that $H_p(0) > H_{c2}(0)$, except for the $Nb_3(Al,Ge)$ ternary compound, which clearly should be paramagnetically limited.

One factor that has been neglected in all of the foregoing is electron spin–flip scattering induced by spin–orbit coupling. The effect of this scattering process is to increase the paramagnetism of the superconducting state and counteract the limiting effect of normal-state paramagnetism. A new paramagnetic critical field can be defined by

$$H_{so}(0) = 1.33\sqrt{\lambda_{so}}H_p(0) \qquad (7)$$

where λ_{so} is the spin–flip scattering frequency parameter. This effect will account for both the less-than-expected paramagnetic limitation in the vanadium-based compounds and the absence of paramagnetic limitation in the $Nb_3(Al,Ge)$ ternary compound. λ_{so} can be estimated from Eq. (7), assuming the observed value of $H_{c2}(0) = H_{so}(0)$. Values of λ_{so} are 0.48 for V_3Ga, 0.34 for V_3Si, and 0.74 for $Nb_3(Al,Ge)$.

Little can be done to raise the upper criticial field of A15 compounds by the manipulation of the normal-state resistivity. Because of their intrinsic brittleness, they cannot be deformed as can the ductile alloys. The addition of alloying elements or impurities in sufficient quantities to have significant effect on ρ_n will almost certainly result in a reduction in T_c. Radiation damage can produce an initial increase in H_{c2} through an increase in ρ_n, but this is ultimately counteracted by the drastic decrease in T_c, as discussed in the chapter by Sweedler, Snead, and Cox, this volume. Examination of Table V suggests that T_c is the most important factor in determining H_{c2} and a significant increase in H_{c2} can only be achieved by an increase in T_c.

IV. Critical Current Density J_c

A. *Microstructure and Critical Currents*

Those defects that pin flux lines and are responsible for the observed high critical current densities in A15 compound superconductors are precipitates, grain boundaries, and radiation-induced defects. The latter are discussed in the chapter by Sweedler, Snead, and Cox, this volume, and will not be dealt with here. Examples of the microstructure of A15 compounds as revealed by transmission electron microscopy are given in the chapter by Pande, this volume.

The effect of refinement of grain size in increasing J_c has been unequivocally demonstrated by many workers; Hanak and Enstrom [66], Caslaw [67], Scanlan *et al.* [68], West and Rawlings [69], and Livingston [70] have varied the grain size of Nb_3Sn by altering growth conditions. The current density increases linearly with decreasing grain size, reaching a maximum value at a grain diameter $a \approx 800$ Å, and then decreases. Livingston [70] has also shown that $J_c \propto a^{-1}$ for V_3Si and V_3Ga, though his grain sizes are not sufficiently small to show the maximum.

The role played by precipitates in raising J_c is ambiguous. The precipitates themselves may act as flux-pinning centers, or they may cause grain refinement, which in turn gives rise to higher J_c values, or both. Nembach and Tachikawa [71] and Nembach [72] have shown that precipitates in V_3Ga and Nb_3Sn refine grain size and increase J_c. J_c is raised when precipitates are introduced into Nb_3Sn. This may be done in diffusion-grown material as ZrO_2 particles in the niobium substrate [73], or as oxides, carbides, or other compounds via the reaction gases in chemically vapor-deposited material [74–78]. Flux pinning in Nb_3Ge prepared by chemical vapor deposition is attributed to precipitates of the Nb_5Ge_3 phase [79, 80]. A quantitative relationship between precipitate size and dispersion and superconducting critical current density has not yet been established. Analysis of the above-mentioned work suggests that maximum flux pinning is obtained with precipitates ~ 50 Å in diameter at a spacing of ~ 200 Å. This is on a finer scale than the most effective grain size, ~ 800 Å.

When results of critical current density J_c versus applied magnetic induction B for Nb_3Sn are plotted as critical Lorentz force $J_cB(=F_p$, the pinning force) versus reduced induction b ($=B/\mu_0H_{c2}$), they generally show a maximum in F_p at 3–7 T ($b \approx 0.2$–0.33) and approach B_{c2} as $(1 - b)^2$. Some results on very-weak-pinning Nb_3Sn show a peak at a higher value of b [81]. J_cB versus b curves for V_3Ga tend to look rather different from those of Nb_3Sn if b is determined by reference to the experimental value of H_{c2}. This is because V_3Ga is strongly paramagnetically limited (see Section III). The value of H_{c2} that occurs in theories of flux pinning is the true, nonlimited upper critical field. When b is recalculated in terms of the true critical field, the peak in F_p versus b curves for V_3Ga falls in the same range as those for Nb_3Sn. Critical current data on other A15 compounds is too limited to determine whether or not this form of behavior is universal to all A15 compounds. Some representative data for Nb_3Sn and V_3Ga are shown in Fig. 20.

Because, as mentioned in Section III, values of H_{c2} are high for these materials and not generally accessible directly, and because

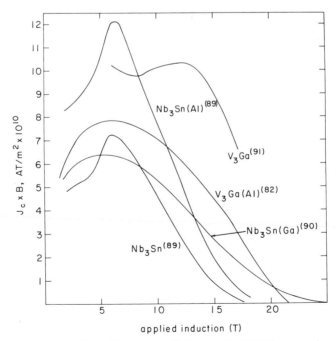

Fig. 20. Critical Lorentz force J_cB versus applied magnetic field H for several samples of Nb_3Sn and V_3Ga grown by the bronze process.

$F_p(b) \propto (1 - b)^2$ as $b \to 1$, it is difficult to make an exact determination of H_{c2} from J_c versus B data alone. This leads to an imprecision in values of b and hence the form of $F_p(b)$ is often not known exactly. However, in the majority of cases for Nb_3Sn and in at least one case for V_3Ga [82] the pinning force can be described by

$$F_p(b) = (\text{const})b^n(1 - b)^2 \qquad (8)$$

where n varies between 0.5 and 1. $n = 0.5$ gives a peak in F_p at $b = 0.2$; for $n = 1$ the peak is at $b = 0.33$. Possible theories to account for these pinning functions are considered in Section IV,B.

B. Theories of Flux Pinning in A15 Compounds

None of the theories of flux pinning described in Section III,B of the introductory chapter by Dew-Hughes provides a really satisfactory explanation of the experimental observations on Nb_3Sn and V_3Ga. Pinning by normal precipitates, ~ 50 Å in diameter, suggests that

$$\begin{aligned} f_p &= \tfrac{1}{2}\mu_0 H_c^2 \pi \xi^2(1 - b) \\ &= (\mu_0 \phi_0 H_{c2}/8 \ \kappa^2)(1 - b) \end{aligned} \qquad (9)$$

If the precipitate spacing L is ~ 200 Å, at inductions greater that 5.75 T, which incidentally is in the vicinity of the maximum in the F_p versus b curve, each precipitate will be able to interact with at least one flux line, and $n_v \approx 1/L^3$. Using direct summation

$$F_p = n_v f_p = \mu_0 \phi_0 H_{c2}(1 - b)/8\kappa^2 L^3 \qquad (10)$$

which clearly does not agree with observation. If the flux-lattice elasticity summation is used, then

$$F_p = \frac{n_v f_p^2}{C} = \frac{\mu_0^2 \phi_0^2 H_{c2}^2(1 - b)^2}{64\kappa^4 L^3 C} \qquad (11)$$

This would give an expression equivalent to Eq. (8) if the appropriate lattice elastic constant C were proportional to either $B^{-1/2}$ or B^{-1}. It is not clear exactly what combination of primary lattice elastic constants should be used for C in this case, but is is unlikely that any reasonable combination would show the required dependence upon B. The field dependence of the elastic constants, and the appropriate constants for different deformation geometries, are discussed by Ullmaier [83].

There are at least three ways in which flux lines might be expected to interact with grain boundaries. The boundary, and a region of adjacent material, will have a shorter electron mean free path, and its value of κ

will be higher than that of the grain interior. It might be thought that the width of the grain boundary, ~3 atomic spacings or 10 Å, would be too small in comparison to the coherence length $\xi(1 \approx 40$ Å, in these materials) to have any effect. However, the effect of scattering from a grain boundary will persist over a distance $\approx$(normal electron mean free path), which is in turn $\approx \xi$, from the boundary. In addition, the grain boundary region in a compound that is normally fully ordered is a region of disorder, and could, irrespective of the change in normal state resistivity, have lower values of T_c and other superconducting parameters as a result of this disorder. To some extent these two effects will be in conflict, and it is not at all clear whether grain boundaries in A15 compounds should be expected to have higher, through increased resistivity, or lower, due to disorder, values of H_{c2}. Nevertheless, it is possible to treat the grain boundary as a sheet of material whose thickness is approximately ξ and whose κ differs by $\Delta \kappa$ from that of the bulk of the material.

A unit length of flux line lying in such a boundary will occupy a volume $\sim \xi^2$, and from Eq. (46) of the introductory chapter by Dew-Hughes the pinning force per unit length will be

$$f_p = \mu_0 H_{c2}^2 \pi \xi \, \Delta \kappa \, (1 - b)/4\kappa^3 \tag{12}$$

Direct summation yields, for the pinning force per unit volume,

$$F_p = f_p \, S_v/d = \mu_0^{3/2} H_{c2}^2 \pi^{\frac{1}{2}} \, \Delta \kappa \, b^{1/2}(1 - b)/4.3\sqrt{2}\kappa^3 \tag{13}$$

where S_v is the surface area of grain boundary per unit volume favorably oriented for pinning. $S_v \propto a^{-1}$, where a is the average grain diameter, which does agree with experiment, up to a diameter ~ 800 Å. This function peaks at $b = 0.33$, also in agreement with much experimental data, but the high-field form of the function is incorrect. Using flux-lattice elastic summation, where the appropriate elastic constant is now $C_{11} \approx \mu_0 H_{c2}^2 b^2$ [83], the pinning force

$$F_p = f_p^2 S_v/C_{11}d = \mu_0 \phi_0^{1/2}\pi \, \Delta \kappa^2 \, (1 - b)^2/34H_{c2}^{1/2}\kappa^6 b^{3/2} \tag{14}$$

Again, this does not agree with experiment.

The second possible interaction arises out of the fact that the superconducting properties of A15 compounds are anisotropic. H_{c2}, and hence κ, varies according to the crystallographic direction along which the magnetic field is directed. H_{c2} along {100} is higher than that along {110}. It has been proposed that the variation in κ can lead to bulk $\Delta \kappa$ pinning [84]. Kramer and Knapp have measured H_{c2} parallel to {100} and {110} in V_3Si at temperatures close to T_c [85]. They extrapolated their results to predict that the anisotropy $\delta = (H_{c2}\{100\} - H_{c2}\{110\})/H_{c2} > 0.16$ at low temperatures. They showed that this effect would be sufficiently large to account for the measured critical current densities in V_3Si. More recently, Foner and McNiff have extended direct measurement of the anistropy in V_3Si down to 4.2 K [86]. They find that $\delta \approx 0.03$ independent of temperature, a value they believe to be too small to account for the observed flux pinning. It should also be mentioned that the V_3Si samples had undergone the martensitic transformation and measured anisotropy is thus that of the tetragonal structure.

The local pinning force per unit length of flux line for $\Delta\kappa$ pinning is given by

$$f_p = \mu_0\phi_0(H_{c2} - H) \, \Delta\kappa/2.32\,\kappa^3 d$$
$$= \phi_0 B_{c2}(1 - b) \, \Delta\kappa/2.32\,\kappa^3 d \qquad (15)$$

Direct summation yields, for the pinning force per unit volume,

$$F_p = f_p\frac{S_v}{d} = \frac{\phi_0 B_{c2}(1 - b) \, \Delta\kappa \, S_v}{2.32\,\kappa^3 d^2} = \frac{B_{c2}^2 b(1 - b) \, \Delta\kappa \, S_v}{2.67\,\kappa^3} \qquad (16)$$

This equation has already been derived for flux pinning by the dislocation cell structure in Nb–Ti alloys [see McInturff, this volume, and Eq. (56) from the introductory chapter by Dew-Hughes in this volume], but does not describe the observed field dependence in A15 compounds. Using flux-lattice elasticity summation (again the appropriate elastic constant is C_{11}) yields

$$F_p = \frac{f_p^2 s_v}{C_{11}} = \frac{\phi_0^{1/2}\mu_0 B_{c2}^{3/2}(1 - b)^2 \, \Delta\kappa \, S_v}{6.6\,\kappa^6 b^{1/2}} \qquad (17)$$

Yet again, there is a discrepancy between this formula and the observed dependence upon b.

The third possible interaction is an elastic one with the elastic strain field of the boundary. Pande and Suenaga have calculated this interaction, assuming the grain boundary to be made up of arrays of dislocations [87]. Their results predict $F_p \propto a^{-1}$ for grain sizes large compared to the inter-flux-line spacing d, but that F_p will decrease for grain diameters less than about five times the inter-flux-line spacing. Critical current density has been found to peak for a grain diameter ~ 800 Å at about 5 T, where $d = 0.25a$ [69], which seems to be in agreement with this prediction, though the field dependence of F_p has not yet been worked out for this theory.

There is often observed a tendency in F_p versus h curves for the peak to shift to lower values of h as the overall pinning force is increased. This led Kramer to propose his theory of flux-lattice shear [88]. It is supposed that, at high fields, flux motion first occurs when the Lorentz force on the flux lattice exceeds the lattice shear strength, allowing some portions of the lattice to move relative to more strongly pinned, stationary, portions. This theory, and a modification of it, is more fully discussed in the chapter by Luhman in this volume in relation to results on bronze-processed Nb_3Sn. The theory does predict that, at high fields, $F_p \propto b^{1/2} (1 - b)^2$, as is observed for many samples of Nb_3Sn. However, if pinning is due to grain boundaries, the critical act is expected to be the passage of a flux line across a boundary. It is not at all clear how lattice shear assists in, and reduces the forces required for, this process.

As with the upper critical field, many of the practices that can be used to enhance J_c in ductile alloys, such as alloying, deformation, and irradiation, cannot be applied to the A15 compounds because of the greater degradation of T_c. The compounds must be prepared in such a manner as to give the smallest possible grain size, or to induce a fine dispersion of a second phase, in order to maximize J_c.

References

1. G. F. Hardy and J. K. Hulm, *Phys. Rev.* **87**, 884 (1953); **93**, 1004 (1954).
2. B. T. Matthias, T. H. Geballe, S. Geller, and E. Corenzwit, *Phys. Rev.* **95**, 1435 (1954).
3. J. J. Hanak, K. Strater, and R. W. Cullen, *RCA Rev.* **25**, 342 (1964).
4. B. T. Matthias *et al., Science* **156**, 645 (1967).
5. P. R. Sahm and T. V. Pruss, *Phys. Lett.* **28A**, 707 (1969).
6. G. W. Webb, L. T. Vieland, R. E. Miller, and A. Wicklund, *Solid State Commun.* **9**, 1769 (1971).
7. J. R. Gavaler, *Appl. Phys. Lett.* **23**, 480 (1973); J. R. Gavaler, M. A. Janocko, and C. K. Jones, *J. Appl. Phys.* **45**, 3009 (1974).
8. L. R. Testardi, J. H. Wernick, and W. A. Royer, *Solid State Commun.* **15**, 1 (1974).
9. A. K. Sinha, *Prog. Mater. Sci.* **15**, 79 (1972).
10. F. Laves, "Theory of Alloy Phases," p. 183. American Society of Metals, Cleveland, Ohio 1956.
11. S. Geller, *Acta Crystallogr.* **9**, 885 (1956); **10**, 380 (1957).
12. G. R. Johnson and D. H. Douglass, *J. Low Temp. Phys.* **14**, 565 (1974).
13. Y. Tarutani and M. Kudo, *J. Less-Common Met.* **55**, 221 (1977).
14. L. D. Hartsough, *J. Phys. Chem. Solids* **35**, 1691 (1974).
15. B. T. Matthias, *Prog. Low Temp. Phys.* **2**, 138 (1957).
16. W. Rehwold, M. Rayl, R. W. Cody, and G. D. Cohen, *Phys. Rev. B* **6**, 363 (1972).
17. L. R. Testardi, W. A. Reed, R. B. Bateman, and V. G. Chirba, *Phys. Rev. Lett.* **15**, 250 (1965).
18. B. W. Batterman and C. S. Barrett, *Phys. Rev.* **145**, 296 (1966).
19. E. Nembach, K. Tachikawa, and S. Takano, *Philos. Mag.* **8**, 869 (1970).
20. E. M. Savitskii, V. V. Baron, Y. V. Efimov, M. I. Bychkova, and L. F. Myzenkova, "Superconducting Materials," p. 115. Plenum Press, New York, 1973.
21. B. N. Kodess, V. B. Kurithzin, and B. V. Tretjakov, *Phys. Lett.* **A37**, 415 (1972).
22. M. Weger and I. B. Goldberg, *Solid State Phys.* **28**, 1 (1973).
23. L. R. Testardi, *Phys. Acoust.* **10**, 194 (1973); *Rev. Mod. Phys.* **47**, 637 (1975).
24. D. Dew-Hughes, *Cryogenics* **15**, 435 (1975).

25. J. J. Hanak, G. D. Cody, P. R. Aron, and H. C. Hitchcock, *in* "High Magnetic Fields" (B. Lax *et al.*, eds.), p. 592. Wiley, New York, 1961.
26. J. J. Hanak, G. D. Cody, J. L. Cooper, and M. Rayl, *Pro. 8th Int. Conf. Low Temperature Physics, London, 1962* (R. O. Davis, ed.), p. 353. Butterworths, London, 1963.
27. T. B. Reed, H. C. Gatos, W. J. LaFleur, and T. J. Roddy, "Metallurgy of Advanced Electronic Materials" (G. E. Brock, ed.), p. 71. Wiley (Interscience), New York, 1963.
28. T. H. Courtney, G. W. Pearsall, and J. Wulff, *Trans. AIME* **232,** 212 (1965); *J. Appl. Phys.* **36,** 3256 (1965).
29. J. F. Bachner and H. C. Gatos, *Trans. AIME* **236,** 1261 (1966).
30. R. D. Blaugher, R. A. Hein, J. E. Cox, and R. M. Waterstrat, *J. Low Temp. Phys.* **1,** 531 (1969).
31. R. A. Hein, J. E. Cox, R. D. Blaugher, and R. M. Waterstrat, *Physica* **55,** 523 (1971).
32. J. Muller, R. Flukiger, A. Junod, F. Heiniger, and C. Susz, *Pro. 13th Int. Conf. Low Temperature Physics, Boulder, Colorado, 1972* (K. D. Timmerhaus *et al.*, eds.) Vol. 3, p. 446. Plenum Press, New York, 1974.
33. A. Müller, *Z. Naturforsch.* **25A,** 1659 (1970).
34. A. Müller, *Z. Naturforsch.* **26A,** 1035 (1976).
35. M. Drys, *J. Less-Common Met.* **44,** 229 (1976).
36. D. Dew-Hughes, Physiques sous champs magnetiques intenses, Colloques Internationaux C.N.R.S., No. 242, p. 44 (1974).
37. D. Dew-Hughes and T. S. Luhman, *J. Mater. Sci.* **13,** 1868 (1978).
38. F. A. Shunk, "Constitution of Binary Alloys," 2nd Suppl., p. 203. McGraw-Hill, New York, 1969.
39. W. G. Moffatt, "Handbook of Binary Phase Diagrams." General Electric Co., Technology Marketing Operation, Schenectady, New York, 1976.
40. F. A. Shunk, "Constitution of Binary Alloys," 2nd Suppl. p. 370. McGraw–Hill, New York, 1969.
41. B. N. Das, J. E. Cox, R. W. Huber, and R. A. Meussner, *Metall. Trans.* **8A,** 541 (1977).
42. J. N. H. Van Vucht, H. A. C. M. Bruning, H. C. Donkersloot, and A. H. Gomes de Mesquita, *Philips Res. Rep.* **19,** 407 (1964).
43. V. N. Sveshnikov, V. M. Pan, and V. I. Latysheva, *Sb. Metallofizi. Fazovye Prevrashcheniya (Phase Transormations)* **22,** 54 (1968).
44. S. Moehlecke, Ph.D. Thesis, Universidad Estudial de Campinas, Brazil (1977).
45. J. L. Jorda, R. Flukiger, and J. Muller, *J. Less-Common Met.* **55,** 249 (1977).
46. R. Flukiger and J. L. Jorda, Phase Diagrams in Metallurgy and Ceramics (G. C. Carter, ed.), Nat. Bur. Stand. Special Publ. No. 496, p. 375 (1978).
47. P. Feschotte and F. L. Spitz, *J. Less-Common Met.* **37,** 233 (1974).
48. E. M. Savitskii, V. V. Baron, Y. V. Efimov, M. I. Bychkova, and L. F. Myzenkova, "Superconducting Materials," p. 286. Plenum Press, New York, 1973.
49. B. T. Matthias, T. H. Geballe, and V. Compton, *Rev. Mod. Phys.* **35,** 1 (1963).
50. B. T. Matthias, T. H. Geballe, R. H. Willens, E. Corenzwit, and G. W. Hull, Jr., *Phys. Rev.* **139,** A1501 (1965).
51. L. R. Newkirk, F. A. Valencia, A. L. Giorgi, E. G. Sklarz, and T. C. Wallace, *IEEE Trans. Magn.* **MAG-11,** 221 (1975).
52. G. H. Roland and A. I. Braginski, *Adv. Cryog. Eng.* **22** (1977).
53. Y. Tarutani, M. Kudo, and S. Taguchi, *Int. Cryogenic Engineering Conf., Tokyo, 5th, 1974,* p. 477. IPC Science and Technology Press, 1974.
54. H. Lutz *et al., Proc. Int. Conf. Low Lying Lattice Vibrational Modes, San Juan, Puerto Rico, 1975, Ferroelectrics* **16,** 259 (1977).
55. D. A. Rogowski and R. Roy, *J. Appl. Phys.* **47,** 4635 (1976).

56. R. Flukiger, Ph.D. Thesis, Université de Genève, Switzerland, 1972.
57. R. E. Somekh and J. E. Evetts, _Solid State Commun._ **24**, 733 (1977).
58. F. Heiniger _et al., Proc. Int. Conf. Low Temperature Physics. Kyoto, 12th, 1970_ (E. Kanda, ed.), p. 331.
59. M. Weger, _Rev. Mod. Phys._ **36**, 175 (1964).
60. A. M. Clogston and V. Jaccarino, _Phys. Rev._ **121**, 1367 (1961).
61. J. Labbé and J. Friedel, _J. Phys._ **27**, 153, 303, 708 (1966).
62. J. Labbé and E. C. Van Reuth, _Phys. Rev. Lett._ **24**, 1232 (1970).
63. K. Hechler, G. Horn, G. Otto, and E. Saur, _J. Low Temp. Phys._ **1**, 29 (1969).
64. D. Dew-Hughes, _Rep. Prog. Phys._ **34**, 821 (1971).
65. K. Maki, _Physics_ **1**, 127, 207 (1964).
66. J. J. Hanak and R. E. Enstrom, _Pro. Int. Conf. Low Temperature Physics, Moscow, 10th, 1966_, Vol. IIB, p. 10, Viniti, Moscow (1966).
67. J. S. Caslaw, _Cryogenics_ **11**, 57 (1971).
68. R. M. Scanlan, W. A. Fietz, and E. F. Koch, _J. Appl. Phys._ **46**, 2244 (1975).
69. A. W. West and R. D. Rawlings, _J. Mater. Sci._ **12**, 1862 (1977).
70. J. D. Livingston, _Phys. Status Solidi (a)_ **44**, 295 (1978).
71. E. Nembach and K. Tachikawa, _J. Less-Common Met._ **19**, 359 (1969).
72. E. Nembach, _Z. Metallkd._ **61**, 734 (1970).
73. M. G. Benz, _Trans. Metall. Soc. AIME_ **242**, 1067 (1968).
74. P. B. Hart, C. Hill, R. Ogden, and C. W. Wilkins, _J. Phys. D Appl. Phys._ **2**, 521 (1969).
75. R. E. Enstrom, J. J. Hanak, and G. W. Cullen, _RCA Rev._ **31**, 702 (1970).
76. G. Ziegler, B. Blos, H. Diepers, and K. Wohlleben, _Z. Angew. Phys._ **31**, 184 (1971).
77. R. E. Enstrom, J. J. Hanak, J. R. Appert, and K. Strater, _J. Electrochem. Soc._ **119**, 743 (1972).
78. R. E. Enstrom and J. R. Appert, _J. Appl. Phys._ **43**, 1915 (1972); **45**, 421 (1974).
79. A. I. Braginski, _et al., IEEE Trans. Magn._ **MAG-13**, 300 (1977).
80. R. V. Carlson, R. J. Barlett, L. R. Newkirk, and F. A. Valencia, _IEEE Trans. Magn._ **MAG-13**, 648 (1977).
81. D. Dew-Hughes, _Philos. Mag._ **30**, 293 (1974).
82. D. Dew-Hughes, _J. Appl. Phys._ **49**, 327 (1978).
83. H. Ullmaier, "Irreversible Properties of Type III Superconductors," pp. 22 et seq. Springer, New York, 1975.
84. A. M. Campbell and J. E. Evetts, _Adv. Phys._ **21**, 199 (1972).
85. E. J. Kramer and G. S. Knapp, _J. Appl. Phys._ **46**, 4595 (1975).
86. S. Foner and E. J. McNiff, Jr., _Appl. Phys. Lett._ **32**, 122 (1978).
87. C. S. Pande and M. Suenaga, _Appl. Phys. Lett._ **29**, 443 (1976).
88. E. J. Kramer, _J. Appl. Phys._ **44**, 1360 (1973).
89. D. Dew-Hughes, _IEEE Trans. Magn._ **MAG-13**, 651 (1977).
90. D. Dew-Hughes and M. Suenaga, _J. Appl. Phys._ **49**, 357 (1978).
91. D. G. Howe and L. S. Weinmann, _IEEE Trans. Magn._ **MAG-13**, 251 (1975).

Superconductivity and Electron Microscopy

C. S. PANDE

Department of Energy and Environment
Brookhaven National Laboratory
Upton, New York

I. Introduction 171
II. Transmission Electron Microscopy 173
  A. Specimen Preparation 173
  B. Liquid Helium Stages 181
  C. Theory of Image Contrast from Lattice Defects 183
III. Electron Microscopy of Niobium and Its Alloys 188
IV. Transmission Electron Microscopy of A15 Superconductors . . . . · . . . 191
  A. The Martensitic Phase Transformation 192
  B. Flux Pinning by Grain Boundaries 194
  C. Stacking Faults 195
  D. Radiation Damage 196
  E. Grain Growth in Multifilamentary Superconductors 198
  F. Electron Diffraction from A15 Compounds 200
  G. Dislocations in A15 Materials 201
V. Observation of Magnetic Flux Lines by Electron Microscopy 203
  A. High-Resolution Bitter Technique 204
  B. Unconventional Techniques 205
  C. Direct Resolution of Flux Lines without Decoration 206
VI. Superconducting Lenses 207
  A. Superconducting Lens Design 209
  B. Advantages and Disadvantages of Superconducting Lenses 209
  C. High-Resolution Electron Microscopy with Superconducting Lenses . . . 213
VII. Conclusion 215
  References 216

I. Introduction

Two mutually related aspects of superconductivity and electron microscopy are briefly considered in this chapter: the application of electron microscopy to the study of fundamental and technological problems connected with superconductivity in general and A15 materials in particular; and the application of superconducting magnet technology to electron microscopy, especially to electron-optical lenses.

Over the past 25 years electron microscopy has developed into one of the most powerful techniques available to materials scientists. It is capable of very high resolution (2–3 Å in modern commercially available instruments). The crystal lattices of many materials can be resolved with these instruments. Alternatively, strains and defects in the lattice can be imaged directly but with a rather poorer resolution (15–20 Å for a dislocation, for example). The microscope can also be used to display electron diffraction information, valuable for determination of crystal structures and for studies of phase transformations such as order–disorder reactions. The recent development of scanning transmission instruments using finely focused electron beams (often as an attachment to the electron microscope) has opened up new possibilities such as the use of energy-dispersive x-ray spectroscopy and electron energy loss spectroscopy to obtain chemical analyses of regions as small as 50–100 Å in diameter. It is certain that electron microscopy will continue to maintain its usefulness as a tool for the investigation of materials.

In the field of superconductivity, however, electron microscopy has played only a minor role in studies of the structure and properties of existing superconductors and in the development of new and improved materials. Most of the published work in this field relates to the ductile superconductors such as Nb, Nb–Zr, and Nb–Ti (see Section III), with less attention paid to the A15 compounds, which have lately begun to replace the ductile materials in technological applications. One possible reason for this disparity is the difficulty of preparing suitable specimens from the A15 compounds. Techniques for specimen preparation, especially for A15 materials, are therefore described in some detail in Section II,A.

Electron microscopical studies of superconducting materials at ambient temperatures can yield useful results, but studies below the superconducting transition temperature require the use of specimen stages cooled by liquid helium. Such stages do exist but are only rarely available commercially. Their design is therefore described in some detail in Section II,B. Theories of image formation in superconductors are very similar to those for normal materials and this topic is therefore treated only very briefly in Section II,C. Many attempts have been made to observe directly or indirectly the magnetic flux lines in type II superconductors because of the role they play in determining critical currents in these materials (see the introductory chapter by Dew-Hughes, this volume). These attempts are described in Section V.

A separate but related topic is the application of superconducting magnet technology to electron microscopy. Superconducting magnet technology has already made some impact in electron microscope lens design. In 1977 an electron microscope using superconducting lenses, comparable in resolution to the best conventional electron microscopes, was developed. Superconducting lenses are therefore briefly treated in Section VI.

It is hoped that the material presented here will be sufficient to enable readers to understand the potential of both electron microscopy and superconducting electron optics. However, a much more thorough understanding in both the fields will be required by anyone wishing to utilize electron microscopy or superconducting optical lenses. The necessary references to selected source material have been provided for this purpose wherever possible.

II. Transmission Electron Microscopy

A. *Specimen Preparation*

In order to be transparent to the electron beam a specimen for electron microscopy must be very thin, less than about 2000 Å for conventional 100-kV electron microscopes. To correlate properties of the bulk specimens with microstructure observed in thin foils it is necessary to be able to thin specimens from relatively thick materials without modifying or destroying the microstructure to be studied. The other possibility is to make the thin films directly (for example, by vacuum evaporation to the desired thickness). This is especially convenient for superconducting materials obtained by coevaporation or sputtering. This section will be concerned only with the techniques for obtaining thin films from the bulk.

In most cases the preparation of thin foils for electron microscopy from bulk occurs in two stages. First, the specimen material is reduced in thickness to 0.1–0.05 mm by techniques such as rolling, grinding, spark erosion, or chemical attack. For many applications it is important that this stage of preparation not introduce mechanical damage. In the author's experience, the most convenient method of initial thinning, for both Nb and its ductile alloys and for A15 compounds, is to mechanically thin the specimen by grinding and then to remove the damaged surface layer chemically, using in many cases hydrofluoric acid. If the specimen thickness is 0.1 mm or more at this stage, no mechanical damage is noticed by electron microscopy.

The final thinning, until recently, was nearly always done by either electrolytic or chemical means. Ion milling procedures have recently been introduced. Ion milling is especially valuable for brittle materials such as A15 compounds. These techniques for final thinning are briefly described below with special reference to Nb, its alloys, and A15 compounds (the brittleness of the last makes them especially difficult to prepare). Further details of the techniques themselves can be obtained from Hirsch *et al.*

[1] and references quoted therein. A book dealing specifically with specimen preparation for electron microscopy is also available [2].

1. ELECTROLYTIC THINNING TECHNIQUES

Electropolishing is most commonly used for final thinning of specimens suitable for electron microscopy. It combines wide applicability with reasonable speed and simplicity. The specimen may initially be polished using a fast electrolyte and subsequently polished with a slower electrolyte. This procedure reduces the need for preliminary mechanical thinning. In all electropolishing techniques (including jet polishing, discussed later) the specimen is introduced into an electrolyte bath as the anode. The cathode is usually of a material not chemically affected by the electrolyte. The best voltage and current for electropolishing depend on the electrolyte, the specimen material, etc., and usually have to be established by trial and error. Table I gives the voltage, electrolytes, and appropriate data for electropolishing niobium and some niobium alloys.

The detailed mechanisms of electropolishing are not yet completely understood. Thinning rarely occurs uniformly over the whole specimen. If the specimen is mounted between flat platelike electrodes, maximum polishing occurs around the edges of the specimens. If the edges are varnished with an insulating material, as is usually the case, the thinning is fastest next to the varnish. Specimens are usually mounted vertically. In this situation thinning is fastest at the top. A viscous layer forms adjacent to the specimen surface and drops downward due to gravity. The opposite occurs if the electrolyte is such that gas formation takes place during electropolishing. Thinning can usually be made more uniform by a gentle stirring. Thin foils suitable for electron microscopy are obtained near areas where perforation occurs. The point where perforation occurs can be controlled to some extent by using special cathode geometry. (For example, in the Bollmann technique this is achieved by using needle-shaped cathodes held close to the specimen surfaces. Perforation occurs near the needle points and the current is then immediately switched off.) In the author's experience, the best results are obtained when the starting specimen is fairly large (~ 0.5 cm^2 or more) and thin (less than 0.1 mm) and the rate of polishing slow. The last condition can usually be achieved by cooling the specimen using liquid nitrogen. A convenient method is to pour small amounts of liquid nitrogen directly into the electrolyte at regular intervals. An important part of specimen preparation is the final washing after electropolishing is complete. Repeated washing in methyl alcohol is a must.

It is common to prepare specimens that can be mounted directly inside

TABLE I

ELECTROPOLISHING TECHNIQUES FOR Nb AND ITS DUCTILE ALLOYS

Material	Technique	Electrolyte[a]	Voltage (V)	Current (A/cm²)	Temperature (°C)	Electrode	Reference
Nb	"Window"	15% HF + 85% HNO_3	8	—	50	Pt or C	3, 4
Nb	—	15% HF + 85% H_2SO_4	8	—			5
Nb	—	17.5% HF 17.5% HNO_3 65% H_2O	4–5	—	40	Pt	6
Nb		10% HF 90% H_2SO_4	12	1.0	0	Stainless steel	7
Nb	PTFE holder	10% HF 90% H_2SO_4	14	6.0		Pt	8
Nb–Zr	—	5% H_2SO_4 2.5% HF 92.5% methanol	25 6.5	1.0	−60	Pt	9
Nb–Ti (wires)		10% Perchloric acid in methanol	6.5		−10	Ti or stainless steel	10

[a] All acids used are in concentrated form. HF strength is 40%. All the polishing electrolytes could usually be used for alloys of Nb also.

175

the electron microscope. A typical size would be a disk 3 mm in diameter and less than 0.1 mm thick. The disk can be punched out or cut by spark erosion and dished in the middle chemically or by spark erosion without perforation. Final polishing can be done by either electropolishing or chemical polishing (see below). The appearance of a perforation is usually observed by putting a light source behind the specimen. This is the method preferred by the author for thinning Nb and its ductile alloys [14].

When a specimen is relatively thick to start with, the electropolishing process just described can be quite time-consuming and tedious. In such cases one can either dish the samples by spark erosion or use what is known as the "jet" technique. In this technique, a jet of electrolyte from a nozzle impinges on the specimen either from one or from both sides. The nozzle also acts as the cathode of the electrolytic cell. Polishing proceeds until perforation or final thinning can be done in the standard electropolishing bath described earlier.

The jet technique is now widely used for specimen preparation and can be performed in highly automated, commercially available machines. The technique is used for specimens that have already been machined to a standard size suitable for the electron microscope specimen holder. When a small hole appears, the polishing current must be switched off very quickly. This is usually accomplished by means of an electronic sensing circuit employing a light source and photocell. The jet technique is not really suitable for brittle materials like A15 compounds because the flow of electrolyte may damage thin areas. Even for superconductors like Nb it has rarely been used. For Nb and its alloys jet polishing can be used under the following conditions: electrolyte 2% HF, 5% H_2SO_4, and 93% methanol; applied voltage, 300 V; current density, 10 A/cm^2; temperature of the electrolyte, $-60°C$ [11].

In spite of many technical improvements, electropolishing remains somewhat of an art, with successful results depending on painstaking attention to detail. For niobium and its alloys electropolishing is especially tricky; it will often be preferable to use ion milling or chemical techniques for the final thinning.

2. Hydrogen-Free Electropolishing

During the process of electropolishing an intake of hydrogen from the electrolyte by the specimen may take place. This is in many cases undesirable. In superconductors, for example, an uptake of hydrogen may result in changes in the superconducting transition temperature. For such specimens a new electrolyte has been introduced by Schober and Sorajic [12]. The electrolyte is 0.05 mole/liter $Mg(ClO_4)_2$ in CH_3OH. The electropolishing voltage is 50–70 V, bath temperature −5°C, and total current ~90 mA for a specimen of Nb in the shape of a 3-mm-diam disk 25 μm thick. Since the electrolyte does not contain a hydrogen radical, no intake of hydrogen is possible.

3. Chemical Thinning Techniques

Chemical thinning is probably less satisfactory than electropolishing simply because it is not possible to stop dissolution quickly when a hole appears. However, this technique is fairly simple and is useful for insulators and other materials for which suitable electropolishing conditions cannot be found. The basic approach is similar to the window technique described previously. Starting with 3-mm disks, one would protect the edges of the disks with varnish and expose the center of the disks to the thinning solution until a hole appeared. In some cases the edges of the hole would then be painted over and the process repeated until a suitable specimen were obtained. Varnish is washed off in acetone. A variation of this technique has been used by the author. Disks 3 mm in diameter and less than 0.1 mm thick, suitable for mounting directly in the electron microscope, are initially obtained by mechanical grinding or spark erosion. The disks are then dished in the middle on both sides by spark erosion (or chemical means using HF and a toothpick). These disks (without protecting the rim with varnish) are then polished chemically using HF in a shallow dish until perforation is detected. Polishing is stopped by pouring methyl alcohol in the polishing solution when perforation occurs. Excellent thin areas are obtained by this technique. For this technique to work successfully it is necessary that the polishing time not be more than about 20 but not less than 5 min. Table II contains a list of chemical polishing solutions that have been used for Nb.

TABLE II

Chemical Polishing Techniques for Nb and Its Alloys

Material	Technique	Initial or final	Solution	Investigators
Nb	Immersion	Initial	30% HF (40% strength) 70% HNO_3	Keh and Weissman [6]
Nb	Immersion	Final	40% HF (40% strength) 60% HNO_3	Van Torne and Thomas [13]
Nb, Nb–Zr	Immersion	Initial	HF (40% strength)	Pande [14]

4. Ion Beam Milling

The ion micromilling technique has been used extensively in recent years for thinning brittle materials such as ceramics, metal oxides, geological specimens, and even lunar rocks. Ion micromilling is indispensible for thinning A15 compounds such as Nb_3Sn. In chemical and electrolytic methods thin sections of these brittle compounds are easily broken by mechanical motion during either the thinning or washing process.

Ion milling equipment is essentially a device for thinning samples by bombardment with inert heavy ions (see Fig. 1). Argon ions are usually used. An ~3-mm-diameter sample is prethinned by spark erosion, mechanical polishing, or chemical etching methods to ≤0.1 mm. Ion milling can then be used to prepare the final transmission specimen. The ion milling chamber containing the holder is evacuated and dry argon gas is introduced through needle valves to a pressure of about 5×10^{-5} mm. The gas is ionized between the anode–cathode gaps with a typical potential of 4–6 kV. A jet of ionized gas ~1 mm in diameter and violet in color impinges onto the specimen at a glancing angle. An ion current of 50–100 μA is used and can be varied by varying the argon pressure or the accelerating voltage. In order to obtain uniform milling the sample is rotated at about 30 rpm about an axis perpendicular to its flat surface. There are usually two independent guns irradiating each face of the specimen. The specimens are sufficiently thin for microscopy when perforation occurs, and this can be detected by observation with a microscope or by using a current-detecting system that periodically checks for perforation and switches off the accelerating voltage when perforation is achieved. Unlike electropolishing or chemical thinning techniques, ion milling at an obtuse

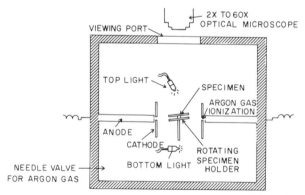

Fig. 1. Schematic diagram of the ion milling chamber.

angle does not require the thinning process to be stopped immediately following specimen perforation.

The parameters involved in the ion micromilling process are (1) accelerating voltage, (2) ion current, and (3) angle of the specimen surface with respect to the direction of ion impingement. The accelerating voltage must be kept at a lower value than that which produces lattice defects in the specimen. For Nb_3Sn and Nb, 5 kV has been found suitable [14]. Too low an accelerating voltage will result in prohibitively long milling time. The rate of milling is roughly proportional to the ion current for small ion currents (less than 100 μA) but rises rapidly at higher ion currents. Thinning is found to be better in the linear range. An ion current of 50–100 μA was found to be suitable for both Nb_3Sn and Nb. Ion currents at these values usually raise the temperature of the specimen to about 100°C [14]. As expected, the milling rates increase with increasing specimen angle and temperature. By experience, an angle of 10–15° was found to give the best specimens, i.e., those with large electron-transparent areas. Typical preparation time for Nb_3Sn specimens ~0.05 mm thick is about 20 h with an ion current of ~100 μA, an ion accelerating voltage of 5 kV, and an angle of 12° [14].

In summary, ion milling is very well suited for preparing brittle materials such as A15 compounds, but care must be taken to minimize radiation damage and surface roughness. It should also be noted that the surface regions of an ion-milled specimen may suffer some loss of crystallinity. Figure 2a shows an example of small, disordered regions produced by radiation damage in Nb_3Sn when the specimen was thinned with argon ions at 6 kV [15]. Figure 2b shows the sort of surface feature that can be obtained if milling is carried out at too high an angle of incidence. For further details concerning the ion milling technique the reader is referred to a review article by Barber [16].

5. DETAILS OF THINNING TECHNIQUES FOR A15 COMPOUNDS

As stated earlier, the micromilling technique is perhaps most suitable for A15 compounds. However, if one wants to avoid the radiation damage introduced in the ion milling, or if the ion milling instruments are not available, the following methods can be used.

Nb_3Sn: The specimen, in the form of a 3-mm-diam disk, is mounted at the center of a Teflon disk and etched in a solution of H_2SO_4 (5 parts by volume), HNO_3 (2 parts), and HF (2 parts). This solution ages while standing in air, with consequent reduction in thinning rate. Thus, the thinning rate can be adjusted by controlling the relative proportion of fresh and aged solution. Thinning is continued until perforation occurs and the specimen should then be cleaned in acetone [17]. The success rate for polycrystalline material is typically 5–10% [14].

V_3Ga: This material can be thinned electrolytically in a mixture of ethanol (100 ml), sulfuric acid (3 ml), and HF or HCl (56%, 5 ml) at $-55°C$ and 55 V [18].

V_3Si: A suitably sized disk is first indented at the center by spark erosion and then thinned electrolytically, at 6–16 V, in a solution containing phosphoric acid (50 ml), acetic acid (50 ml), and hydrofluoric acid (15 ml) using a graphite cathode. Best results are obtained by switching off the polishing current just before perforation so that thinning can be completed by the slower chemical attack [19].

Thinning conditions for other superconductors are given in Tables I and II. Many of the solutions recommended make use of hydrofluoric acid.

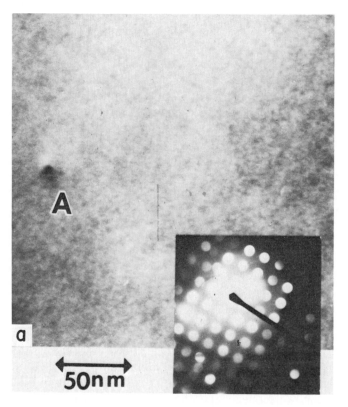

Fig. 2a. Artifacts introduced by ion milling. Disordered regions of size ~10 Å.

Fig. 2b.   Precipitate at A is not due to ion milling. Surface features (From Pande [14].)

This is a highly corrosive substance that should be handled with great care. HF burns should be treated immediately with a saturated paste of magnesium sulfate, a container of which should always be at hand.

Having discussed various techniques for obtaining thin superconducting specimens, we shall describe special stages for observing these specimens at liquid helium temperatures.

B.   Liquid Helium Stages

Development of a liquid helium stage capable of providing, at liquid helium temperatures, all the manipulations possible in a modern conventional electron microscope has progressed rapidly in recent years. This work was stimulated partly in the hope of observing superconductivity phenomena, and partly to study phase transformations at low temperatures. Liquid helium stages have been designed for scanning microscopes and higher-voltage electron microscopes as well as for conventional transmission microscopes. The main problem with such stages is that of vibration produced by circulation of the coolant. This vibration limits the resolution of the system to about 20 Å, compared to about 2–5 Å for a conventional stage in a 100-kV electron microscope. The angle by which the specimen can be tilted in such stages is also relatively small ($\sim \pm 10°$). Temperature around the specimen is typically ~ 10 K, though lower temperatures have sometimes been achieved. It appears from the work of Valdrè and Goringe and others [20–22] that many initial problems with liquid helium microscope stages have now been solved. Hobbs [23], for example, has obtained micrographs consistently showing resolutions of ~ 25 Å with exposures of 15 sec at a temperature of 10 K and utilizing the full $\pm 20°$ tilt available. The stage has been operated for over an hour at a time with liquid helium consumption at 1000 ml an hour. A stage similar to that designed by Hobbs is now commercially available.† Details concerning the construction of helium stages will not be given here, but are available in Hobbs's thesis [24] and in the literature [25–27]. Other designs are available. A design suitable for Phillips 100-kV microscopes is given by Chlebek and Curzon [28] and is shown in Fig. 3.

For the study of magnetic properties of superconductors, it is necessary

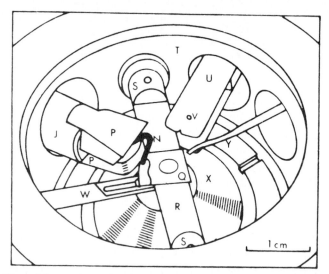

Fig. 3 View of the specimen chamber in a Phillips 300 microscope with microscope column split at the objective lens level showing the liquid helium stage. Copper conductors P are soldered to tube J, which is at 77 K. Specimen holder Q is clamped around a piece of mica R screwed onto Celeron posts S. These in turn are glued into the object-positioning ring T located inside the stage. Copper wires N are soldered to specimen holder Q at one end and to copper end plug on the other side maintained at liquid helium temperature. U is the cooling blade, in withdrawn position, V the hole for the electrons, X the pole piece, Y the gas inlet, and W the objective aperture holder. (From Chlebek and Curzon [28], copyright The Institute of Physics, London.)

† From Oxford Instruments, England.

to apply a suitable magnetic field to the specimen. To some extent the objective lens of the electron microscope can be used for this purpose. By placing the specimen at different angles to the horizontal plane and altering its height by means of a lifting stage, the magnetic field acting on it can be changed almost independently of the excitation of the lens. It is also possible to obtain fields up to 0.6 T using three pairs of Helmholtz coils, the central one being movable horizontally relative to the other pairs [29, 30].

In summary, specimen holders capable of cooling the specimens to about 10 K do exist. However, much work remains to be done before liquid helium stages combining high resolution, tilt, and magnetic field capability become available.

Whether a liquid helium stage or a conventional stage is used for observing specimens of superconductors, certain basic imaging techniques have to be followed. These techniques are not special to superconductors, but are common to all materials. In most cases it is the lattice defects that are studied by these techniques. The theory of image contrast in electron microscopy from these lattice defects will now be discussed.

C. Theory of Image Contrast from Lattice Defects

Lattice defects play an important role in determining irreversible superconducting properties. Lattice defects pin fluxoids and hence influence the critical current density. (For details see the chapters by Dew-Hughes, by McInturff, and by Luhman, this volume.) In transmission electron microscopy (TEM) five main mechanisms of image formation can be utilized to observe and analyze lattice defects. They are

(1)   direct resolution of the crystal lattice,
(2)   moiré patterns,
(3)   out-of-focus contrast,
(4)   absorption contrast, and
(5)   diffraction contrast.

1. LATTICE RESOLUTION

The best electron microscopes have resolution limits in the range of 2–3 Å and are thus able to resolve the crystal lattices of many materials. An example is given in Fig. 4, which shows the first direct observation of 110 lattice planes in an A15 compound (Nb_3Sn) [31]. Lattice resolution clearly makes the study of lattice defects possible, but for this purpose it has a number of serious difficulties:

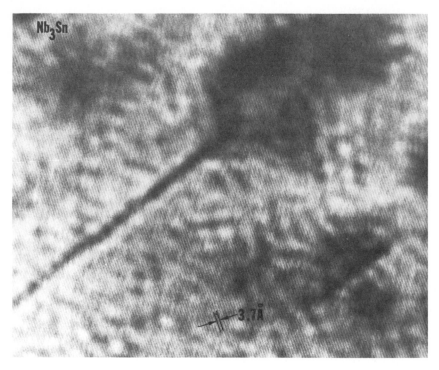

Fig. 4. Direct lattice image of 110 planes in Nb_3Sn. (D. R. Clarke and C. S. Pande, unpublished work.)

(i)   The specimen thickness is even more limited than for conventional electron microscopy, perhaps to about 300 Å.

(ii)   Interpretation of the images obtained is not straightforward.

(iii)   In order for the crystal lattice to be resolved, very high magnifications are used and this precludes study of large defects and long-range interactions between defects.

(iv)   Very high resolution is required, which is difficult to obtain routinely.

2. MOIRÉ FRINGES

A technique that overcomes some of these problems but still provides an indirect view of lattice defects consists of overlapping two crystal lattices, with slightly different spacings and/or with a small misorientation between them, and observing them in the electron microscope. The images under such conditions form what are known as moiré fringes. Defects in either lattice show up in the moiré fringes and are considerably magnified. Figure 5 shows moiré fringes in irradiated Nb_3Sn [14].

3. Out-of-Focus Contrast

Out-of-focus contrast is useful for studies of lattice defects such as small strain-free voids or magnetic domain walls, which alter the phase of the electron beam without changing the absorption coefficient or causing much scattering. By careful control of defocus, fairly detailed information about the defect can be obtained. This technique has been used to study radiation damage in Nb [32] and to attempt an observation of magnetic flux lines in other type II superconductors.

4. Absorption Contrast

Absorption contrast depends on variations of the absorption coefficient for high-energy electrons in different parts of the specimen. For this purpose any electron that fails to reach the final image plane is considered to have been absorbed by the specimen. This would include electrons scattered through large angles, beyond the limiting aperture of the imaging system, and electrons that suffer appreciable energy loss, so that they are not correctly focused by the imaging lenses. Absorption contrast can be used equally well for crystalline and amorphous materials. Unfortunately, most lattice defects do not greatly alter the absorption coefficient and therefore this technique is very limited in this application.

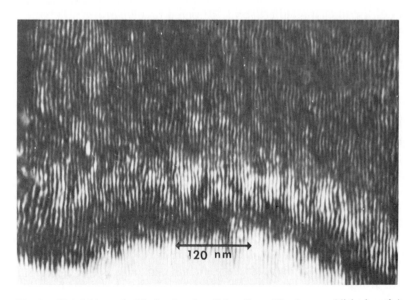

Fig. 5. Moiré fringes in Nb_3Sn showing dislocations. (Pande, unpublished work.)

5. DIFFRACTION CONTRAST

To produce diffraction contrast from a crystalline material usually only one of the beams diffracted by the specimen is allowed to reach the screen, the others being stopped by an aperture in the focal plane of the objective lens. Either the undiffracted beam ("bright-field" image) or any one of the diffracted beams ("dark-field" image) may be used. Imperfections in a nearly perfect lattice produce contrast by causing local variations in the diffracting conditions and so changing the intensities of the diffracted beams. Relating the image contrast to the nature of the defect usually requires complex calculations that can only be carried out by computer; however, some simple guidelines are useful [1]. (See Table III.) If the crystal is oriented so that the Bragg conditions are almost exactly satisfied, most defects will produce rather broad images, about 100 Å. Better resolutions can be achieved by tilting the specimen far away from the Bragg position, the so-called weak-beam technique [33]. In this condition the images become much narrower, a few tens of angstroms. In image contrast theory it is simplest to assume that only two sets of Bragg planes are operating so that only two beams (the undiffracted beam and one diffracted beam) are present. Such conditions can usually be obtained in practice by appropriately tilting a single crystal but the grain size of A15 materials is usually small and is thus a real handicap in such work.

The equations describing the passage of two beams through a crystal are a pair of coupled first-order differential equations relating the amplitudes of the undiffracted (ψ_0) and diffracted (ψ_g) beams, viz. [1].

$$\frac{d\psi_0}{dz} = \frac{\pi i}{\xi_0}\, \psi_0 + \frac{\pi i}{\xi_g}\, \psi_g\, \exp(2\pi i s Z + 2\pi i g_{hkl} \cdot R) \qquad (1)$$

$$\frac{d\psi_g}{dz} = \frac{\pi i}{\xi_0}\, \psi_g + \frac{\pi i}{\xi_g}\, \psi_0\, \exp(- 2\pi i s Z - 2\pi i g_{hkl} \cdot R) \qquad (2)$$

In these equations, g_{hkl} is the diffracting vector and is by definition normal to the diffracting planes with a magnitude equal to the reciprocal of the interplanar spacing, i.e.

$$|g_{hkl}| = 1/d_{hkl} \qquad (3)$$

ξ_0 and ξ_g are characteristic lengths, ξ_0 represents the mean refractive index of the crystal and ξ_g the extinction distance (being a reciprocal measure of the strength of the diffraction). R is the displacement vector of the lattice defect and is normally a function of position in the crystal. Z is the distance in the crystal parallel to the main beam and s is a measure of the deviation from the exact Bragg orientation. Table III gives the nature of

TABLE III

NATURE OF ELECTRON MICROGRAPHS AND ELECTRON DIFFRACTION PATTERN FOR VARIOUS TYPES OF DEFECTS

Type of defect	Elastic displacement R	Image contrast	Criterion for invisibility	Diffraction pattern	Example or reference
Dislocations	R = function of space coordinates in the film	Visible as a line ~20 Å wide in weak beam and ~100 Å in bright field. Components of the superdislocation may be revealed in weak beam	$g \cdot b = 0$. In ordered materials normal invisibility is not usually possible; b = Burgers vector	No effect	See Fig. 17
Disordered regions	$R \propto 1/r^2$; r = distance from the defect	Visible as black dot in superlattice (dark field) or fundamental reflection (bright field)	—	No effect for low density of defects	See Fig. 13
Antiphase boundary	R = lattice vector of the basic lattice but *not* of superlattice	Visible as fringes ($\alpha = g \cdot R$) in superlattice reflection with long depth period	$g \cdot R = \alpha = 0$ or integer	No effect except some fine structure in selected area diffraction	Hirsch et al. [1]
Stacking fault	$R \neq$ lattice vector.	Visible as α fringes again, but in fundamental reflections and with smaller depth period	Same as above	Same as above	See Fig. 12
Domain boundaries	$R = CZ\tau$ where τ is the twinning vector, Z the distance from the interface, and C a small constant	Visible as δ fringes at the interface in fundamental reflections with a smaller depth period than α fringes	$\delta = 0$ for imaging reflection, $g \cdot R = 0$ in general	Superposition of two slightly different patterns. High-order spots could be split	Hirsch et al. [1]
Twin boundaries	Same as above	Interface imaged by thickness fringes	$g \cdot R = 0$	Superposition of two diffraction patterns with splitting	Hirsch et al. [1]

the electron micrograph and electron diffraction image for various values of **R**. To a great extent the two coupled equations (1) and (2) form a basis of the theory of image interpretation in modern electron microscopy as applied to crystalline materials. These equations together with **R** given in Table III almost completely determine the nature of the image of a given lattice defect observed in diffraction contrast. However, these equations have not been solved analytically except in some simple cases. Computer solutions are available in most cases. A more recent trend is the method of "computer matching" [34] in which the equations are used to predict the observed image in the electron microscopes, and an actual computer image is obtained by plotting the resulting intensities. Theoretical micrographs are thus computed for a range of possibilites for the unknown defect and for a range of values of **g** and *s*, and identification is made by visual matching of the computer and observed (experimental) micrographs.

Studies of A15 superconductors using diffraction contrast, especially the weak-beam method, appear very promising. The technique can resolve superdislocations in ordered A15 materials and reveal the associated antiphase boundaries. It should also be helpful in understanding the nature of radiation damage in these materials. Such work will not be easy because of the small grain size in these compounds, which makes it difficult to excite various Bragg reflections by tilting.

Several examples of diffraction contrast in superconducting materials are given in the following sections. A much more detailed description of all the mechanisms of contrast (except the weak-beam technique) is given in Hirsch *et al.* [1].

In summary, the detailed interpretation of the images obtained in the electron microscope is not usually straightforward and at times the observed image of an defect may bear little physical resemblance to the actual defect. Correct interpretation of the images is thus as important as the techniques of obtaining these images. The applications of these techniques to superconductors are considered next.

III. Electron Microscopy of Niobium and Its Alloys

Transmission electron microscopy of ductile, type II superconductors such as Nb, Nb–Ti, and Nb–Zr has been primarily directed toward understanding the interactions between the Abrikosov vortices (present when these alloys are in the superconducting state) and lattice defects, such as dislocations and precipitates. It is these interactions that control the nonreversible superconducting properties (see the introductory chapter by Dew-Hughes). The role of electron microscopy has been to characterize the defect microstructures as illustrated in the following examples.

Narlikar and Dew-Hughes [35] studied the effect of dislocation microstructure on superconducting properties in niobium and its alloys and showed that uniform dislocation density has only a weak pinning effect on the flux lines. Much stronger pinning is obtained if the dislocations are clustered (as for example in Fig. 6) into walls surrounding relatively dislocation-free areas. Such structures tend to occur during cold working. This result was later extended to included drawn superconducting filaments. Neal *et al.* [36] studied the effects of deformation and annealing on the superconducting properties of copper-clad Nb 44% Ti wires and concluded that the highest critical current densities were obtained in samples in which the dislocation cell structure was as small as could be achieved.

Baker [37] studied the effects on J_c of small second-phase precipitate particles in Nb–Ti alloys and showed that the highest values of J_c were obtained by using the finest possible precipitate dispersions. Similar results were obtained for Nb 25% Zr by Milne [38] and for Nb 65% Ti by Witcomb and Dew-Hughes [39]. Walker *et al.* [9, 40] related the flux-pinning properties of Nb 25% Zr alloy to the microstructure of precipitates formed between 500 and 800°C. The precipitate particles were found to inhibit long-range rearrangement of dislocations and thus led to the formation of a fine, equiaxed cell structure when the material was deformed and annealed. Milne and Finlayson [41] showed that the value of J_c in these Nb–Zr alloys depended upon the volume fraction of β_{Zr} precipitated during cold rolling and subsequent annealing. Hillmann and Hauck [42], using electron microscopy, determined the density and size of α precipitates in a Nb 50% Ti alloy to be 10^{10} cm^{-2} and 200 Å, respectively. The critical current density for these alloys at 5 T was 2×10^5 A/cm^2. This high value was accounted for by a "matching effect" between the spacing of the flux-line lattice and the distribution of pinning sites.

Electron microscopy has also proved very useful in the study of multifilament superconducting composites both because of its high resolution compared with optical microscopy and because of its ability to give information about lattice defects. For example, it is known that the critical current density for multifilament niobium superconductors is increased by decreasing the filament size [43]. Transmission electron microscopy showed this improvement to be due to increased cold work (and consequently, a smaller dislocation cell size) in the finer-drawn composites. Figure 7 shows a scanning electron micrograph [44] of the cross section of

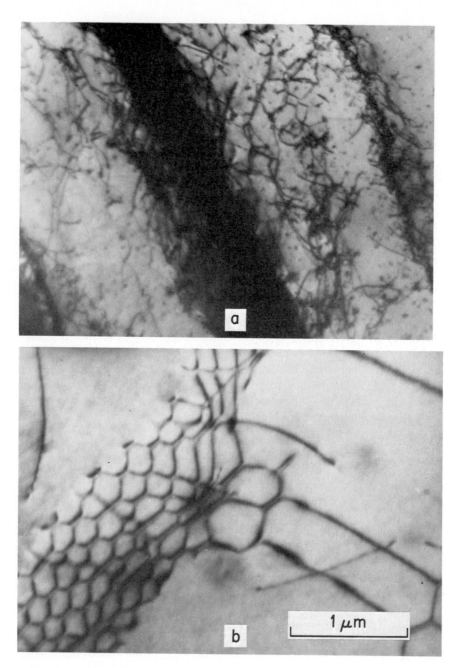

Fig. 6. Electron micrographs showing (a) cell walls in cold-worked niobium and (b) low-angle boundary in cold-worked and annealed niobium. (Pande, unpublished work.)

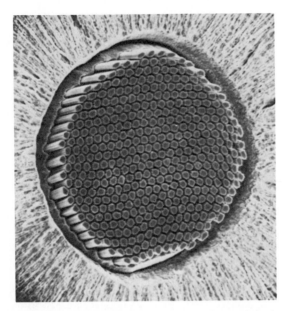

Fig. 7. Scanning electron micrograph of a transverse cross section of multifilament niobium wire. The filament diameter is 7 μm. Copper matrix has been etched away. (From Santhanam [44].)

a multifilament superconductor with a filament diameter of 7 μm and demonstrates the better resolution of this technique as compared with optical microscopy. The transmission electron micrographs in Fig. 8 show dislocation cell structures from four such differently drawn multifilament specimens [44]. Comparison of the critical current densities from these specimens with the results obtained by Conard *et al.* [45] for alloys with much smaller grain sizes shows that as flux-pinning agents the cell boundaries behave in a way similar to grain boundaries.

These examples serve to illustrate the usefulness of electron microscopy in relating superconducting properties to microstructures in Nb and its ductile superconducting alloys.

IV. Transmission Electron Microscopy of A15 Superconductors

In contrast with the large body of work on the ductile superconductors such as Nb–Ti, relatively few electron microscopical studies of

Fig. 8. Transmission electron micrographs of transverse sections of multifilament niobium wire. Filament diameters: (a) 7.4 μm, (b) 10 μm, (c) 13.5 μm, and (d) 18 μm; cell size, 1000–12,000 Å. (From Santhanam [44].)

A15 compounds have been reported. In part this has been due to the difficulty of preparing suitable specimens from these materials. This is particularly true for the multifilamentary superconductors because of the many different constituent materials they contain. Since ion milling has recently proved to be a successful technique, there will probably be an increase in the application of electron microscopy to A15 compounds in the near future. Here are described some of the electron microscope observations in A15 superconductors.

A. The Martensitic Phase Transformation

One of the earliest uses of electron microscopy in the study of A15 compounds was the direct observation of the martensitic phase transformation (see the chapter by Dew-Hughes on the physical metallurgy of A15 compounds, this volume) in V_3Si. The transformation occurred as a thinned specimen was cooled below 20 K in a liquid helium stage [19]. Twin-related lamellae of martensite with thicknesses as low as 100 Å are observed. These lamellae are shown in Fig. 9. A similar observation was later made in V_3Ga [18] but no further work has been done on any other A15 compound. With the advent of liquid helium stages and improved thinning techniques, direct observation of the martensitic transformation in other A15 compounds is expected.

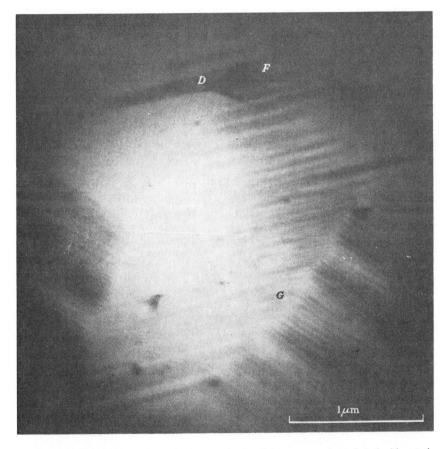

Fig. 9. Transmission electron micrograph showing contrast from interlocking twin boundaries on {110} plane in tetragonal V_3Si. Temperature of observation ~20 K. Note association with a dislocation at D, overlapping sets of fringes in different directions at F, and junction of such sets as G. (From Goringe and Valdrè [19], copyright Royal Society, London.)

B. Flux Pinning by Grain Boundaries

The value of J_c obtained in A15 materials is relatively high and has been ascribed to the fine grains formed in these materials, possibly stabilized by a dispersion of precipitates. The first workers to obtain an experimental correlation between grain size and critical current on the basis of electron microscopy were Nembach and Tachikawa [46] in V_3Ga and Scanlan *et al.* in Nb_3Sn [47]. Scanlan *et al.* speculated that the fine grain structure in their Nb_3Sn–Zr material was stabilized by the presence of second-phase precipitates in the grain boundaries. Figures 10a and 10b show grain boundaries in Nb_3Sn with and without ZrO_2, respectively. Pande [14], by electron microscopy on Nb_3Sn tapes, showed that the grain boundaries contain no visible precipitates. The dislocation density in these tapes was found to be less than 10^4 cm^2; very few other defects were found. However, the grain size d in these materials was found to be as small as 400 Å. It was therefore concluded that the pinning in these materials is mostly due to grain boundaries. A theory based on this idea by Pande and Suenaga [48] gives the experimentally observed J_c versus $1/d$ dependence, and shows that J_c is a maximum for a certain grain size

Fig. 10. Transmission electron micrographs showing grain boundaries in Nb_3Sn; (a) commercial Nb_3Sn tape containing ZrO_2; (b) filamentary Nb_3Sn.(From Scanlan *et al.* [47].)

in a given field (see the chapter by Dew-Hughes on the physical metal-lurgy of A15 compounds). Direct evidence for such a peak has now been obtained by using scanning electron microscopy [49] and TEM [50].

A method for determining the grain size of brittle fine-grained materials that avoids the difficulty of preparing the thin highly polished specimens needed for transmission electron microscopy is to use a scanning electron microscope to study an intergranular fracture surface [14]. Intergranular fracture can be obtained in A15 compounds simply by deforming to failure after etching off any ductile coating. This technique does not have as good a resolution as transmission electron microscopy but can deal with grain sizes down to about 100 Å. It has been routinely used for this purpose by the author (see Fig. 11).

C. Stacking Faults

In A15 structures, stacking faults on {111} planes are expected [51]. These faults break the linear chains of A atoms (in A_3B compound) and hence are expected to affect the superconducting critical temperature (see

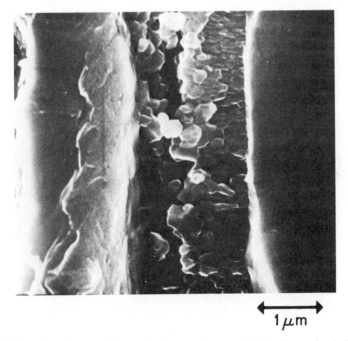

1 μm

Fig. 11. Scanning electron micrograph (in secondary mode) showing grains in Nb_3Sn. (Pande, unpublished work.)

the chapter by Dew-Hughes on the physical metallurgy of A15 compounds, this volume). Because of their planar nature, they would also be expected to increase the flux-pinning force. The most promising way of observing such stacking faults is by transmission electron microscopy using diffraction contrast. Under favorable imaging conditions, stacking faults give rise to oscillations in intensity in the image. Such fringes have recently been obtained by Uzel and Diepers [52] in Nb_3Sn containing a small amount of carbon (see Fig. 12). However, in pure Nb_3Sn stacking faults are very rare: only one such fault was observed by the author after examining over 100 specimens. This presumably implies that interstitial impurities are needed to form these faults.

D. Radiation Damage

One of the controversial subjects in the study of A15 materials has been the nature of the defect introduced in these materials by high-energy nuclear irradiation such as with α-particles and fast neutrons with $E > 1$ MeV. These defects significantly affect both the J_c and T_c in these materials (see the chapter by Sweedler, Snead, and Cox, this volume). Two types of defects have been proposed: (1) antisite defects homogeneously distributed on the atomic scale, and (2) "unknown universal" defects containing a large static displacement resulting in bond bending. A systematic electron microscope study of neutron-irradiated Nb_3Sn has now been completed to ascertain the nature of this defect [15, 53, 54]. Figure 13 shows a transmission electron micrograph, imaged in superlattice re-

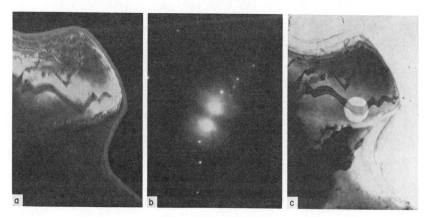

Fig. 12. Transmission electron micrograph showing: (a) dark field; (b) electron diffraction; and (c) bright field image for Nb_3Sn foil containing 0.35% (atomic) carbon showing stacking faults, $g = 200$. (From Uzel and Diepers [52].)

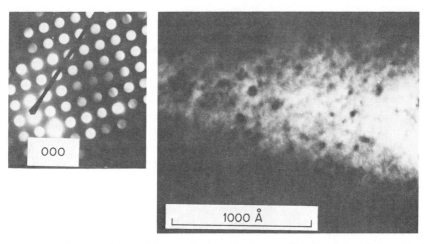

Fig. 13.   Electron micrograph in superlattice reflection g = 011, showing small disordered regions of size ~40 Å in fast neutron irradiated Nb_3Sn. The electron diffraction for the same region is shown on top left corner. (From Pande [53].)

flection, showing disordered regions in neutron-irradiated Nb_3Sn (size ~20–60 Å) [53]. These regions were interpreted from detailed contrast calculations to be regions in which local values of S, the long-range parameter, is much less than in the surrounding matrix. These regions retain the A15 structure and therefore probably contain a high density of antisite defects. Some specimens when imaged in fundamental reflections showed that the specimen contained localized regions of strain, again of size 20–60 Å. Simultaneous use of the two techniques showed [15] that most of the localized regions of disorder have an associated region of strain (which is to be expected since these regions are coherent). (A summary of the techniques used and the results obtained are given in Table IV.) Thus, both the antisite defects and the large static displacements are present in these disordered regions. The seemingly opposite points of view about the nature of the defect are thus reconciled. However, some reservations have to be made here. First, it should be stated that virtually any defect will lead to static displacement; thus, the disordered regions, or their associated strain, are not necessarily the "unknown universal defect." On the other hand, the antisite defects are not uniformly distributed on the atomic scale (there was no solid justification for this assumption). They are in fact clustered, as is expected from the radiation damage models when it is remembered that long-range replacement sequences seldom lead to antisite disorder in the A15 structure [15]. Unless low-angle scattering is employed, x-ray or neutron diffraction cannot distinguish

TABLE IV

ELECTRON MICROSCOPIC TECHNIQUES USED AND RESULTS OBTAINED IN STUDYING
RADIATION DAMAGE IN NEUTRON-IRRADIATED Nb_3Sn

Technique used	Nature of the image	Results [15] observed
Dark-field superlattice reflection	Disordered regions with local regions of low S imaged	Small disordered regions of size 20–60 Å (average ~35 Å) randomly distributed
Bright-field fundamental reflection	Localized regions of strain	Small regions again of size 20–60 Å randomly distributed
Techniques (1) and (2) used on the same area	Identification of the two regions, i.e., their relationship	Most of the disordered regions have an associated region of localized strain
Fundamental reflection with $s = 0$ and s large (s denotes deviation from Bragg angle)	Nature of the strain	The disordered regions are not small dislocation loops, i.e., vacancies or interstitials in them have not rearranged substantially

between the two cases. Details about these observations can be found in Pande [15, 53, 54]. A detailed theory of T_c degradation based on these observations, given by Pande, is in good agreement with experiment [53, 54].

In summary, the electron microscope techniques have led to a detailed knowledge of the nature of the defects in irradiated A15 compounds; these results are in accord with the defects expected from radiation damage models. These observations have, in turn, led to a quantitative theory of T_c degradation. It should be stated that early electron microscopic observations (ca. 1965) [55–57] had already given some indication of the existence of the disordered regions in neutron-irradiated Nb_3Sn (although their exact nature could not be ascertained because the superlattice reflection technique had not been developed) (see Fig. 14).

E. Grain Growth in Multifilamentary Superconductors

Multifilamentary superconductors have found increasing use in recent years (see the chapter by Dahl, this volume). The kinetics and nature of grain growth of the A15 superconductor in bronze-processed material (see the chapter by Luhman, this volume) is important since the grain size de-

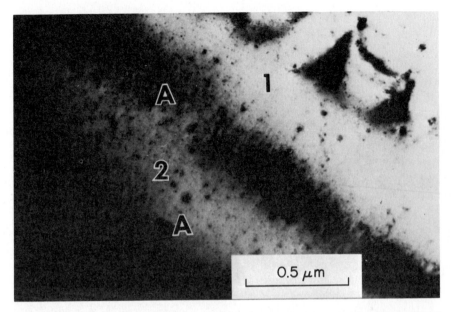

Fig. 14. Transmission electron micrographs of Nb_3Sn single crystal foil after irradiation (top) and before irradiation (bottom). Notice the black dots in the irradiated foils; 1 and 2 denote thickness fringes. Contamination spots near A. (Cullen and co-workers, unpublished work.)

termines the value of the critical current J_c. Specimen thinning of this composite material for electron microscopic investigation is very difficult. As pointed out earlier, ion milling is successful in thinning multifilamentary superconductors. Ion-milled specimens were used in the observation of grain growth in Nb_3Sn as a function of temperature. Electron micrographs (see Fig. 15) have also been obtained showing the grain structure at the interfaces between Nb and Nb_3Sn [14]. As pointed out earlier, another technique is to use the scanning electron microscope. Here, no special specimen preparation is necessary. However, the resolution is not sufficiently high to permit observation of the initial stages of grain growth.

F. *Electron Diffraction from A15 Compounds*

For many superconducting metals, such as Nb and its alloys, electron diffraction is routinely used. However, hardly any A15 compounds have been analyzed by electron diffraction. In the case of some A15 materials,

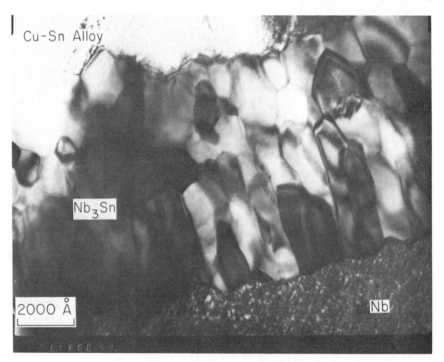

Fig. 15. Electron micrograph of a multifilamentary superconductor showing Nb–Nb_3Sn and Sn–Nb_3Sn interfaces as well as the individual grains. Note the shape of the grains near the interface. (Pande, unpublished work.)

the grain size is large enough (>1000 Å) to give characteristic spot patterns in selected area diffraction, specimen preparation being the limiting factor.

The A15 structure is well known and the corresponding reciprocal lattice points can be calculated. Not all the reciprocal points are associated with Bragg reflections, however. Some of the predicted reflections do not appear because of various symmetries in the crystal. Table V gives the expected intensities for various reflections in terms of the order parameter S and scattering factors f. All actual intensities, of course, may also depend on other factors, such as double diffraction. Forbidden spots such as 100 may appear by double diffraction. In the case of ring patterns, the multiplicity factor must be taken into account. Notice that some intensities in Table V involve $(fA - fB)^2$ for the A_3B compound. These are the superlattice reflections, which are especially sensitive to disorder (see Section IV,D). Based on an electron diffraction study, Schmidt et al. [58] postulated the existence of charge density waves in A15 materials. However, detailed [14] electron diffraction study of Nb_3Sn and Nb_3Ge failed to reveal such waves in these materials, although some diffraction patterns in disordered specimens were similar to those obtained by Schmidt et al.

G. Dislocations in A15 Materials

Most of the A15 materials are brittle. Polycrystalline Nb_3Sn under deformation usually fails by intergranular cracking along grain boundaries

TABLE V

FUNDAMENTAL AND SUPERLATTICE SPOTS IN A15 STRUCTURE (A_3B)

hkl	S Contribution to F^a	hkl	S Contribution to F^a
110	$2S(fB - fA)$	420	$S(fB - fA) + (3fA + fB)$
200	$S(fB - fA) \pm (3fA + fB)$	421	$S(fB - fA) - (3fA + fB)$
210	$S(fB - fA) - (3fA + fB)$	332	$S(fB - fA) + (3fA + fB)$
211	$S(fB - fA) + (3fA + fB)$	422	$2S(fB - fA)$
220	$2S(fB - fA)$	510, 431	$2S(fB - fA)$
310	$2S(fB - fA)$	520, 432	$S(fA - fB) + (3fA + fB)$
222	$3S(fB - fA) - (3fA + fB)$	521	$S(fB - fA) + (3fA + fB)$
320	$S(fA - fB) + (3fA + fB)$	440	$2(3fA + fB)$
321	$S(fB - fA) + (3fA + fB)$	530, 433	$2S(fB - fA)$
400	$2(3fA + fB)$	600, 442	$S(fB - fA) + (3fA + fB)$
411, 330	$2S(fB - fA)$	610	$S(fB - fA) - (3fA + fB)$
		611, 532	$S(fB - fA) - (3fA + fB)$

^{a} S is the long-range order parameter, f the scattering amplitude, and F the structure factor. The intensity will be the square of F. For polycrystals, multiplicity factors must be included.

(see Fig. 16). Hence it is difficult to produce dislocations by mechanical deformation. Some ingrown dislocations are observed in Nb_3Sn and V_3Si. Figure 17a shows an ingrown dislocation in V_3Si with a Burgers vector $b = [100]$ [59]. Pande has observed [15] superdislocations in Nb_3Sn (see Fig. 17b). More detailed study of the dislocations in A15 materials is needed to better understand the deformation process in A15 materials, especially at high temperatures.

These examples illustrate the usefulness of electron microscopy in the study of A15 superconductors. The problem of obtaining thin films from these materials has been more or less overcome by ion milling techniques. Initial studies on these materials have already yielded valuable information and it is expected that electron microscopy will be increasingly used in characterizing and developing technologically important A15 compounds.

Fig. 16. Transmission electron micrograph of a crack propagating along grain boundaries in Nb_3Sn. (Pande, unpublished work.)

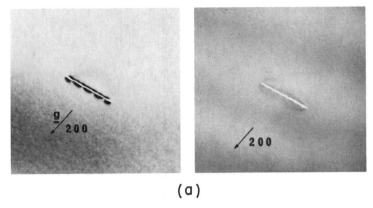

(a)

Fig. 17a. Ingrown dislocation in V_3Si with $b = [100]$. (From Mahajan *et al.* [58].)

V. Observation of Magnetic Flux Lines by Electron Microscopy

By far the greatest efforts in electron microscopy of superconductors have been devoted to observing magnetic flux lines. The situation is in some respect similar to the time when methods using electron microscopy for observing dislocations were not available and various "decoration techniques" were used. The introduction of the technique of diffraction

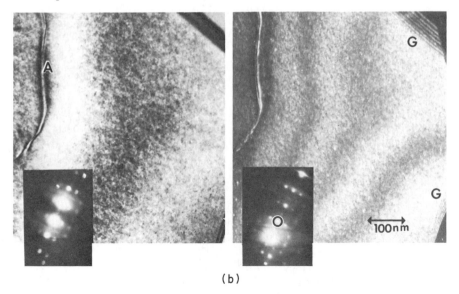

(b)

Fig. 17b. Superdislocations in Nb_3Sn at A. Diffraction condition is also shown. Black and white dots are disordered regions due to neutron irradiation. (From Pande [15].)

contrast provided a simple and versatile means of observing dislocations and made transmission electron microscopy indispensible for materials research. A direct method of observing magnetic flux lines by electron microscopy is not yet available. This section briefly discusses some indirect methods of observing flux lines and indicates some possible new techniques that might result in their direct observation. Table VI summarizes the various electron-optical methods that have been used or can potentially be used for direct or indirect observations of flux lines.

A. High-Resolution Bitter Technique

It was shown by Hutchinson *et al.* [64] that when a ferromagnetic material such as cobalt or iron is evaporated in an inert atmosphere (helium ~0.5 Torr) small ferromagnetic single crystals ranging in size from 50 to 200 Å are formed. These crystals can then be deposited preferentially on regions of a sample having high flux density gradients. Essmann and

TABLE VI

Methods Used or Available for Investigating Flux-Line Lattice

Method	Type of information available or possible	Resolution	Investigators or reference
High-resolution Bitter technique (surface replica)	Decorated flux lines	~100 Å	Sarma and Moon [60], Essmann and Traüble [61]
High-resolution Bitter technique (thin films)	Decorated flux lines *and* lattice defects	~100 Å	Herring [62]
Shadow electron microscopy	Leakage field	~10^4 Å	Blackman *et al.* [65], Pozzi and Valdrè [29]
Mirror electron microscopy	Leakage field	~10^4 Å	Frost and Wende [69]
Electron interferometry	Leakage flux	~10^{-7} G/cm²	Merli *et al.* [71], Wahl [72]
Out-of-focus TEM and STEM	Flux lines (?) *and* lattice defects	~	Darlington and Valdre [75]
Scanning electron microscopy	Decorated flux lines, leakage field, and crystal defects	50–200 Å	Singh, Curzon and Koch [63]
High-resolution electron microscopy or HVEM with field emission gun and image enhancement	Direct visibility of flux lines (?) and lattice defects		Pande [14]

Traüble [61] and Sarma and Moon [60] used this idea to mark the points at which quantized flux lines intersect the surface of a superconducting sample in its mixed state. By observing the particle distribution on the surface, the distribution of the flux lines could be determined with a resolution of about ~ 100 Å. The magnetic particle distribution can be observed in a number of different ways:

(a) Replica technique: After depositing the ferromagnetic particles, a replica of the surface is made by deposition of a carbon film that is subsequently removed and examined in the electron microscope. This was the method by which the flux-line lattice was first observed (see Fig. 18a).

(b) Thin-film technique: After depositing ferromagnetic particles on a thin film (≤ 2000 A) of a superconductor, usually prepared by evaporation, the magnetic particles are observed directly, without replication [62]. An advantage of this technique is that in addition to the flux lines, other lattice defects are also visible, and hence the interaction between the flux lines and lattice defects can be studied directly (Fig. 18b). It should be mentioned that the behavior of flux lines and of dislocations in a thin-film specimen can be somewhat different from that in the bulk.

(c) Use of scanning electron microscopy: The surface on which ferromagnetic particles have been deposited can be observed directly in a scanning electron microscope. Replication is not needed and the flux lines are observed at the surface of a bulk specimen rather than in a thin film. Singh et al. [63] have observed the flux lines in an A15 compound with this technique (Fig. 19).

B. Unconventional Techniques

Many unconventional methods have been tried in attempts to image magnetic flux lines. In shadow electron microscopy the electron beam is kept near the edge of the specimen. A "shadow" is thus formed, the position and shape of which depend on magnetic fields experienced by the electron beam [29, 65, 66]. The method has a resolution of about 1 μm and is obviously too crude for observing individual flux lines but rather can be used [29, 65–67] to image areas of flux penetration. The method has been applied to the study of the intermediate state of lead [29] and of the mixed state of Nb [22]. In mirror electron microscopy [68–70] the electrons directed toward the sample surface are made to reverse their paths by a magnetic field before they hit the sample surface; they are then imaged. These electrons carry the information about any magnetic fields in the sample and this can be made visible in the image. Although the method is suitable for observing dynamic effects, such as a sample with slowly changing magnetic structure, the resolution is again poor (>1 μm) [74]. Other methods include electron interferometry [71–73] and out-of-focus TEM and scanning transmission electron microscopy (STEM) [75].

C. Direct Resolution of Flux Lines without Decoration

As stated earlier, the direct resolution of individual flux lines in thin films using TEM, STEM, or (in bulk specimens) scanning electron microscopy has not yet been achieved, in spite of many ingenious experiments, Valdrè and his co-workers [76] have shown theoretically that individual flux lines in relatively thick specimens would produce visible images in lightly defocused electron micrographs because of their associated magnetic fields. Their magnetic contrast calculations show that sufficient contrast may be available using instrumentation required for detecting low-contrast changes. Pande [14] has recently proposed that instead of trying to observe the fluxoids by out-of-focus techniques, using

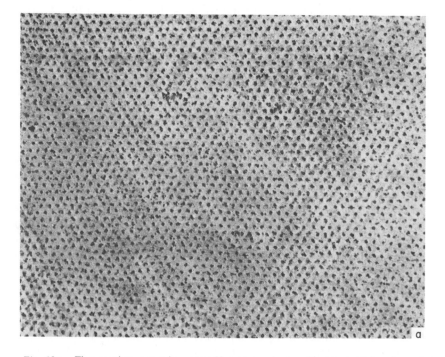

Fig. 18a. Flux vortices array in a type II superconductor, Pb 6.3 at. % In, decorated with ferromagnetic particles. Transmission electron micrograph of the replica. (From Essman and Traüble [61].)

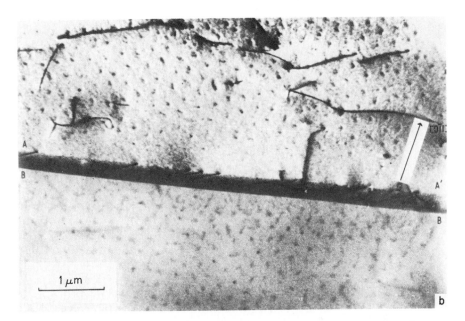

Fig. 18b. Transmission electron micrograph of a niobium foil about 2000 Å thick showing lattice dislocations near a grain boundary as well as decorated flux lines. (From Herring [62], copyright The Institute of Physics, London.)

magnetic contrast, it may be easier to observe the fluxoids using contrast arising from the elastic stress field associated with each fluxoid. This stress field is known quite accurately, and has been used to estimate J_c in superconducting materials [48]. Contrast calculations done using the two-beam case [Eqs. (1) and (2)] show that the contrast observed will be very faint, but within range of detection, provided high-voltage electron microscopy (HVEM) with a field emission gun is used. One advantage of this method is that if successful it will provide a means of experimentally observing many flux-line lattice configurations, such as screw dislocations in the flux-line lattice and flux-line cutting that have recently been proposed to explain flux pinning.

VI. Superconducting Lenses

So far the discussion has been limited to the application of electron microscopy in the study of various superconductors. Superconducting magnet technology has now reached a stage where it is being used to design lenses for electron microscopes. The rest of this chapter discusses

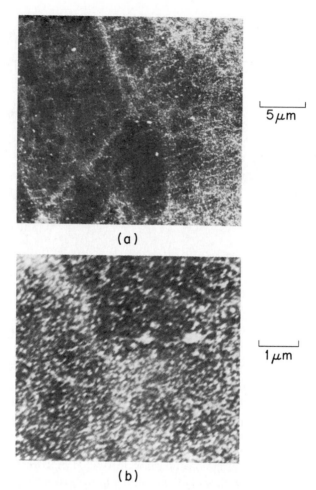

Fig. 19.   Flux distribution in V 25.4 at. % Ga showing flux pinning by grain boundaries. Parts (a) and (b) show the same area in different magnification. (White dots indicate fluxoids. Field cycled from 15 to 8 kG at 4.2 K.) (From Singh *et al.* [63], copyright The Institute of Physics, London.)

such lenses and their use in electron microscopes instead of conventional electron lenses.

Conventional electron lenses used in electron microscopes have reached a high degree of excellence in design and performance. Current requirements call for improvements in the following areas: (1) higher resolution—this means resolution on the atomic scale, with the possibility of routinely resolving individual atoms; (2) observation in thicker specimens, usually obtainable only by increasing the accelerating electron voltage; and (c) analytical and in situ electron microscopy. These goals could in principle be achieved with high-voltage electron microscopy using superconducting lenses, since the latter can satisfy the demand for small focal lengths, small lens aberrations, and extremely steady lens excitation currents. Type II superconductors cooled below T_c are able to support overall current densities several orders of magnitude greater than pure copper. Therefore superconducting coils are in principle able to create (with smaller cross sections) higher flux densities with small half-widths and shorter focal lengths. Until recently, however the only gain appeared to be in obtaining higher peak densities ($\sim$80 kG compared to about 20 kG) than in conventional lenses, and superconducting lenses are not yet in routine use.

A. Superconducting Lens Design

The first superconducting lenses were simply short solenoids made of Nb–Zr or Nb–Ti wire. They were first used by Fernández-Morán in 1965 [77] to obtain electron microcope images. Since then the field has developed at a relatively slow pace. Electron microscopes using only superconducting lenses are, however, now available. The main achievement has been in reducing the lens aberrations. Because of its high aberrations, the simple Fernández-Morán type of superconducting lens is no longer used unless a large space is needed around the focal plane for special specimen stages. The superconducting coils are now usually enclosed in a ferromagnetic case or used in conjunction with a ferromagnetic pole piece, as in conventional magnetic lenses. Figure 20 gives a brief sketch of various general types of superconducting lenses that have been developed. Table VII gives further details of these lenses. More details can be found in a recent review by Bonjour [79]. A more thorough account of superconducting magnet lens design is given by Iwasa and Montgomery [80].

B. Advantages and Disadvantages of Superconducting Lenses

The main advantages of superconducting lenses are

(1) Better electron-optical performance, i.e., small focal length and reduced spherical and chromatic aberration, especially at high voltages. This is so because in superconducting lenses peak flux densities can be made higher and half-widths of the field distribution can be reduced. It

TABLE VII

SUPERCONDUCTING LENSES (AFTER VALDRÈ AND HAWKES[110])

Type	Description	Highest peak flux density B_m obtained (T)	Highest experimental resolution reported	Magnetic circuit	Superconducting winding	Remarks
Ironless lenses	Coils without any pole pieces (unshrouded)	33	~10–20 Å	Free space	Nb–Zr [77, 78, 81, 83]; Nb–Ti [82, 84]; 65 BT [85]; Nb$_3$Sn [86]	Field not axially asymmetric; astigmatism present; rarely used
Shrouded (shielded) lens	Coils of the type above enclosed in a ferromagnetic case, reducing flux leakage	78		Fe, FeCo	Nb–Zr [87]; Nb$_3$Sn [86]; Nb–Ti [88]	As above; field symmetry defect dependent on coil defects
Lenses with conventional pole pieces	Lenses of classical design, excited by superconducting coil	52	3–4 Å	Fe, rare earth metals dysprosium, and holmium pole pieces can be used to increase B_m	Nb–Ti [89–98]	Relatively small half-width field
Superconducting magnetic shielding lenses	Diamagnetic properties of superconductors used to shape the magnetic field and prevent flux leakage	70	<3 Å	Free space or Nb$_3$Sn [101]	Nb$_3$Sn tube [99–102]	Nb$_3$Sn tube needs precise alignment; otherwise transverse field component results, aberrations relatively smaller

Trapped flux lenses (rings)	Superconducting currents induced in ring by cooling it below T_c in a field B_m and slowly reducing B_m to zero	56		Fe, Cu [103] free space	Nb [103] Nb–Zr [103] Nb$_3$Sn [104]	Impractical to operate; difficult to change field; flux trapping takes hours
Trapped flux lenses (disks)	Massive Nb$_3$Sn is replaced by a stack of about 20 disks, each having 6 rings of Nb$_3$Sn disks placed in the gap of a magnetic circuit	22	10 Å	Free space [86] Permendur [105]	Nb$_3$Sn [86, 105, 106]	See above

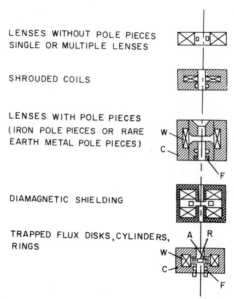

Fig. 20. Types of superconducting lenses: z axis shows electron beam direction; P, pole pieces; W, superconducting exciting coil; C, ferromagnetic shroud; A, stack of annuli; R, ring; F, focusing coils.

should be mentioned that the reduction of chromatic aberration is especially welcome since in many cases, like observation in thick specimens, chromatic aberration becomes the resolution-limiting factor. When superconducting lenses are operated in the persistent current mode, instability is virtually eliminated. Expensive highly stable lens power supply units are no longer necessary [107].

(2) Reduced size   For high-voltage electron microscopes (in the megavolt range) the weight of a superconducting lens could be two orders of magnitude smaller than that of a conventional lens. This would enable such microscopes to be housed in smaller rooms. For conventional microscopes (100 kV) the two types of lenses are comparable in size.

(3) New improved geometry   Space is very restricted in the pole piece of a conventional lens. This is a problem when special specimen stages are required. It also limits the space available for additional in situ experimental facilities. With specifically designed superconducting lenses more space can be provided.

(4) Indirect advantages   Cryogenic pumping of the helium cryostat improves the vacuum in the system, with a reduction in contamination. The low-temperature systems already present for superconducting lenses can also be utilized in the study of materials at low temperatures.

The main disadvantages are

(1) Axial asymmetry   This is the main disadvantage in a supercon-ducting lens and can be quite serious. Open coil type superconducting lenses (without pole pieces), the first superconducting lenses to be used, have poor axial symmetry. This problem besets the shrouded supercon-ducting lenses also and hence in spite of their high peak fields and reason-ably low half-width they are rarely used. The only solution is to go back to the use of pole pieces, or in the case of diamagnetic lenses, which operate by shielding, to improve the homogeneity of flux-squeezing bodies. The asymmetry can also be corrected to some extent by the use of stigmators.

(2) Instability   Mechanical and thermal instability can hinder the per-formance of superconducting lenses. Such instabilities can be reduced through use of liquid helium II. This improves heat transfer and makes the cryostat structure more vibration free.

We have given only a brief account of superconducting lenses. A de-tailed review of the literature on superconducting lenses up to about 1970 has been given by Hardy [108]. Reference can also be made to a brief re-view by Septier [96] for the early work on superconducting lenses. An ex-cellent account of the later work is now available in reviews by Dietrich [109], Bonjour [79], and Hawkes and Valdrè [110].

C. High-Resolution Electron Microscope with Superconducting Lenses

As stated previously, the high expectations for improvement in electron microscopes by the introduction of superconducting lenses have not been realized on a wide scale. In 1977, however, an electron microscope using superconducting lenses was introduced [111] with a resolution of 1.7 Å, better than most of the conventional electron microscopes and ap-proaching the theoretical resolving power of an electron-optical system. The experimental setup is shown in Fig. 21. It consists of a 400-kV elec-trostatic generator (with $\Delta V/V \approx 3 \times 10^{-6}$) and an accelerator with three tube lenses, a double condenser, a cryostat with four superconducting lenses (Fig. 22), a normal intermediate lens, a projector, and an image-recording unit. The cryostat consisted of three chambers suspended from a thick-walled steel casting. The liquid nitrogen reservoir and other He chamber are supported by glass epoxy rods. The outer He chamber en-closes the inner He chamber. This arrangement permits the use of super-fluid He in the inner chamber. The objective lens is a shielded lens and is equipped with twofield coils (Fig. 22). The fine focusing is performed with the beam voltage because of a strong field hysteresis in this kind of lens.

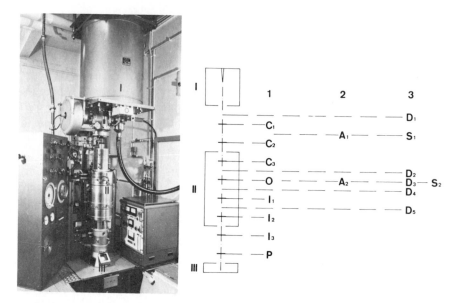

Fig. 21. Experimental setup of the 400-keV electron microscope with superconducting lenses, I, accelerator; II, cryostat; III, image registration. (1) Lenses; $C_{1,2,3}$, condensers; O, objective lens; $I_{1,2,3}$, intermediate lenses; P, projective lens; (2) $A_{1,2}$, apertures; (3) correction systems; $D_{1,2,3,4,5}$, deflectors: $S_{1,2}$, stigmators. (From Dietrich *et al.* [111].)

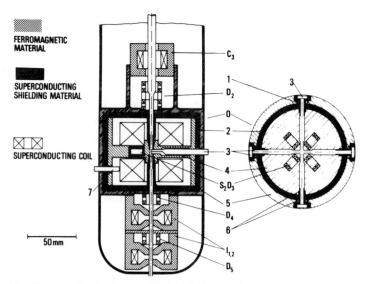

Fig. 22. Superconducting lens system with helium-cooled specimen stage and transverse section of objective lens: 1, double flange sealing; 2, shielding casing; 3, channel for specimen and aperture holders; 4, support disks; 5, wall of inner He chamber; 6, indium seals; and 7, adjustment screws. (From Dietrich *et al.* [111].)

Characteristic lens data are focal length $f = 1.8$ mm; spherical aberration $C_s = 1.45$ mm chromatic aberration $C_c = 1.3$ mm. The specimen stage is the side-entry type. After positioning the specimen, the connections on the holder to room temperature are released, thus minimizing the specimen drift.

The intermediate and the condenser lenses (Fig. 22) are similar to conventional lenses. An iron circuit surrounds the superconducting excitation coil. All the superconducting lenses are excited one after another by the same power supply using the persistent-current mode in the superconductor. Changing of the excitation current is accomplished by heating the persistent-current switch. Deflectors are small superconducting coils used for adjusting the electron beam. The current drift $\Delta I/I$ is maintained at less than 10^{-6}. Since the inductance of these coils is relatively small, the contact resistance of the superconducting connection has to be $< 10^{-11}$ Ω.

Cooling of the cryostat and aligning of the microscope takes about 2 h and the working time determined by helium consumption in the inner helium chamber is about 5 h. Images of amorphous carbon foils indicated a point resolution well below 2 Å. Dietrich *et al.* [111] showed that their system can also be applied to microprobe analysis without strongly reducing the theoretical resolution limit. This is accomplished by increasing the lens gap of the shielding lens. The system can also in principle be used with an energy-dispersive x-ray analyzer, Auger spectroscopy, and energy-loss spectroscopy.

Other electron microscopes equipped with superconducting lenses but with somewhat poorer resolution are or were available at the following institutions: Cornell University, Ithaca [106], Enrico Fermi Institute, Chicago [77, 78], Collège de France, Paris [112], Faculté des Sciences, Rheims, France [89], Hitachi, Tokyo [87], Laboratoire d'Optique Electronique de CNRS, Toulouse, France [93], Oak Ridge National Laboratory, Oak Ridge [89], Institute fur Experimentalle Kernphysik, Karlsruhe Germany [113], and Academy of Sciences, USSR [114]. The accelerating voltages of these electron microscopes range from 50 to 5000 kV [113], the consumption of helium is typically ~1 liter/h, and in most cases the superconducting lenses, operate in the persistent mode. Focusing, unlike conventional lenses, is usually done by adjusting accelerating voltage. Further details can be found in the references cited.

VII. Conclusion

In this brief review an attempt has been made to describe the great potential of electron microscopy and related techniques in the study of superconducting materials. Standard methods of transmission electron microscopy are now beginning to be used extensively in observing crystal defects, which are then correlated with bulk superconducting properties. Very little use has been made so far of scanning microscopes in either reflection or transmission modes. The study of flux pinning has been hampered by the inability to resolve flux vortices directly without decoration. It is hoped that further developments will change this situation. With regard to superconducting lenses, their development is still far from complete and they have yet to routinely surpass conventional lenses in quality. They do, however, offer certain advantages, especially in high-voltage (>1 MV) electron microscopes. With continued improvement, their use is expected to increase.

ACKNOWLEDGMENTS

The author is grateful to various workers who supplied the micrographs of their work. Figures 3, 7–10, 12, 13, 17b, 18a, 18b, 19, 21, and 22 are taken from published work and are reprinted with permission; for sources see the figure legends.

The author is also grateful to Dr. A. T. Winter for commenting extensively on the manuscript.

References

1. P. B. Hirsch, A. Howie, R. B. Nicholson, D. W. Pashley, and M. J. Whelan, "Electron Microscopy of Thin Crystals." Butterworths, London, 1965.
2. I. S. Brammer and M. A. P. Dewey, "Specimen Preparation for Electron Metallography." Blackwell, Oxford, 1966.
3. A. Fourdeaux and A. Wronski, *J. Less-Common Met.* **6,** 11 (1964).
4. J. O. Stiegler, C. K. H. Dubose, and C. J. McHargue, *Acta Metall.* **12,** 263 (1964).
5. D. P. Gregory, A. N. Stroh, and G. H. Rowe, *Trans. Am. Inst. Min. Metall. Pet. Eng.* **227,** 678 (1963).
6. A. S. Keh and S. Weissman, "Electron Microscopy and Strength of Crystals" (G. Thomas and J. Washburn, eds.), p. 231. Wiley (Interscience), New York, 1963.
7. A. V. Narlikar and D. Dew-Hughes, *Phys. Status Solidi* **6,** 383 (1964).
8. G. W. Briers, D. W. Dawe, M. A. P. Dewey, and I. S. Brammer, *J. Inst. Met.* **93,** 77 (1964–1965).
9. M. S. Walker, R. Stickler, and F. E. Werner, *Z. Metallk.* **54,** 331 (1963).
10. J. A. F. Gidley and P. N. Richards, *J. Sci. Instrum.* **2,** 297 (1969).
11. R. Stickler and R. J. Engle, *J. Sci. Instrum.* **40,** 518 (1963).
12. T. Schober and V. Sorajic, *Metallography* **6,** 183 (1973).
13. L. I. Van Torne and G. Thomas, *Acta Metall.* **11,** 881 (1963).
14. C. S. Pande, Unpublished work.
15. C. S. Pande, *Phys. Status Solidi* (1979) (in press).
16. D. J. Barber, *Proc. Eur. Congr. Electron Microsc., 5th* p. 293, Institute of Physics, London, 1972.
17. Superconducting Phenomena, Phase III, RCA Report (1965).

18. E. Nembach and K. Tachikawa, *Philos. Mag.* **21,** 869 (1970).
19. M. J. Goringe and U. Valdrè, *Proc. R. Soc. London Ser A* **295,** 192 (1966).
20. U. Valdrè and M. J. Goringe, *J. Sci. Instrum.* **42,** 268 (1965).
21. J. A. Venables, *Rev. Sci. Instrum.* **34,** 582 (1963).
22. H. Watanabe and I. Ishikawa, *Jpn. J. Appl. Phys.* **6,** 83 (1967).
23. L. W. Hobbs (1971), quoted by Valdrè [25].
24. L. W. Hobbs, D. Phil. Thesis, Univ. of Oxford (1972) (unpublished).
25. U. Valdrè, *Proc. Eur. Congr. Electron Microsc., 5th* p. 317. Institute of Physics, London, 1972.
26. U. Valdrè and M. J. Goringe, "Electron Microscopy in Materials Science," p. 207. Academic Press, New York, 1971.
27. L. W. Hobbs and M. J. Goringe, *Proc. Int. Congr. Electron Microsc. 7th, Grenoble,* Vol. 2, p. 239. Société Francaise de Microscopie Electronique, Paris, 1970.
28. H. G. Chelbek and A. E. Curzon, *J. Phys. E.* **6,** 1105 (1973).
29. G. Pozzi and U. Valdrè, *Philos. Mag.* **23,** 745 (1971).
30. J. H. Bennett and R. P. Ferrier, *Proc. Int. Congr. Electron Microsc., 7th, Grenoble* Vol. 2, p. 607. Société Francaise de Microscopie Electronique, Paris, 1970.
31. D. R. Clarke and C. S. Pande, Unpublished work.
32. J. B. Mitchell and W. L. Bell, *Acta Metall.* **24,** 147 (1976).
33. D. J. H. Cockayne, I. L. F. Ray, and M. J. Whelan, *Philos. Mag.* **20,** 1265 (1969).
34. A. K. Head, P. Humble, L. M. Clarebrough, A. J. Morton, and C. T. Forwood, "Computed Electron Micrographs and Defect Indentification." North-Holland Publ., Amsterdam, 1973.
35. A. V. Narlikar and D. Dew-Hughes, *J. Mater. Sci.* **1,** 317 (1966).
36. D. F. Neal, A. C. Barber, A. Woolcock, and J. A. F. Gidley, *Acta Metall.* **19,** 143 (1971).
37. C. Baker, *J. Mater. Sci.* **5,** 40 (1970).
38. I. Milne, *J. Mater. Sci.* **7,** 413 (1972).
39. M. J. Witcomb and D. Dew-Hughes, *J. Mater. Sci.* **8,** 1383 (1973).
40. M. S. Walker and M. J. Fraser, *AIME Annu. Meeting. Supercond. Mater. Magn., New York February, 1962,* (unpublished).
41. I. Milne and T. R. Finlayson, *Appl. Supercond. Conf.* p. 425. Institute of Electrical and Electronics Engineers, New York, 1972.
42. H. Hillmann and D. Hauck, *Appl. Supercond. Conf.* p. 429. Institute of Electrical and Electronics Engineers, New York, 1972.
43. M. P. Mathur, M. Ashkin, and D. W. Deis, *J. Appl. Phys.* **45,** 3627 (1974).
44. A. T. Santhanam, *J. Mater. Sci.* **11,** 1099 (1976).
45. H. Conard, L. Rice, E. L. Fletscher, and F. Vernon, *Mater. Sci. Eng.* **1,** 360 (1967).
46. E. Nembach and K. Tachikawa, *J. Less-Common Met.* **19,** 359 (1969).
47. R. M. Scanlan, W. A. Fietz, and E. F. Koch, *J. Appl. Phys.* **46,** 2244 (1975).
48. C. S. Pande and M. Suenaga, *Appl. Phys. Lett.* **29,** 443 (1976).
49. B. J. Shaw, *J. Appl. Phys.* **47,** 2143 (1976).
50. A. W. West and R. D. Rawlings, *J. Mater. Sci.* **12,** 1962 (1977).
51. U. Essmann and G. Zerweck, *Phys. Status Solidi (b)* **57,** 611 (1973).
52. Y. Uzel and H. Diepers, *Z. Phys.* **258,** 126 (1973).
53. C. S. Pande, *Solid State Commun.* **24,** 241 (1977).
54. C. S. Pande, *J. Nucl. Mater.* **72,** 83 (1978).
55. G. W. Cullen, *Proc. Summer Study Supercond. Devices and Accelerators Part II Brookhaven National Laboratory, Upton, New York* p. 437 (1968) (unpublished).
56. G. W. Cullen and G. D. Cody, *J. Appl. Phys.* **44,** 2838 (1973).

57. G. D. Cody, Private communication.
58. P. H. Schmidt, E. G. Spencer, D. C. Joy and J. M. Rowell, *Conf. Supercond. d- and f-band Met., 2nd* (D. H. Douglass, ed.) Plenum, New York, 1976. p. 431.
59. S. Mahajan, S. Nakáhara, J. H. Wernick, and G. Y. Chin, Unpublished.
60. N. V. Sarma and J. R. Moon, *Philos. Mag.* **16**, 433 (1967).
61. U. Essmann and H. Traüble, *Phys. Lett.* **24A**, 526 (1967); See also H. Traüble and U. Essmann *J. Appl. Phys.* **39**, 4052 (1968).
62. C. Herring, D. Phil. Thesis, Univ. of Oxford (1973) (unpublished); *Phys. Lett.* **47A**, 105 (1974); *J. Phys. F.* **6**, 99 (1976).
63. O. Singh, A. E. Curzon, and C. Koch, *J. Phys. D.* **9**, 611 (1976).
64. R. I. Hutchinson, P. A. Lavin, and J. R. Moon, *J. Sci. Instrum.* **42**, 885 (1965).
65. M. Blackman, A. E. Curzon, and A. T. Pawlowic, *Nature (London)* **200**, 157 (1963).
66. M. J. Goringe and U. Valdrè, *Proc. Eur. Regional Conf. Electron Microsc., 3rd, Prague* p. 305. Czechoslovak Academy of Sciences, Prague, 1964. p. 305.
67. R. Aoki, U. Kräegeloh, T. Miyazaki, and B. Shinozaki, *Phys. Lett.* **45A**, 89 (1973).
68. A. B. Bok, J. B. Lee Poole, J. Roos, and H. de Long, *Adv. Opt. Electron Microsc.* **4**, 161 (1971).
69. G. Forst and B. Wende, *Z. Angew. Phys.* **17**, 479 (1964).
70. O. Bostanjoglo and G. Siegel, *Cryogenics* **7**, 157 (1967).
71. P. G. Merli, G. F. Missiroli, and G. Pozzi, *J. Phys. E.* **7**, 729 (1974).
72. H. Wahl, *Optik* **28**, 417 (1968–1969).
73. H. Wahl, *Ber. Bunsenges. Phys. Chem.* **74**, 1142 (1970).
74. J. P. Jakubovics, "Electron Microscopy and Materials Science" (E. Ruedl and U. Valdrè, eds.), Part IV, p. 1303. The Commission of the European Communities, Luxembourg, 1976.
75. E. H. Darlington and U. Valdrè, *J. Phys. E.* **8**, 321 (1975).
76. C. Capiluppi, G. Pozzi, and U. Valdrè, *Philos Mag.* **26**, 865 (1972).
77. H. Fernández-Morán, *Proc. Nat. Acad. Sci. U. S.* **53**, 445 (1965).
78. H. Fernández-Morán, *Proc. Nat. Acad. Sci. U. S.* **56**, 801 (1966); *Proc. Int. Conf. Electron Microsc., 6th, Kyoto* Vol. 1, p. 147. Tokyo Manizen Co., Tokyo, 1966.
79. P. Bonjour, *Proc. Eur. Congr. Electron Microsc., Jerusalem* (D. G. Brandon, ed.) Vol. 1, p. 63. Tal, Jerusalem, 1976.
80. Y. Iwasa and D. B. Montgomery, "Applied Superconductivity" (V. L. Newhouse, ed.), Vol. II, p. 387. Academic Press, New York, 1975.
81. A. Laberringue, P. Levinson, G. Bergeot, and P. Bonhomme, *Ann. Univ. ARERS* **5**, 50 (1967).
82. P. G. Merli and U. Valdrè, *Proc. Eur. Congr. Electron Microsc., Rome* (D. S. Bocciarelli, ed.) Vol. 1, p. 197. Tipografia Poliglotta Vaticana, Rome, 1968.
83. J. Trinquier and J. L. Balladore, *Proc. Eur. Congr. Electron Microsc., Rome* (D. S. Bocciarelli, ed.) Vol. 1, p. 191. Tipografia Poliglotta Vaticana, Rome, 1968.
84. P. G. Merli, *Proc. Int. Congr. Electron. Microsc. 7th, Grenoble* Vol. 2, p. 100. Société Francaise Microscopie Electronique, Paris, 1970.
85. G. V. Der-Shvarts and I. S. Makarova, Izv. Akad. Nauk SSSR Ser. Fiz **32**, 932 (1968) [*Engl. transl.: Bull Acad. Sci. USSR Phys. Ser.* **32**, 866 (1968)]
86. N. Kitamura, M. P. Schulhof, and B. M. Siegel, *Appl. Phys. Lett.* **9**, 377 (1966).
87. S. Ozosa, S. Katagiri, H. Kimura, and B. Tadano, *Proc. Int. Conf. Electron Microsc. 6th, Kyoto* Vol. 1, p. 149. Tokyo Manizen Co., Tokyo, 1966; S. Ozasa, S. Katagari, N. Kitamura, and H. Kimura, *J. Electron Microsc.* **17**, 240 (1968); S. Ozasa et al., *Proc.*

Int. Conf. Electron Microsc. 7th, Grenoble Vol. 1, pp. 121, 123. Société Francaise de Microscopic Electronique, Paris, 1970.
88. D. Génotel, A. Laberrigue, P. Levinson, and C. Séverin, *C. R. Acad. Sci. Paris B,* **265,** 226 (1967).
89. A. Laberrigue, P. Levinson, and J-C. Homo. *Rev. Phys. Appl.* **6,** 453 (1971).
90. P. Bonhomme and A. Laberrigue, *J. Microsc.* **8,** 795 (1969).
91. P. Bonhomme, A. Beorchia, and A. Laberrigue, *Optik* **39,** 39 (1973).
92. A. Laberrigue *et al.,* "High Voltage Electron Microscopy" (P. R. Swan, C. J. Humphreys, and M. J. Goringe, eds.), p. 108. Academic Press, New York, 1974.
93. J. Triniquier, J. L. Ballodore, and J. P. Martinez, *Proc. Eur. Congr. Electron Microsc., 5th* p. 118. Institute of Physics, London, 1972.
94. T. Tringuier and J. L. Balladore, *Proc. Int. Congr. Electron. Microsc., Grenoble* Vol. 2, p. 97. Société Francaise de Microscopic Electronique, Paris, 1970.
95. P. Bonjour and A. Septier, *Proc. Eur. Congr. Electron Microsc., Rome* (D. S. Bocciarelli, ed.) Vol. 1, p. 189. Tipografia Poliglotta Vaticana, Rome, 1968.
96. A. Septier, "Electron Microscopy in Material Science" (U. Valdré, ed.), p. 37. Academic Press, New York, 1971; A. Septier, *Eur. Cong. Electron Microsc., 5th* p. 104. Institute of Physics, London, 1972.
97. P. Bonjour, *J. Microsc.* **20,** 219 (1974).
98. R. E. Worsham, J. E. Mann, and E. G. Richardson, *Proc. Annu. EMSA Meeting, 29th, Boston* (C. J. Arcencaux, ed.) p. 10. Claitor, Baton Rouge, 1971.
99. I. Dietrich *et al., Optik* **45,** 291 (1976).
100. I. Dietrich, R. Weyl, and H. Zerbst, *Cryogenics* **7,** 178 (1967).
101. I. Dietrich, A. Koller, and G. LeFranc, *Optik* **35,** 468 (1972).
102. I. Dietrich, G. LeFranc, R. Weyl, and H. Zerbest, *Optik* **38,** 449 (1973).
103. H. Boersch, O. Bostanjoglo, and B. Lischke, *Optik* **24,** 460 (1966–1967).
104. G. Berjot, P. Bonhomme, F. Payen, A. Beorchia, J. Mouchet, and A. Laberrigue, *Proc. Int. Conf. Electron Microsc.* Vol. 2, p. 103. Société Francaise de Microscopic Electronique, Paris, 1970.
105. H. Kawakatsu, F. H. Plomp, and B. M. Siegel, *Proc. Eur. Conf. Electron Microsc., Rome* (D. S. Bocciarelli, ed.) Vol. 2, p. 193. Tipografia Poliglotta Vaticana, Rome, 1968.
106. M. Hibino, D. F. Hardy, F. H. Plomp, H. Kawakatsu, and B. M. Siegel, *J. Appl. Phys.* **44,** 4743 (1973).
107. K. Kobayashi and N. Uyeda, *Int. Conf. Electron Microsc., Canberra* (J. V. Sanders and D. J. Goodchild, eds.) Vol. 1, p. 264. Australian Academy of Sciences, Canberra, 1974.
108. D. F. Hardy, *Adv. Opt. Electron. Microsc.* **5,** 201 (1973).
109. I. Dietrich, "Superconducting Electron Optic Devices." Plenum, New York, 1976.
110. P. W. Hawkes and U. Valdrè, *J. Phys. E. Sci. Instrum.* **10,** 309 (1977).
111. I. Dietrich *et al., Ultramicroscopy* **2,** 241 (1977).
112. M. Pluchery, *J. Microsc.* **8,** 771 (1969).
113. C. Passow, *Optik* **44,** 427 (1976).
114. P. A. Stoyanov, T. V. Artemova, E. V. Susov, and A. A. Talashev, *Izv. Akad. Nauk SSSR Ser. Fiz* **32,** 928 (1968).

Metallurgy of A15 Conductors

THOMAS LUHMAN

Department of Energy and Environment
Brookhaven National Laboratory
Upton, New York

I. Introduction	221
II. A15 Compound Stability	224
III. Conductor Processing Methods	231
A. Liquid Solute Diffusion	232
B. Chemical Vapor Deposition (CVD)	233
C. Physical Vapor Deposition (PVD); Electron Beam and Sputtering Deposition	234
D. Solid State Diffusion (Bronze Process)	235
E. Variations on the Solid State and Liquid Solute Diffusion Techniques	238
IV. Relationships between Superconducting Properties and Microstructure in Bronze-Processed Conductors	240
A. Introduction	240
B. Compound Growth and Its Effect on T_c	240
C. Critical Current Density and Microstructure	244
D. Phenomenological Flux-Pinning Theories	250
V. Superconducting Critical Currents, Temperatures, and Magnetic Fields of Nb$_3$Sn Wire Conductors under Tensile Strain	254
A. Introduction	254
B. Stress–Strain Relationships	254
C. Degradation of Superconducting Properties under Mechanical Strain	256
References	263

I. Introduction

This chapter reviews the metallurgical aspects of fabricating and testing A15 conductors. The nomenclature A15 refers to a series of compounds having the cubic Cr$_3$Si (A$_3$B) structure. Several of these compounds have relatively high superconducting transition temperatures T_c, namely, those compounds for which the A atom is either Nb or V and the B atom is a nontransition metal from group III or IV of the periodic table. These particular A15 compounds also exhibit high upper critical field values H_{c2}, of the order of several tens of teslas. Their technological importance stems from a combination of high T_c, H_{c2}, and current carrying capacity J_c. Close packed in structure, brittle, and in some cases metastable at the stoichiometric composition, these compounds present a variety of challenging problems in their utilization as conductors for magnetic field generation and power transmission. The physical metallurgy of A15 compounds has been discussed by Dew-Hughes in an earlier chapter in this volume.

Of particular importance to the development of A15 conductors was the observation in 1960 by Bozorth et al. [1] that the critical magnetic field for Nb_3Sn exceeded 7.0 T at 4.2 K. The following year, Kunzler et al. [2] produced the first wires of Nb_3Sn by compacting Nb and Sn powders in a Nb tube and then drawing the composite down to a wire. The Nb_3Sn compound could be formed by reaction at a temperatture above 930°C. This process was also originally used for V_3Ga and later tried for $Nb_3(Al, Ge)$ [3, 4] and Nb_3Al [5, 6] powders. (See, for example, discussions in Section III,E.) A variation of this technique developed by Martin et al. [7] and refined by Saur and Wurm [8] involved Sn coating of Nb wire followed by a similar high-temperature heat treatment. The necessity of using a reaction temperature above 930°C for Nb_3Sn, combined with the inherent brittleness of the A15 compound, put severe limitations on the manufacture and use of these conductors. Nevertheless, by using appropriate coil former and insulating materials experimental laboratory magnets were constructed. Using this method, known as "wind and react," the conductor is first wound into a solenoid and then converted to Nb_3Sn by heat treatment of the entire solenoid. Although fields as high as 10 T were generated, these early wind and react A15 wire conductors were unacceptable commercially since they suffered from severe flux jumping and damage during quenching to the normal state.

The problem of fabricating brittle A15 compounds into useful conductor shapes was first considered by Hanak et al. [9] and Stauffer [10], who used a chemical vapor deposition technique to deposit micron-thick layers of Nb_3Sn onto ribbon substrate. The thin layers were placed in compression by deposition onto a substrate whose expansion coefficient exceeded that of the compound, thus minimizing the compound's sensitivity to mechanical damage. In this manner somewhat flexible tape conductors were manufactured. Nb_3Sn tapes were also developed by Benz and co-workers [11, 12], by coating a Nb ribbon with liquid Sn and reacting it at 930°C or higher.

Just as for the early wire conductors, however, these tape conductors suffered from premature quenching and burnout. Quenching was due primarily to geometrical design considerations that were not fully recognized

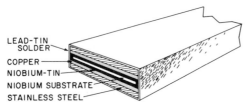

LEAD-TIN SOLDER
COPPER
NIOBIUM-TIN
NIOBIUM SUBSTRATE
STAINLESS STEEL

Fig. 1.   Schematic of tin-dipped Nb₃Sn tape conductor.

at the time (see the chapter by Dahl, this volume). To prevent conductor burnout during quenching a parallel current path of high-conductivity Cu or Al was needed. Figure 1 presents a schematic view of an early Nb₃Sn tape conductor containing individual layer components of Nb₃Sn, pure Cu, Nb ribbon substrate, and a stainless-steel reinforcer. In this composite tape, the Cu provides protection against burnout during quenching to the resistive normal state while the stainless steel provides a high elastic modulus to minimize strain in the brittle compound. Similar utilization of composite technique was seen in the early cabled wire conductors. Figure 2 illustrates the mechanical and electrical reinforcement components of a bundled seven-strand wire cable known commercially as "Cryostrand" [11]. This cable was manufactured by tin dipping the unreacted, bundled, niobium strands.

Toward the latter half of the 1960s it was shown both theoretically and experimentally that flux jumping is absent in superconductors that are less than about 0.05 mm in diameter [13–15]. (See the introductory chapter by Dew-Hughes, and the chapter by Dahl, this volume.) This work provided the incentive for Kauffman and Pickett (Nb₃Sn) [16] and Tachikawa and Tanaka (V₃Ga) [17, 18] to develop a solid state diffusion technique that eventually led to the commercial production of fine-filament A15 con-

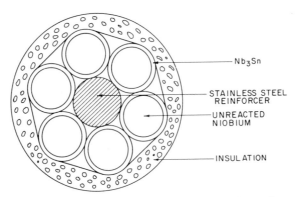

Nb₃Sn

STAINLESS STEEL REINFORCER

UNREACTED NIOBIUM

INSULATION

Fig. 2.   Bundled seven-strand cable known commercially as Cryostrand.

ductors (see Section III,D). This new technique, referred to as the bronze process, takes advantage of the fact that Nb_3Sn and V_3Ga are formed by a solid state reaction at an interface between Nb(V) and a Cu–Sn(Ga) bronze. Conductors containing small A15 filaments (1–10 μm) were produced by drawing ductile composites of Nb(V) rods, in appropriate bronze matrices, to wire dimensions. Heat treatment at 575–750°C formed the A15 compound and could be implemented either prior to magnet construction or after, using a wind and react technique.

Presented in Fig. 3 is a cross-sectional view of a bronze-processed Nb_3Sn multifilament conductor. Note the cylindrical A15 compound that has formed at the bronze–Nb interfaces. Filaments of pure Cu are often incorporated into present-day conductors, as are filaments of tungsten or stainless steel. High electrical conductivity Cu in bronze-processed composite wires must be protected against contamination by the bronze solute during compound formation heat treatments. Figure 4 presents a cross-sectional view of Nb_3Sn bronze-processed multifilament wire containing pure Cu; the Cu is protected from Sn contamination by a Ta tube that acts as a barrier to Sn diffusion during reaction heat treatments. With the advent of practical A15 conductors such as that in Fig. 4, higher magnetic fields and wider temperature ranges between operating temperatures and the material's T_c were achievable. These improvements in the conductor's superconducting properties, when coupled with the lower refrigeration costs that accrue from the use of higher operating temperatures, make the A15 multifilament wires very appealing. Indeed, they are receiving serious consideration for future projects such as plasma confinement magnets in controlled thermal nuclear reactors [19], high-energy particle accelerators [20], and power transmission lines [21]. It seems that realization of the full potential of superconductivity will be further enhanced by the continued successful development of usable A15 conductors.

II. A15 Compound Stability

Of the many known superconducting A15 compounds, only two, V_3Ga and Nb_3Sn, have as yet been developed as commercial conductors. Three binary compounds, Nb_3Al, Nb_3Ga, and Nb_3Ge, and one ternary compound, $Nb_3(Al, Ge)$, have higher values of T_c and H_{c2} than do Nb_3Sn and V_3Ga, but none of these has yet been successfully developed into a commercial conductor. The reason for this failure is that the latter compounds cannot be produced either by directly reacting solid niobium with the appropriate liquid or by a solid state interfacial reaction between Nb and an appropriate bronze.

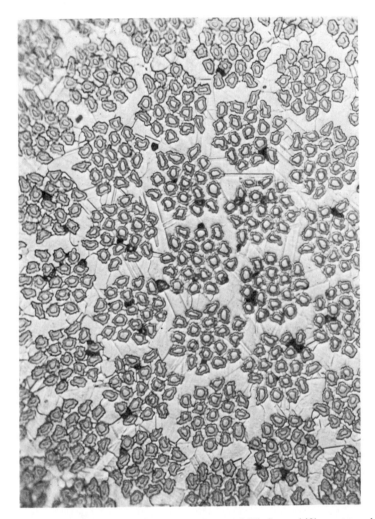

Fig. 3. Cross-sectional view of a bronze-processed Nb₃Sn multifilament conductor. Note the cylindrical shape of the A15 compound located at the interface between the Cu–Sn matrix and the Nb filaments (500×). [From D. Dew-Hughes, *Cryogenics,* August, 450 (1975).]

In the Nb–Sn and V–Ga binary systems the A15 compounds are the only compounds stable above 930 and 1300°C, respectively. This high-temperature thermodynamic stability permits direct compound formation by heat treatment of elemental diffusion couples. For example, reaction of Sn-dipped Nb tape above 930°C results in Nb_3Sn formation at the Nb–Sn surface. Sn-coated Nb heat treated at temperatures less than 930°C produces a spectrum of interfacial compounds, including Nb_3Sn, Nb_6Sn_5, and $NbSn_2$ [22]. Heat treatment of V–Ga couples below

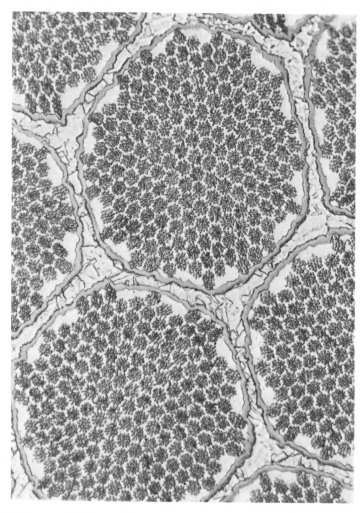

Fig. 4. Cross-sectional view of unreacted composite showing Nb filaments in a bronze matrix. The bronze matrix is separated from the pure Cu constituent by Ta diffusion barriers (500×). [From D. Dew-Hughes, *Cryogenics,* August, 450 (1975).]

~ 1300°C produces V_3Ga_2 and VGa_2 [23]. The presence of these additional compounds restricts A15 compound formation by consuming the available Sn or Ga, thereby degrading the superconducting properties. Furthermore, the requirement of a high-temperature heat treatment inhibits control of metallurgical structure, including, for example, grain size. Thus control over structure-sensitive superconducting properties such as critical current density is also inhibited.

The reason why stoichiometric high-T_c Nb_3Al, Nb_3Ga, and Nb_3Ge are not formed by direct reaction between constituent elements is related to their relative stability compared with competing compounds in the respective phase systems. In each case one or more solute-rich compounds are more stable, having higher melting temperatures, and are therefore preferred to form. (See the discussion in Section II,B of the chapter by Dew-Hughes on the physical metallurgy of A15 compounds, this volume.) In theory it ought to be possible to produce a conductor of $Nb_3(Al, Ge)$ or perhaps Nb_3Al, (see the phase diagram in Dew-Hughes's chapter on the physical metallurgy of A15 compounds, this volume), by dipping a Nb tape or wire into a melt. At high temperatures of $\sim 1800°C$, the ternary A15 compound is the system's only stable compound. A discussion of a variation on this approach, utilizing a solid Al–Ge alloy core in a Nb tube, is given in Section III,D. When heated to $\sim 1800°C$ the molten alloy forms the ternary compound by reaction with the Nb tube.

The ability or inability of the bronze process to produce single-phase A15 compounds can only be understood by reference to the relevant Nb–Cu–X ternary phase diagram. Formation of phases by diffusion between elements or other phases is governed by thermodynamic considerations. In multicomponent systems, diffusion follows the two-phase tie lines [24]. The only phases that will form in a diffusion couple are those phases that lie on the most direct route, following the two-phase tie lines, between the two initial components of the couple. Of the phases that can form, the ones most likely to form are those with the highest relative thermodynamic stability [25]. The most stable phases will generally have larger homogeneity ranges and will be the termini for more tie lines than will be the less stable phases. This tends to ensure that the more stable phases are most likely to be encountered on the tie-line routes.

Three types of phase diagrams appropriate to the formation of A15 compounds by the bronze process may be recognized [26]:

(a) Those in which the A15 phase is the only relevant stable phase, other than the terminal Nb(V) and Cu-based solid solutions. The Nb–Sn–Cu diagram, of which the preferred version is that due to Hopkins *et al.* [27], is an example of this type. Figure 5 is an estimate of the 700°C isotherm for this system, deduced from the 1100°C section of Hopkins *et al.* This diagram shows that the diffusion path from the Cu–Sn solid solution to the Nb–Sn solid solution passes through the A15 Nb_3Sn phase field. V_3Ga is formed with the same readiness as Nb_3Sn, and the V–Ga–Cu ternary diagram is similar to that of Nb–Sn–Cu [28].

(b) Those in which the most stable phase is one in the Nb(V)–X bi-

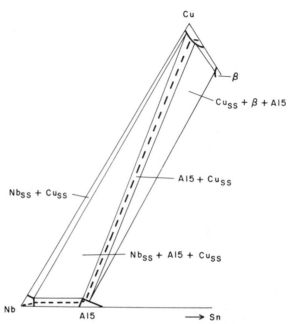

Fig. 5. Estimate of the Nb–Sn–Cu ternary phase diagram at 700°C. (After Hopkins *et al.* [27].)

nary, with a lesser Nb(V) content than the A15 phase. There is no direct tie line to the A15 phase, and the A15 may not even lie on the direct tie-line route from the most stable phase to the Nb–X solid solution. An example of this type is the Nb–Ge–Cu diagram, a section of which, again from the work of Hopkins *et al.* [27], is shown in Fig. 6. There are in fact two stable phases: Nb_5Ge_3 and Nb_3Ge_2. Two possible diffusion patterns are indicated, depending upon the initial concentration of germanium in the bronze. In both cases the resultant reaction layer is expected to be Nb_5Ge_3, as is found experimentally [25, 27].

(c) In the third type of system, the most stable phase is a new, ternary phase that bears no relation to, and whose existence cannot be predicted from, any of the three constituent binary systems. The example here is Nb–Al–Cu [29]; a 1000°C isothermal section of this diagram is shown in Fig. 7. Two ternary compounds, χ phase and a Laves phase $Nb(Al_xCu_{1-x})_2$, block the tie-line routes between the Cu–Al bronze and the Nb–Al solid solution. Which one of these actually forms can be seen to depend upon the starting composition of the bronze. The wide homogeneity range of the Laves phase suggests that it is very stable. This is the phase found by Luhman *et al.* [25] for their Nb/Cu–Al diffusion couple.

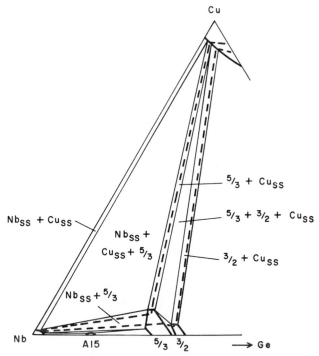

Fig. 6. Isotherm of the Nb–Ge–Cu ternary phase diagram at 700°C.

The foregoing explains why Nb_3Ge and Nb_3Al cannot be formed by the bronze process. The Nb–Cu–Ga ternary diagram has not been investigated, but it is probably of a type intermediate between (a) and (c) [26]. The products of a Nb–Cu(Ga) diffusion couple are nonstoichiometric, low-T_c Nb_3Ga and a series of other Nb–Ga compounds richer in gallium [25]. Laves and χ phases similar to those found in Nb–Al–Cu have

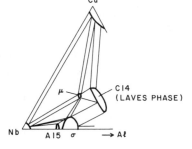

Fig. 7. Isotherm, 1000°C, of the Nb–Al–Cu phase diagram. (After Hunt and Raman [29].)

been reported. An attempt to form $Nb_3(Al,Ge)$ by diffusion between pure niobium and $Cu(Al,Ge)$ ternary bronzes produced only Nb_3Ge_2 [25].

Livingston has illustrated the effect of bronze composition on solute activity in a $V-Cu + Si$ bronze diffusion couple [30]. The activity of Si in the bronze is a function of the Si composition, decreasing with decreasing Si content. Two compounds, V_3Si and V_5Si_3, are normally observed in bronze diffusion couples heat treated at $\sim 700°C$. After initiating compound formation by an appropriate heat treatment, Livingston diluted Si in the bronze by cladding the samples with pure Cu. The resulting alloy was thus depleted in Si and the thermodynamic activity of Si reduced accordingly. Decreasing the Si content in this manner had the effect of reducing the volume fraction of the $5:3$ phase while increasing the layer thickness of the $3:1$ A15 phase. This approach to A15 compound formation should also work in other systems. It has been attempted in the Nb–Ga system, where it was found that kinetic factors inhibited layer growth when the Ga concentration of the bronze was lowered [31].

Another approach to forming A15 compounds via the solid state technique is to marry the stability of either Nb_3Sn or V_3Ga with the more desirable superconducting properties of the less stable compounds. Ternary additions of Al, Ga, Ge, etc. to the binary bronzes has met with some success, for it has been found that ternary Cu–Sn–Ga bronzes react with Nb to form a pseudo-binary $Nb_3(SnGa)$ compound with a $T_c \approx 0.5$ K above Nb_3Sn and J_c at high magnetic fields ($H_a > 12$ T) increased by a factor of 2 [32]. H_{c2} can also be increased from 22 to 25 T. Unfortunately, in order to retain the ductile properties associated with the bronze matrix, ternary additions must be limited to the Cu solid solution range. Thus, raising the ternary alloy content requires a concomitant decrease in the Sn concentration and an associated decrease in the superconducting properties of the A15 phase. The upper limit for an increased Ga concentration is approximately 8 wt. % Ga.

The author has found that ternary additions of Ge to Cu + Sn bronzes result in formation of the phase Nb_3Ge_2 when Ge concentrations in the bronze exceed 1 wt. % Ge. This phase acts as a diffusion barrier, preventing formation of the $3:1$ compound. For lesser Ge contents, thin layers of Nb_3Ge_2 ($\sim 1-2 \mu m$) permit limited Sn diffusion through the interface and some $Nb_3(SnGe)$ compound formation. Although electron microprobe analysis indicated nearly 0.5 wt. % Ge present in the A15 compound, no enhancement of T_c or J_c was observed.

In summary, it appears that the six high-T_c A15 compounds can be listed according to decreasing stability as Nb_3Sn, V_3Ga, V_3Si, Nb_3Al, Nb_3Ga, and Nb_3Ge. Generally, the first three compounds can easily be made at the stoichiometric composition by arc melting and annealing. These three compounds are formed via the bronze diffusion method, V_3Si forming with an additional compound V_5Si_3. A common feature pertaining to the stability of these three A15 compounds is their characteristic martensitic transformation to tetragonal structures on cooling below room temperature. This transformation is structure and sample sensitive. Some evidence exists that indicates even the commercially available composite A15 conductors may undergo such transformations [33].

Compound instability is reflected in Nb_3Al by the equilibrium compositions being off the 3:1 stoichiometry, toward Nb-rich compositions for all temperatures below ~1750°C. In order to prepare high-T_c stoichiometric Nb_3Al, anneals at high temperature are required in combination with rapid quenching and ordering heat treatments. Nb_3Al does not form when solid state bronze techniques are employed. Fabricating conductors of Nb_3Al has been limited to powder, thin foil, and physical vapor deposition techniques (see Section II,D for references and further discussion). Second-phase material is often present following heat treatments and although reasonably good T_c and J_c values have been obtained, filament integrity has always been inferior to Nb_3Sn and V_3Ga produced by the bronze technique.

High-T_c Nb_3Ga can only be produced by physical vapor deposition or rapid quenching from the melt with post ordering anneals. To date no effective means of fabricating Nb_3Ga multifilament conductors has been devised. Finally, it is virtually impossible to make bulk samples of high-T_c (>21 K) Nb_3Ge by any combination of melting and quenching. Nb_3Ge is metastable at the stoichiometric composition and has only been produced with T_c's greater than 21 K by chemical or physical vapor deposition under stringent experimental conditions. These techniques approximate ultra-high quenching rates and some evidence exists that in addition to the high quenching rates achieved by evaporation, the presence of oxygen is also necessary [34–36]. Chemical vapor deposition (CVD) and physical vapor deposition (PVD) techniques offer promise for production of several A15 materials in tape form, such as that used in ac superconducting power transmission. Considerably more technical development is required to adapt them to the fabrication of multifilament composite wires, although single strands of a W 1 wt. % ThO_2 wire have successfully been coated with Nb_3Ge using a chemical vapor deposition technique [37].

III. Conductor Processing Methods

There are several processing techniques for making Nb_3Sn and V_3Ga tape and wire conductors. In this section, each technique is summarized and the main advantages and disadvantages are pointed out with respect to individual applications. Nb_3Sn has received more attention than V_3Ga primarily due to its higher T_c (18.2 K as compared with 15.0 K) and the lower cost of tin than gallium.

A. *Liquid Solute Diffusion*

Liquid solute diffusion is one of the initial processes that were used commercially to produce Nb_3Sn tapes for high-field magnets. The process consists simply of introducing a thin niobium tape ($\sim$25 μm thick) into a molten tin bath. The tin-coated tape is then heated at 930–1100°C in vacuum or an inert atmosphere to form Nb_3Sn layers on both sides of the tape.

There are several metallurgical parameters controlling the morphology of these Nb_3Sn layers. The most important one, with regard to ac losses, is the amount of Sn left on the Nb tape after passing through the tin bath. When the amount of tin on the tape is limited so that all the tin is used in forming thin Nb_3Sn layers, these layers will be dense throughout and their surfaces smooth. When there is an excess of tin on the tape, thicker layers can be grown but the Nb_3Sn layers are porous near their surfaces and the surfaces are rough. This is due to the penetration of excess liquid tin into the Nb_3Sn grain boundaries. For tapes made using excess tin, further surface treatments are required to reduce ac losses. At present both of the approaches above are used by individual producers. Kawecki Berylco Industries has been producing Nb_3Sn by the limited tin process while Intermagnetics General Company has developed the excess tin process, with subsequent surface treatment, for their magnet and ac transmission line applications.

In order to enhance critical current densities of liquid tin diffused Nb_3Sn, it is now common practice to form ZrO_2 in the niobium tapes and to add copper or lead into the tin bath. The ZrO_2 precipitates minimize the size of Nb_3Sn grains while the copper or lead additions increase the growth rate of the Nb_3Sn layers. The exact roles of these additives in improving the ac loss characteristics of Nb_3Sn are not yet clear. The kinetics of this process have been discussed in a review paper by Echarri and Spadoni [38].

An advantage of the liquid solute process is that it can easily be developed commercially for producing long lengths of tape for magnet fabrication. Although some modifications, such as surface etching, have had to be made for use of these tapes in low-field ac application, these are not major obstacles. Copper can be soldered directly to the reacted ribbon for electrical stabilization, while the entire tape can be encased in stainless steel for protection and mechanical strength (see Fig. 1). Nb_3Sn tape made by the liquid solute method is flexible, providing the compound layer is sufficiently thin $(< \sim 2 \ \mu m)$ and close to the tape's neutral axis. Typical commercial tapes can be bent around a radius of ~ 1 cm without suffering deterioration of their superconducting properties.

The original V_3Ga tapes made by Tachikawa and co-workers [39] involved Ga coating of V ribbon. In this case an additional layer of Cu was clad onto the tapes to form a Cu–Ga bronze, thereby suppressing formation of V_3Ga_2 and VGa. This also provided for a lower reaction temperature (600–750°C).

As discussed in Section II, direct diffusion of other solute elements into Nb to form A15 conductors is limited by the relative stability of the A15 compound. In the cases of Nb tapes with Ga, Ge, or Al, formation of other nonsuperconducting compounds is favored. Only Nb_3Al can be formed by the liquid solute method, and this requires reaction temperatures in excess of 1400°C where control over microstructural parameters is very difficult [40, 41].

B. Chemical Vapor Deposition (CVD)

CVD is accomplished by the hydrogen reduction of a mixture of the chlorides of the constituent elements. Hanak and co-workers, working on Nb_3Sn, developed the first successful CVD deposits [42, 43]. By passing chlorine over the metallic elements at 800–900°C the gaseous chlorides are produced. In the presence of HCl, which suppresses formation of solid $NbCl_3$, the chlorides are reduced by hydrogen and deposited onto heated substrates, normally niobium, stainless steel, or Hastelloy. A15 layer thickness is controlled by the rate at which the substrate is passed through the reaction chamber. Variations on the CVD process have been reviewed [38]. Recently, the CVD process was developed for Nb_3Ga [44] and Nb_3Ge [45–47]. The superconducting properties of these compounds are very sensitive to factors like deposition temperature and flow rates, indeed more sensitive than was seen for CVD Nb_3Sn. In all cases, various impurity gases such as O_2, CO, and CH_4 have been added to improve the critical current densities. Braginski et al. have achieved self-field J_c's of $\sim 10^7$ A/cm^2 at 4.2 K in coarse-grained CVD Nb_3Ge by incorporating a fine dispersion of second-phase particles [48]. Deposits of Nb_3Ge have been formed on copper substrates with T_c's close to 22 K and critical current densities of 1.8×10^6 A/cm^2 at 13.8 K by Newkirk et al. [47].

Several experimenters have demonstrated the feasibility of applying the

CVD process to the preparation of continuous high-strength tape and wire conductors [37, 48]. Some difficulty in producing tapes by the CVD process comes with the application of the normal metal cladding needed for flux jump stability. In the past this has been done by electrolytic deposition of a nickel flash followed by a silver or copper deposit. The resulting normal metal layers of Cu or Ag have had poor residual resistance ratio values (~ 10), while the nickel flash has contributed large ferromagnetic losses [49]. Airco has developed a method for electron beam evaporation of high-purity copper cladding that appears to provide a solution to this problem [50].

Another aspect of some CVD clad tapes is the possibility of diffusion of nickel from the substrate into the Nb_3Sn. This results in degraded superconducting properties at the Nb_3Sn–substrate interface. At present, there are limited data on this point, although measurements have shown high losses [51]. It is to be expected that low loss values, comparable with those for other processes, would result if a similar development effort were applied in the case of the CVD process. Further successful development of CVD Nb_3Ge will certainly advance the potential of this compound as a viable commercial superconducting material.

C. Physical Vapor Deposition (PVD); Electron Beam and Sputtering Deposition

Electron beam vapor deposition (EBD) and sputtering deposition (SD) methods are possible candidates for commercial-scale production of A15 conductors. In contrast to CVD, EBD and SD are adaptable to independent control of the composition, individual atoms being placed onto the substrate. The substrate temperature is also easily varied. Hammond has reviewed the accomplishments and prospects of fabricating A15 compounds by EBD [52]. Among these, EBD, through control of the substrate temperature, provides for grain size control as well as synthesis of compounds stable only at low temperatures. Other features include control over compound thickness and the capability of producing ternary compounds.

Electron beam techniques have been developed extensively for fabrication of superconducting power transmission cables. One technique, which has been used successfully to fabricate low-loss Nb_3Sn, is coevaporation of elemental Nb and Sn from separately controlled electron beam (EB) sources onto a substrate [53]. Although it was shown that suitable Nb_3Sn can be made with this technique, it should be pointed out that this method is rather "technologically intensive." For example, as many as four evaporation monitors, which control EB currents through feedback loops, are required to produce the desired Nb_3Sn. When these sensitive monitors are required for fabrication of long-length conductors, long-term reliability of such a system may become troublesome. This method offers some definite advantages, however, over more conventional techniques since it permits a variation in composition of deposited materials and freedom in choosing substrate materials. Fine layering of Nb_3Sn with nonsuperconducting materials has been used to achieve high current densities in Nb_3Sn [52, 54]. Coevaporation by EB techniques also allows deposition in controlled environments. Oxygen levels of $\sim 5 \times 10^{-7}$ Torr, for example, are found to stabilize high-T_c (>21 K) Nb_3Ge over a wide range of Nb/Ge flux ratios [35]. This is an important consideration for the long-term reliability requirement in commercial-scale applications.

There is a variation on this technique, pursued by Airco, based on the use of EB evaporation [50]. Here a Cu–Sn alloy is deposited onto a Nb substrate and the composite is heat treated to form Nb_3Sn. In this particular case, the evaporation rate monitoring is not as critical as in the coevaporation method since the Cu–Sn alloy composition is not as critical. The Cu–Sn can be evaporated from a casting.

Typical EB deposition rates are ~ 0.1 μm/min. Dahlgren has successfully used a high-rate (>1μm/min) sputtering technique to obtain high-J_c deposits of Nb_3Al and $Nb_3(Al,Ge)$ [55–60]. J_c for thick deposits of $Nb_3(Al_{0.75}Ge_{0.25})$ is greater than that for Nb_3Sn at magnetic fields in excess of 10 T [60]. The A15 phase is obtained by heat treatment of a bcc phase deposited at low temperatures (10–400°C). Using the high-rate SD technique usually produces T_c values somewhat lower than the best reported values for the same compounds produced by other techniques. However, the highest known critical temperature to date, 22.3 K for Nb_3Ge, was produced by Gavaler with special SD conditions, that is, low deposition rates (<0.1 μm/min) and high deposition temperatures (>650°C) [61]. SD is a very important technique and has considerable potential for the manufacture of commercial conductors.

D. Solid State Diffusion (Bronze Process)

This process has been developed extensively in the last few years for Nb_3Sn and V_3Ga multifilamentary conductors [62]. Figure 8 schematically illustrates the solid state (bronze) method for fabricating multifilament Nb_3Sn and V_3Ga wires. Holes are drilled into one end of the starting Sn or

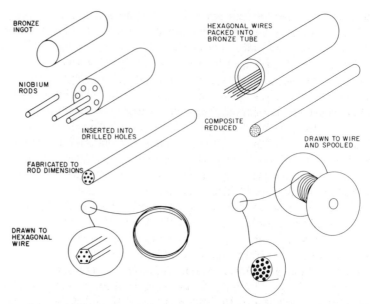

Fig. 8.   Schematic illustration of the bronze-process multifilament conductor production route.

Ga bronze ingot and Nb or V rods are inserted to make a starting composite. The composite is initially fabricated to a rod shape by drawing or ex-
truding. At this stage, the rod may either be drawn directly to wire dimensions or bundled with several rods inside a bronze tube and then drawn.
Often the individual rods are drawn through a hexagonal die to permit
convenient packing in the bronze tube. Depending upon usage, the composite wire may receive a final mechanical twisting so that the filaments
take up a helical path and thereby ensure electrical stability in changing
magnetic fields (see the chapter by Dahl, this volume). Formation of the
A15 compound occurs when the composite is heat treated for several
hours, usually at ~700 and 600°C for Nb_3Sn and V_3Ga, respectively.
Compound formation under these solid state conditions involves the se-
lective diffusion of the bronze solute to the filament–bronze interface and
a subsequent reaction with the filaments to form the A15 compound (see
Figs. 3 and 4). The heat-treatment procedure may occur either prior to or
after winding of the conductor into a solenoid shape. Twelve-mil wires
with more than 10,000 filaments, each less than 1 μm in diameter, are
commercially available today [63].

  The source of Sn or Ga in these composite wires can be either external, a coating on a pure Cu matrix, or internal, a Cu-based bronze matrix [63–65]. In the former instance the use of pure Cu offers the advantage of creating a ductile composite that may be drawn to wire dimensions without intermediate stress relief anneals. The chief disadvantages of this method arise from nonuniformities in coating thickness along long lengths of conductors. Variations of Sn(Ga) content within the Cu-based matrix affect the local growth and superconducting properties of the A15 compound. Insufficient solute tends to starve innermost filaments of Sn(Ga). If the alloy content exceeds the solid solubility limit in Cu, a Cu-based compound is formed in the alloy matrix and, owing to the induced matrix embrittlement, mechanical manipulation of the wires is restricted. Furthermore, when alloy additions to the bronze exceed its limit of solid solubility, A15 compound formation is restricted at those areas where the Cu-based compound contacts the filament surface. It is desirable to have the maximum Sn(Ga) content in the bronze be 9.1 and 19.9 at. % Sn and Ga, respectively [66]. A high-Sn(Ga) content ensures rapid compound growth, which in turn aids in producing large superconducting critical current densities (see Section IV).

An important consideration in the manufacture of A15 multifilament conductors using the bronze process is the overall wire critical current density. This value represents the intrinsic superconducting critical current density of the A15 compound diluted by such conductor components as the bronze matrix, nonsuperconducting metal reinforcers, and any pure copper added for electrical stability. Incorporation of pure Cu necessitates a further addition to the composite, namely, the barrier to prevent contamination and alloying of the Cu during working and heat treatment (see Fig. 4). The maximum A15-to-normal-metal ratio achieved in practical A15 multifilament conductors is approximately 1:3. The amount of Cu added to magnet windings must increase with stored energy in the system and therefore the superconductor-to-normal-metal ratio in large magnet systems is often less than 1:3.

Early observations of very low ac losses in cylindrical samples of bronze-processed Nb_3Sn at Brookhaven National Laboratory encouraged the development of bronze-processed tape conductors for power transmission [67]. Figure 9 illustrates the fabrication of Nb_3Sn tapes by this technique. Fabrication begins with the preparation of a rectangular composite billet in which a Nb sheet is placed in the central plane of a Cu–Sn alloy matrix. The matrix encapsulation can be achieved in a number of ways including alloy casting, electronic beam welding, extrusion, or explosive bonding. After rolling the billet to the desired final thickness and slitting it to width, the resulting tape is heated at an elevated temperature

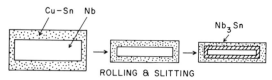

Fig. 9. Schematic illustration of Nb$_3$Sn composite tape fabrication by the bronze method.

(700–750°C) to form Nb$_3$Sn. Thickness of the Nb$_3$Sn layers is determined primarily by heat treating conditions.

The advantages of this process are that only conventional fabrication equipment is required and that heat-treatment temperatures are lower than for liquid processes. Since the formation reaction of Nb$_3$Sn occurs in the solid state and at low temperatures, Nb$_3$Sn surfaces are very smooth and grain growth is minimized. These properties of the material are thought to be the main reasons for the low ac losses of tape produced by the bronze process.

E. Variations on the Solid State and Liquid Solute Diffusion Techniques

An interesting variation on the bronze method has been developed by Tsuei [68] and refined by Roberge, Bevk, and co-workers [69]. In this method multifilamentary wires are fabricated by rapidly quenching (~1000°C/sec) a Nb–Sn–Cu alloy from the liquid state. Nb filaments result both from dendritic growth and from precipitation out of the supersaturated Cu–Sn alloy. This composite can be drawn to a wire and heat treated at temperatures between 600 and 800°C to produce Nb$_3$Sn filaments. In its final form, the composite consists of fine, multiply connected filaments of Nb$_3$Sn in a Cu and Sn matrix. Two critical parameters in this approach are the niobium content and quenching rate. The former must exceed 15 volume percent in order to give sufficient volume fraction of Nb filaments to ensure continuity. The latter must be uniform and sufficiently rapid to produce a fine and uniform structure. It has been demonstrated that the critical current densities of these materials are comparable to the best commercial multifilamentary Nb$_3$Sn wires [69].

A second alternative to achieving multifilament Nb$_3$Sn wire conductors involves a powder metallurgical approach. In this process, developed by Pickus et al. [70], niobium powder is initially cold isostatically compacted to a cylindrical shape. After compacting, the powder is vacuum sintered at ~2000°C to achieve a minimum porosity of ~25%. The compact is

then immersed in a molten tin bath at $\sim 750°C$ to infiltrate the interconnected pore network with tin. To draw wire from this tin-infiltrated cylinder requires double encapsulation with Nb and monel or copper tubing. A
combination of form rolling and wire drawing is then employed. The final
wire is heat treated for ~ 2 min at $950°C$ to form the Nb_3Sn compound.
These wires have shown critical current densities as high as for bronze-
processed material and T_c values that are even slightly higher, ~ 18 as
compared with ~ 17.5 K. The high current densities are almost certainly a
result of higher critical fields, which in turn result from the better T_c values. As a rule of thumb, the minimum bending radius for single-strand
multifilament A15 conductors is ~ 60 times the wire diameter. This critical
radius appears to be reduced by a factor of 2 in wires produced by infiltration of niobium powder. The infiltration method results in an interconnected network of superconducting paths rather than a geometrical
predetermined configuration. Presumably the better properties reflect the
availability of new electrical paths when original paths are damaged by
straining, i.e., a redundancy of some electrical paths. The infiltration
method is well suited to producing long lengths of superconductor and
since the current densities obtained thus far are as good as those achieved
in bronze-processed materials, this approach offers considerable promise.

Luo [70] has fabricated wire and tape conductors using a powder metallurgical technique to achieve a Tsuei-type composite structure. Employing a sintered mixture of the ductile alloy powders Cu–Sn and
Nb–Al, T_c values reaching 17.7 K and $J_c(4.2$ K) self-field values of the
order of 10^4 A/cm^2 were obtained. Both the powder and quenching techniques are in preliminary stages of development and their full potential
cannot yet be accurately evaluated.

It was pointed out in Section II that at high temperatures ($\sim 1750°C$)
Nb_3Al can be formed directly from elemental constituents. Several investigators have attempted to fabricate multifilament Nb_3Al wire and tape
using Al powder in niobium tubes [6, 71, 72] or niobium rods in an Al matrix [73]. The high reaction temperatures have limited the success of this
approach since they restrict control over such metallurgical parameters as
grain size (for critical current density) and Kirkendall porosity (for filament integrity). Lower reaction temperatures have been achieved by
Ceresara and co-workers when interdiffusion distances were reduced
between Nb and Al [74–76]. Using a "Swiss-rolled" foil composite,
Nb_3Al was produced by reaction at temperatures below $1000°C$. A critical
current density of 1.8×10^5 A/cm^2 at 6.4 T is reported after reaction at
$850°C$ for 6 h. T_c was only 15.5 K. Similar results were obtained in Nb_3Al
by Luhman *et al.* [66] who used a process called mechanical alloying to
achieve the short interdiffusion distances required.

IV. Relationships between Superconducting Properties and Microstructure in Bronze-Processed Conductors

A. Introduction

This section reviews the metallurgical parameters that control the superconducting properties of bronze-processed A15 conductors. Losses in alternating fields and their relationships to microstructural parameters are reviewed in the chapter by Bussière, this volume.

As discussed in the chapter by Dew-Hughes on the physical metallurgy of A15 compounds, this volume, the critical temperatures of Nb_3Sn and V_3Ga are very closely related to their stoichiometry, degree of crystallographic ordering, and levels of impurities such as oxygen. These compounds are thermodynamically very stable at their stoichiometric composition and therefore form nearly on stoichiometry. Atomic ordering is a function of heat treatment time, and the impurity levels can be adjusted through use of starting material and atmosphere control during formation.

While T_c depends on atomic ordering and impurity levels, critical current density is a function of microstructure and grain size in particular. The bronze process lends itself well to minimizing grain growth and maximizing grain boundary area. Through the low reaction temperatures, 550–750°C, fine-grained compounds are produced (< 1000 Å) with J_c values $\sim 10^6$ A/cm² at 5 T.

Inherent in bronze-processed compounds is a compressive strain on the superconductor [33]. This strain results from the differential thermal contraction between the bronze and filament materials. As a consequence of the strain, T_c and J_c values are somewhat depressed. (See Section V for further discussion.)

B. Compound Growth and Its Effect on T_c

All of the commercially available bronze-processed conductors are heat treated at temperatures ranging from 550 to 750°C. Compound growth rate under these solid state conditions depends upon heat treatment temperature and bronze solute concentration. Illustrated in Fig. 10 is the dependence of Nb_3Sn layer thickness on the atomic percentage of Sn in a single-filament bronze matrix [66]. With increasing Sn concentration in the matrix, the Nb_3Sn layer thickness increases rapidly for similar heat treatment times at 700°C. In conductors having large superconducting-to-normal-metal ratios, depletion of Sn from the bronze during compound formation can lead to lower growth kinetics and degra-

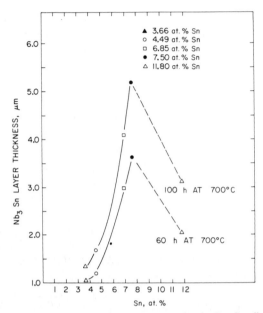

Fig. 10. Nb$_3$Sn layer thickness versus Sn concentration in Cu–Sn alloys heat treated at 700°C for 60 and 100 h.

dation of superconducting properties [77]. To minimize this effect, Sn may be alloyed with the Nb filament, thereby providing an additional Sn source during compound formation. Figure 11 demonstrates how an alloy addition of 2 wt. % Sn to a single core wire boosts the layer growth rate beyond that achieved from solid solution binary bronze alloys [78]. Similar effects were noted earlier by Howe and Weinman for bronze-processed V$_3$Ga, illustrated in Fig. 12 [79–81]. Alloying vanadium cores with up to 9 at. % Ga allows reaction temperatures as low as 525°C.

Layer growth of these compounds frequently follows a power law whose time exponent t^m varies from 0.5, the value expected for bulk diffusion through the existing layer. Figure 13 illustrates the observed Nb$_3$Sn layer thickness versus heat treatment time at 700°C for different matrix Sn concentrations in single-filament wires [66]. The layer growth rate follows approximately a 0.5 dependence for lower Sn concentrations (3.66 and 4.49 at. %), but is somewhat faster for the high Sn concentrations (6.85 and 7.50 at. %). Values exceeding $m = 0.5$ for Sn-rich matrices indicate that an additional mechanism is involved. The additional mechanism may be the formation rate of Nb$_3$Sn at the Nb–matrix interface. Although this point remains uncertain, it has been shown that grain boundary diffusion is an important factor contributing to the growth of Nb$_3$Sn in these wires

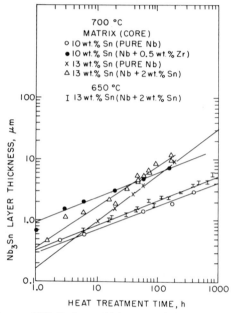

Fig. 11. Dependence of Nb$_3$Sn layer thickness on heat treatment time for various Sn contents in the bronze and Nb filaments.

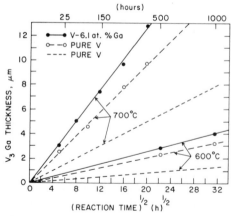

Fig. 12. Isothermal growth of V$_3$Ga on pure V and V 6.1 at. % Ga filaments in a Cu 15.4 at. % Ga matrix. (After Howe, Weinman, and co-workers [79]–[81].)

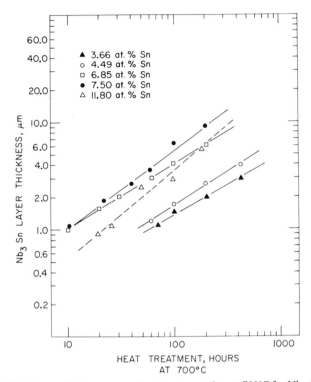

Fig. 13. Nb₃Sn layer thickness versus heat treatment time at 700°C for Nb–(Cu,Sn) alloy composite wires with various Sn concentrations in the alloy.

and such grain boundary diffusion is often associated with a time exponent exceeding 0.5 [82]. Experimental results obtained with conductors having higher superconducting-to-normal-metal ratios usually have $m < 0.5$. Low exponent values are attributed to solute depletion from the bronze during layer growth. It is not presently known how each of these factors interacts in determining overall layer growth rates.

Layer growth rate is rapidly increased when layer thickness becomes nearly equal to the filament radius. This is likely due to a large volume of bronze feeding a decreasing volume of Nb. An observation related to diffusion is the appearance of Kirkendall porosity resulting from Sn diffusion toward the interface. Such porosity in the bronze matrix can degrade the mechanical strength of conductors; however, it is usually only observed after extensive layer growth and seldom occurs for the 1–2-μm-thick layers typically used in commercial multifilament conductors [83].

Figure 14 illustrates how the superconducting transition temperature T_c varies with both reaction heat-treatment time and temperature

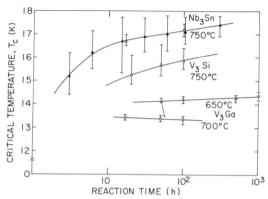

Fig. 14. Examples of the variation of the critical temperature T_c with heat treatment for Nb$_3$Sn, V$_3$Ga, and V$_3$Si composite wires.

for bronze-processed Nb$_3$Sn, V$_3$Ga, and V$_3$Si [84]. The symbols represent resistive transition midpoints (50% transformation) and the total temperature width is expressed through the use of error bars. The superconducting-to-normal transition is very broad (1–2 K) for short reaction times, whereas the transition midpoint increases and the transition width decreases as reaction time lengthens. In each system the onset remains somewhat below the T_c value usually quoted for well-annealed arc-melted material. These characteristics of the superconducting transition in bronze-processed Nb$_3$Sn have recently been shown to be influenced by a stress on the compound arising from thermal contraction of the bronze matrix [33, 85]. Figure 15 illustrates the dependence of T_c on reaction time at 700°C for single-filament Nb$_3$Sn conductors, both with the bronze matrix intact and after having had the matrix removed by etching. Elimination of the matrix-imposed stress by etching off the matrix decreases the temperature width of the transition and significantly raises the inductively measured T_c, particularly for shorter reaction times. T_c values for samples whose matrix has been removed reach those for arc-melted material, ~ 18.2 K, after only 24 h at 700°C. The remaining small decrease in T_c for low-ratio samples has been accounted for in terms of a residual plastic strain in the unreacted niobium core [86]. Similar strain-induced matrix effects have been observed for V$_3$Ga, although the maximum observed decrease in T_c is only ~ 0.25–0.5 K [27].

C.   Critical Current Density and Microstructure

While T_c depends on atomic ordering and composition, critical current density is primarily a function of microstructure. In these high-J_c mate-

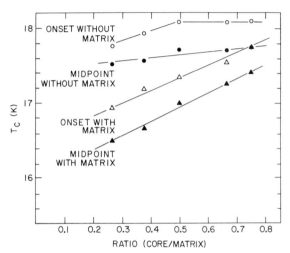

Fig. 15. Effect of matrix-to-core-diameter ratio on the superconducting transition of a Cu 13 wt. % Sn monofilament Nb_3Sn wire heat treated for 24 h at 700°C.

rials there are strong flux-pinning interactions at structural defects within the material. From the electron microscope investigations of Scanlan *et al.* [88] (Nb_3Sn), West and Rawlings [89] (Nb_3Sn), and Nembach and Tachikawa [90] (V_3Ga) it may be concluded that grain boundaries are the only defects present in sufficient quantity to account for the high critical current densities observed in the bronze-processed conductors. Recently Curzon *et al.*, using a Bitter pattern technique and scanning electron microscope, directly observed the flux-lattice–grain-boundary pinning interaction in the mixed state of V_3Ga [91].

Much of the critical current data generated on bronze-processed Nb_3Sn and V_3Ga has been interpreted in terms of flux-pinning interaction with grain boundaries. It has been established that J_c of Nb_3Sn at high magnetic fields is inversely proportional to its grain size until the grains become less than 300–500 Å in diameter. Below this grain size J_c decreases [89, 92]. Although a similar relationship has not been established at low magnetic fields, which are typically used in transmission line operation, it is probably safe to assume that a similar dependence of J_c on grain size will exist.

The grain size in Nb_3Sn and V_3Ga tends to be several thousands of angstroms or even larger unless special efforts are made to ensure minimum grain growth. Owing to this, it is necessary to maximize grain boundary area in order to obtain high critical current densities. The bronze process provides a technique for forming Nb_3Sn and V_3Ga at low temperatures, well suited to minimizing grain growth and maximizing grain boundary area and current densities. Figure 16 illustrates the critical current den-

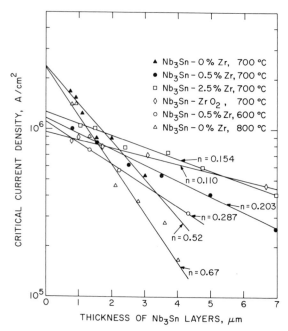

Fig. 16. Superconducting critical current densities at 4.0 T of Nb₃Sn and Zr-doped Nb₃Sn multifilament wires as a function of reaction layer thickness.

sities of several Nb₃Sn wires as functions of compound layer thicknesses at 4.0 T [93]. It is observed that with increasing compound layer thickness the critical current density decreases for all heat treatment conditions. This decrease is attributed to grain growth during layer formation which leads to an overall decrease in grain boundary area per unit layer thickness. Such changes in overall grain boundary density during layer growth have been observed in electron microscope studies on bronze-processed Nb₃Sn [88, 89, 94].

The situation with regard to V₃Ga is similar. Figure 17 illustrates the critical current density (and temperature) of bronze-processed V₃Ga as a function of heat treatment temperature [95]. Here, grain boundaries are also the major source of flux pinning and interpretations similar to those for the Nb₃Sn case are frequently made. That is, higher temperatures tend to increase the grain boundary area per unit layer thickness and consequently decrease J_c.

A common approach to achieving higher critical current densities involves metallurgically increasing the layer growth rate relative to the rate of grain growth. Figure 18 presents the critical current density of Nb₃Sn as a function of layer thickness for several samples, including Cu 10 wt. %

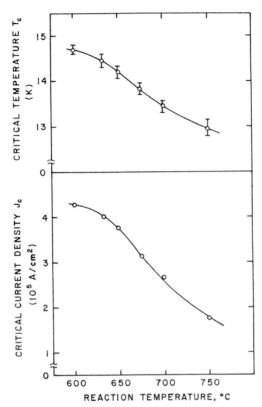

Fig. 17.   The critical current density at 4.0 T and critical temperature of V_3Ga as a function of reaction temperature.

Sn with a pure Nb core, Cu 13 wt. % Sn with a pure Nb core, and Cu 13 wt. % Sn with a Nb 2 wt. % Sn core [78]. The reaction rate for layer growth is increased by increasing the amount of Sn available. The high layer growth rates for Sn-rich samples are reflected in a decreasing dependence of J_c on d. For samples having Sn alloyed with the Nb filaments, critical current densities remain nearly constant at $\sim 2 \times 10^6$ A/cm^2 (4 T) as layer thicknesses approach 12 μm. For reasons of electrical stability bronze-processed tapes with high current densities and larger layer thicknesses have an important application in power transmission cables. Here high current densities are required throughout the entire compound layer thickness to provide security against fault currents that penetrate the surface layer. Howe and Weinman have added up to 9 at. % Ga to vanadium cores and produced V_3Ga with critical current values at 2.5×10^6 A/cm^2 at 4 T [81].

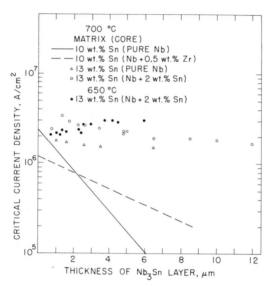

Fig. 18. Nb₃Sn critical current density at 4.0 T as a function of layer thickness.

A common metallurgical technique used to restrict grain growth involves doping with a grain boundary pinner such as ZrO_2. Referring back to Fig. 16, we can see the critical current densities versus layer thickness for a ZrO_2-doped multifilament conductor. Doping with ZrO_2 restricts grain growth in Nb_3Sn and effectively produces less dependence of J_c on d. Unfortunately the critical current densities of the initial nucleated layers have considerably lower current densities than do those for unalloyed Nb. This lowered current density is attributed to preferred heterogeneous nucleation in the doped samples. Such nucleation provides fewer and larger Nb_3Sn grains, which cause the lower J_c values. Consequently, although the doped compound layer grows with less accompanying grain growth, at small layer thicknesses the initial grain size is large enough to lower J_c values below that for the unalloyed material.

Figure 19 presents a comparison of the dependence of critical current density on the magnetic field for bronze-processed Nb_3Sn and V_3Ga [78]. In these critical current measurements, J_c was taken as the sample current in a perpendicular field required to produce 1 μV across approximately 3 cm of 5-cm-long samples. As seen in the figure, V_3Ga possesses a large J_c value at all applied magnetic fields but has significantly better critical currents than does Nb_3Sn for magnetic fields exceeding ~12.0 T. Due to the higher cost of Ga, and the lower T_c of V_3Ga relative to Nb_3Sn, the latter has been chosen over V_3Ga for most applications below 12.0 T.

An attempt has been made to increase the high-field performance of

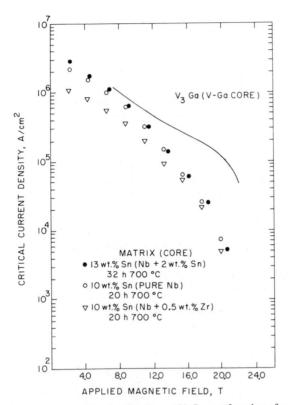

Fig. 19. Critical current densities for Nb$_3$Sn and V$_3$Ga as a function of applied magnetic field. (Data, in part, from Howe, *et al.* [80, 81].)

Nb$_3$Sn by raising its upper critical field H_{c2} [78]. Small additions of Ga (~2 wt. %) to the Cu–Sn bronze lead to increases in the critical current densities of Nb$_3$Sn at fields above ~12.0 T (see Fig. 20). To further increase the Ga concentration it is necessary to correspondingly decrease the accompanying Sn concentration of the bronze, keeping the total alloying content below the solid solubility limit (see Section III,D.). This decrease in Sn concentration lowers the layer growth rate and degrades the critical current density in the Nb$_3$Sn at low fields. A matrix composition of 5 wt. % Sn and 8.5 wt. % Ga leads to approximately 3 at. % Ga in the compound, i.e., Nb$_3$(Sn$_{0.9}$Ga$_{0.1}$).

A relatively sharp decline in J_c is noted for V$_3$Ga as the applied magnetic field in Fig. 19 approaches H_{c2}. Figure 21 presents the variation of the upper critical field with temperature for V$_3$Ga and Nb$_3$Sn bronze-processed composites [96]. The rapid increase in dH_{c2}/dT as T increases for V$_3$Ga is indicative of strong paramagnetic limiting in this material (see

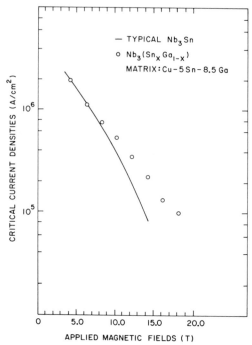

Fig. 20. Critical current densities as a function of applied magnetic fields for Nb_3Sn and $Nb_3(Sn_xGa_{1-x})$ composite wires.

the introductory chapter by Dew-Hughes). H_{c2} of Nb_3Sn is not paramagnetically limited. One approach to increasing the high-field J_c of V_3Ga would be to offset the paramagnetic limiting by inducing spin–orbit scattering. This perhaps can be achieved by alloying vanadium with uranium, an element having a high atomic number Z. Since spin–orbit scattering increases as Z^4, a substantial decrease in the stability of the normal paramagnetic state would ensue and an increase in J_c near the H_{c2} value might occur.

D. Phenomenological Flux-Pinning Theories

The critical current–magnetic field dependence of A15 compounds may be expressed by plotting the pinning force density F_p as a function of the reduced magnetic field h ($= H/H_{c2}$). The pinning force density $F_p(h)$ represents an upper limit of a material's capability to restrict flux flow under the Lorentz force $J_c \times B$. All high-field superconductors, including the bronze-processed A15 compounds, have a maximum (F_p^{max}) in the plot of

F_p versus h. The shape and maximum value of this peak depend on metal-lurgical treatment. Kramer has developed a model that predicts these fea-tures and the changes that accompany metallurgical treatment [97]. $F_p(h)$ is represented as a composite function containing two parts

$$f_p(h) = K_p h^{1/2}/(1 - h)^2 \tag{1}$$

and

$$f_s(h) = K_s h^{1/2}(1 - h)^2 \tag{2}$$

$f_p(h)$ is the pinning force derived from a calculation based on the assump-tion that pinning interactions are broken when the critical Lorentz force is exceeded, K_p being the strength of these pins. Equation (1) accounts for pinning forces between zero field and the peak (F_p^{max}). The function $f_p(h)$ is an increasing function of h. The function $f_s(h)$ corresponds to the plastic shear of the flux-line lattice (FLL) around local strong pinning sites. Equation (2) describes a decreasing function of h near H_{c2}, while K_s is a proportionality constant independent of pinning strength. In the simplest condition where all pins have the same strength K_p the pinning force $F_p(h)$ is given by the smaller of $f_p(h)$ or $f_s(h)$ and the peak in $F_p(h)$, that is, F_p^{max}, occurs at a reduced field h_p, where $f_p(h) = f_s(h)$. The influence of metal-lurgical treatment on F_p^{max} and h_p may be understood through its effect on the pinning strength K_p. Metallurgical treatments that produce small val-ues of K_p, that is, weak or widely spaced pins, result in $f_p(h) = f_s(h)$ at

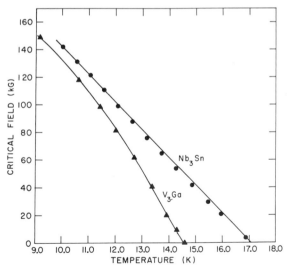

Fig. 21. Upper critical field H_{c2} versus temperature for Nb_3Sn and V_3Ga multifilament wires.

increasingly high values of h. On the other hand, treatments that yield large values of K_p produce $f_p(h) = f_s(h)$ at lower values of h and, correspondingly, higher values of F_p^{max}. Thus, the model predicts an increase in F_p^{max} and its shift to lower h_p values with heat treatments that increase the pinning strength K_p.

Figure 22 illustrates the pinning force density as a function of h for several bronze-processed Nb₃Sn superconducting wires [98]. For F_p^{max} values less than $\sim 3.0 \times 10^9$ dyn/cm³ (i.e., low values of K_p) a decrease in F_p^{max} accompanies a shift toward higher h_p values. However, for the optimally heat-treated samples in which the grain boundary density is maximized (i.e., high values of K_p), F_p^{max} is observed to increase without appreciable changes in h_p. Luhman *et al.* have modified Kramer's model to include these high-K_p A15 materials [98]. The field dependence of the FLL shear contribution to $F_p(h)$ near $h_p = 0.2$ is rewritten as $h^{3/2}(1 - h)^2/(h^{1/2} - n)^2$, where $n = (\phi\rho/H_{c2})^{1/2}$, ρ is the grain boundary density, and ϕ the flux quantum. This function takes into account the field variation of the flux-line lattice spacing and its relationship to sample grain size. For high grain boundary densities (high K_p) the modified function predicts that only small shifts in h_p values will accompany the increases in F_p^{max} near $h = 0.2$–0.3. Figure 23 illustrates the reduced pinning force density as a function of h, including the new function $h^{3/2}(1 - h)^2/(h^{1/2} - n)^2$. Figure 23 is plotted for the values $n = 0.1$, 0.2, and 0.3, which correspond to grain sizes of 1000, 500, and 333 Å, respec-

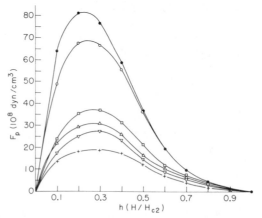

Fig. 22. Pinning force density as a function of reduced magnetic field for a series of bronze-processed Nb₃Sn wire conductors. Matrix (core): ●, 13 wt. % Sn (Nb 2 wt. % Sn), 32 h at 700°C; ○, 10 wt. % Sn (pure Nb), 20 h at 700°C; □, 10 wt. % Sn (Nb 2.5 wt. % Zr), 20 h at 700°C; △, 10 wt. % Sn (pure Nb), 160 h at 700°C; ▽, 10 wt. % Sn (Nb 2.5 wt. % Zr), 150 h at 700°C; +, 10 wt. % Sn (Nb 0.75 wt. % ZrO₂), 161 h at 700°C.

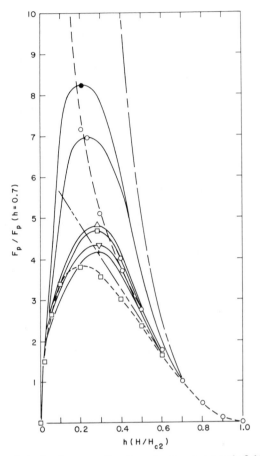

Fig. 23. Reduced pinning force as a function of reduced magnetic field. Included are the functions $h^{1/2}(1 - h)^2$ ($\square$) and $h^{3/2}(1 - h)^2(h^{1/2} - n)^2$ with $n = 0.1$ (————) $n = 2$ (———), and $n = 0.3$ (— — —).

tively. A qualitative estimate of how this magnetic field dependence for $f_s(h)$ affects the determination of $F_p(h)$ can be seen from Fig. 23. The predicted values of $F_p^{\max}$ represent the intersections of the function $h^{3/2}(1 - h)^2/(h^{1/2} - n)^2$ with Eq. (1), not shown in the figure. The intersections approximate the $F_p^{\max}$ values for the bronze-processed Nb$_3$Sn samples. Different intersections between Eq. (1) and the individual function $n = 0.1$, 0.2, or 0.3 produce $F_p^{\max}$ values that decrease as K_p decreases but do not shift appreciably in h_p. The magnitude and field dependence of $F_p(h)$ for bronze-processed A15 conductors are thus fully described by these phenomenological equations.

What about the possible flux-pinning mechanisms that determine K_p values? Pande and Suenaga have developed a theory that predicts a maximum in F_p when the grain size is five times the magnetic flux-line spacing [99]. An electron microscopy investigation of filamentary Nb_3Sn conductors demonstrated a maximum in F_p at a grain size of approximately four times the magnetic flux-line spacing [100]. This is good agreement between theory and experiment and points to the important role of grain boundaries in determining the critical current density for Nb_3Sn conductors.

V. Superconducting Critical Currents, Temperatures, and Magnetic Fields of Nb_3Sn Wire Conductors under Tensile Strain

A. Introduction

Two important considerations regarding the practical application of bronze-processed Nb_3Sn multifilament conductors are their mechanical strengths and the degradation under strain of their superconducting properties. Due to the brittleness of the compound, allowable strain limits must be specified for handling, braiding, cabling, and winding operations. In addition, operational strain limits must be set to meet strains from electromagnetic and thermal forces. In the following discussion the effects of mechanical strains on conductors are divided into two categories: stress–strain relationships and degradation of superconductivity.

B. Stress–Strain Relationships

Deformation of high-modulus fibers in ductile matrices is characterized by three stages on a stress–strain plot: an initial elastic range, a plastic matrix–elastic fiber range, and an all plastic stage. Ideally, the modulus at each stage is a volume average of the individual constituents. Several investigators report observing these three stages in bronze-processed Nb_3Sn conductors tested at 4.2 K [33, 85, 101–103]. Figure 24 presents typical data for single-filament Nb_3Sn conductors. The filament size is approximately the same for each conductor; the amount of matrix differs [104]. The first stage is represented by a common elastic modulus of ~ 1.1 GN/m^2 and extends to $\sim 0.2\%$ in each case. The moduli of stage two vary with bronze-to-niobium ratios, showing higher values for lower-ratio samples. It is difficult to mark the onset of stage three. If, however, the

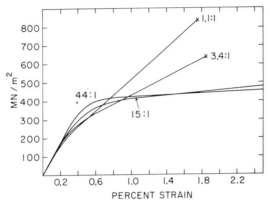

Fig. 24.   Stress–strain relationships for several monofilament Nb₃Sn conductors with differing bronze-to-niobium ratios.

volume fraction of the filament is high, i.e., low-ratio samples, initiation of stage three and the yielding of the fibers result in composite fracture; see, for example, ratios 1.1:1 and 3.4:1. For higher-ratio material fracturing of the filaments does not produce composite failure. In these cases enough matrix material is present to maintain mechanical integrity in the composite. Present designs for large magnet systems could set practical strain limits to approximately 1%. This would necessitate load limits for unreinforced wires less than ~450 MN/m².

Anomalous behavior is exhibited at intermediate strains (stage two) between approximately 0.2 and 0.8%. Here, a simple scaling according to the law of mixtures is obviously not operating. Within this range low-ratio samples appear not to exhibit the reinforcement expected from the filaments. This effect is caused by internal prestrains in the matrix. Due to the relatively higher thermal contraction coefficient for bronze, internal stresses, which arise during cooling from ~700°C to 4.2 K, produce a tensile prestrain in the matrix and a compressive prestrain on the compound. In these composites, tensile prestrains in the matrix have the effect of lowering the apparent yield strength of the bronze. This tensile prestrain increases with decreasing ratios [104, 105]. Lower-ratio samples therefore exhibit, under tensile loading, an onset of plastic deformation in the matrix at progressively lower applied strains. The mechanical strength available from the bronze component is thus limited and different for each sample. A decrease in the yield strength of the bronze with decreasing ratios in this way lowers the overall composite modulus. Evidently such effects from matrix prestrain may occur in multifilament conductors since similar behavior has been observed for multifilament Nb₃Sn conductors with differing bronze-to-niobium ratios [10]. The effects of temperature,

cryostabilizing material, filament size, number, cabling, and bundling geo-
metrics on mechanical properties are only now being investigated.

C. Degradation of Superconducting Properties under Mechanical Strain

As stated above, the compound in bronze-processed conductors is subjected to a compressive strain originating from thermal contraction of the
outer bronze matrix. These compressive strains lower the compound's
superconducting temperature and critical current density, whereas removal of the compressive strains through etching away of the bronze matrix restores T_c and J_c [104, 105]. The existence of these compressive
strains has a very favorable effect. They enhance the tensile strain required to eventually degrade the superconducting properties of composite
conductors.

The changes observed in T_c for stressed and unstressed samples when
the outer bronze matrix is removed by etching [33] are presented in Fig.
25. With increasing bronze-to-niobium area ratios, an increase in ΔT_c is
observed up to ~ 1.2 K at a ratio of $\sim 10:1$. For ratios exceeding $\sim 10:1$
the effect saturates and little increase in ΔT_c occurs. This saturation point
represents the bronze-to-niobium ratio for which all the internal prestrain
is taken up in the niobium and Nb$_3$Sn core. For lesser ratios, a balance is
struck between compression of the core and tensile strain in the matrix. It
is evident in Fig. 25 from the small ΔT_c observed for the stressed samples
that tensile straining at 4.2 K effectively removed the prestresses.

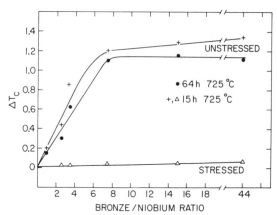

Fig. 25. The change ΔT_c observed in T_c, for stressed and unstressed monofilament
Nb$_3$Sn wires when the outer bronze matrix is removed by etching.

The data presented in Fig. 25 can be used to confirm the thermal-contraction hypothesis. The elastic strain in the superconducting filament is easily calculated by matching the strains and forces in each constituent after cooling separately from 700°C to 4.2 K:

$$\epsilon_s = \left(-\frac{Br}{Nb}\cdot\frac{E_b}{E_s}\cdot\Delta\right)\bigg/\left(1+\frac{Br}{Nb}\cdot\frac{E_b}{E_s}\right) \tag{3}$$

where $\Delta = (\alpha_b - \alpha_s)\,\Delta T$, ϵ_s is the strain in the superconductor, Br/Nb is the bronze-to-niobium ratio and E_b/E_s is the ratio of elastic moduli. Letting $\epsilon_s^2 = \Delta T_c/k$ [106], we have

$$\frac{Br}{Nb}\bigg/\Delta T_c^{1/2} = \frac{E_s}{E_b}\bigg/k\Delta + \frac{Br}{Nb}\bigg/k\Delta \tag{4}$$

Equation (4) describes the dependence of ΔT_c on the bronze-to-niobium ratio and is plotted in Fig. 26 as a straight line through the data of Fig. 25. This representation serves to illustrate how the composite nature of bronze-processed Nb_3Sn conductors influences their superconducting properties.

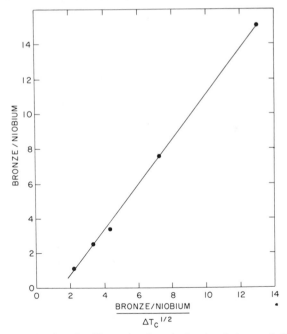

Fig. 26. Straight-line function illustrating the calculated and observed effect of compressive strains on T_c for bronze-processed Nb_3Sn monofilament wires.

Figure 27 presents the superconducting transition temperatures for a series of single-filament wires as a function of the applied tensile strain. The data in Fig. 27 are for a 15-h heat treatment that produced a ~ 2.5-μm layer of Nb_3Sn. The T_c values presented represent the temperatures corresponding to 80% of the total inductive change to the normal state. Each specimen's curve exhibits an increase in T_c for initial strains, reaches a maximum value, and eventually decreases at larger strains. Such behavior reflects, first, removal of compressive prestresses, then application of tensile strain. The observed peak T_c value is related to that observed when the bronze matrix is etched off an unstrained wire. Figure 28 shows the linear relationship found when the increase in T_c under tensile loading, ΔT_c^*, is plotted against the increase in T_c found when the bronze is removed from a unstrained sample, ΔT_c.

The maxima of Fig. 27 represent the strains at which internal prestrains are removed by tensile loading. Thus the magnitude of each prestrain can be estimated as the strain value associated with the T_c maximum ϵ^*. The ϵ^* or prestrain values are approximately 0.2, 0.5, 0.60, 0.80, and 0.95% for the ratios 1.1:1, 2.5:1, 3.4:1, 7.6:1, and (15:1, 44:1), respectively.

Several investigators have reported increases in J_c of bronze-processed conductors during mechanical straining [83, 107–111]. Figure 29 illustrates how the critical current densities for monofilament conductors vary as a function of strain [111]. The general behavior of $J_c(4.0\ T)$ is similar to that observed for T_c. The strain values associated with the maxima in T_c correspond well with those for J_c (see Fig. 30). This observation demonstrates that J_c changes under tensile loading are incurred because of the induced T_c variations.

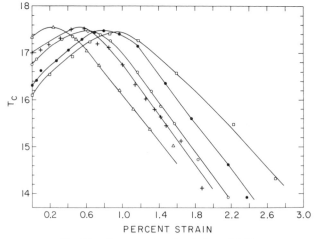

Fig. 27. T_c versus tensile strain at 4.2 K.

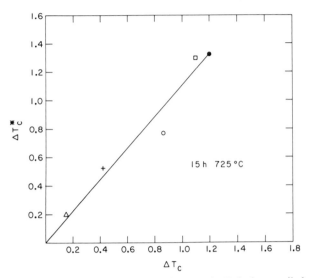

Fig. 28. Relationship between the maximum increase in T_c during tensile loading (ΔT_c^*) and removal of the bronze matrix by etching (ΔT_c).

The dependence of H_{c2} on the bronze-to-niobium ratio is illustrated in Fig. 31. H_{c2} decreases with increasing ratios. Variations in H_{c2} follow the behavior of those seen for T_c and J_c, that is, a saturation occurs for ratios exceeding $\sim 10 : 1$. It should be noted that the magnitudes of the changes in T_c ($\sim 8\%$) are less than those observed for H_{c2} ($\sim 25\%$). Plotting H_{c2}/T_c reveals this difference; see Fig. 32. Calculations by Welch *et al.* show that this variation in H_{c2}/T_c is expected from theory and it is concluded therefore that the observed changes in H_{c2} are also incurred because of the strain-induced changes in T_c [111].

Differences often arise between critical current relationships obtained by various investigators on similar types of conductors. Such differences

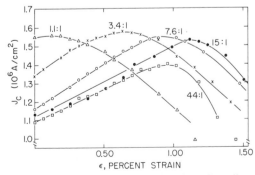

Fig. 29. Critical current density (4.0 T) as a function of tensile strain at 4.2 K.

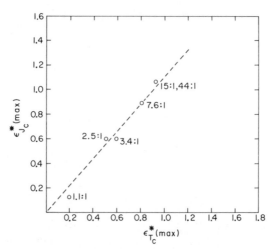

Fig. 30. Correspondence between the strains at maxima in current density and critical temperature plots.

result when the amount of compressive prestrain on the superconductor varies as conductors are strained during handling and mounting. Ekin has shown that variations in published critical current data can be fit to a single curve that is normalized to the peak values of critical current; see Fig. 33 [110]. The assumption is made in Fig. 33 that the peak condition in critical current represents a position of similar strain, presumed to be zero, on the superconducting compound. In this way it is seen that the

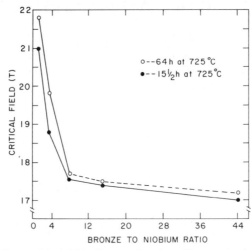

Fig. 31. Upper critical fields H_{c2} as a function of bronze-to-niobium ratios.

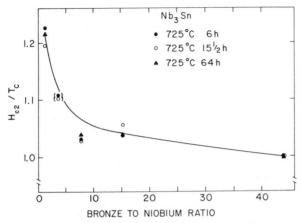

Fig. 32. The ratio H_{c2}/T_c as a function of bronze-to-niobium ratios.

process of critical current degradation under tensile straining depends on the amount of compressive prestress present initially in a conductor. The prestress varies, then, with factors that include the bronze-to-niobium ratio, Sn concentration in the bronze, compound thickness, matrix porosity, and prior strain from handing operations.

Only limited fatigue data exist on mechanical and associated electrical properties of Nb_3Sn conductors. Fisher *et al.* have reported a Nb_3Sn composite fractured during tension–compression cycling, while no damage occurred in load–unload cycling at strains up to 0.2% [112].

The foregoing information suggests that superconducting magnets be designed so their hoop stresses balance the conductor's internal compres-

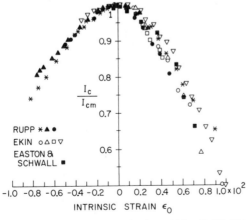

Fig. 33. Normalized I_c versus strain data. (After J. W. Ekin [110].)

sive prestresses. In this way a maximum critical current is evident. Furthermore, perhaps a strain margin of a few tenths of a percent tensile and compressive strain would exist before significant critical current degradation occurred.

Several possibilities exist to improve the reliability of Nb₃Sn conductors by decreasing their sensitivity to strain degradation. Increasing the bronze-to-niobium ratio of low-ratio samples increases the allowable strain limits but at the expense of a decrease in the conductor's overall critical current density as the relative amount of superconductor decreases. It has been shown by the author that a significant increase in the allowable strain limits of low-ratio samples is attained when small additions of Be are made to the Cu–Sn bronze matrix. Strengthening from the Be reduces the accommodating plastic strain in the matrix that occurs during cooling from the heat-treatment temperature, thus increasing, for a given ratio, the imposed compressive strain on the A15 compound. Figure 34 serves to illustrate the improved properties for monofilament wires with bronze-to-niobium ratios of approximately 2.5:1. Here it is seen that an alloy addition of 0.2 wt. % Be increases both the strain values for the peak critical current and the strain associated with the onset of irreversible critical current changes. These increases amount to ~0.4% strain.

In summary, the composite nature of the A15 conductor leads to compressive strains on the superconducting compound. These compressive strains can typically provide strain margins of ±0.2–0.3% at the maximum critical current value. These strain values can be significantly increased with small alloy additions of Be. Magnet systems certainly can be

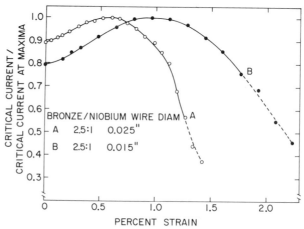

Fig. 34. Improvement gained in the strain limits of a Nb₃Sn monofilament wire through the allow addition of ~0.2 wt. % Be to the matrix.

designed with construction and operational strain limits within such ranges. It appears therefore that A15 multifilament conductors will find extensive use in future generations of power transmission systems and high-field superconducting magnets.

ACKNOWLEDGMENTS

The author would like to express his appreciation to Dr. M. Suenaga for his support and assistance with the author's research presented here. A special thanks to Dr. D. O. Welch for his discussions elucidating the mechanical properties of composite superconductors. Finally, I wish to express my appreciation to A. D. Luhman for her talented reading of this book's entire manuscript.

References

1. R. M. Bozorth, A. J. Williams, and D. D. Davis, *Phys. Rev. Lett.* **6,** 148 (1960).
2. J. E. Kunzler, E. Buehler, F. S. L. Hsu, and J. H. Wernick, *Phys. Rev. Lett.* **6,** 89 (1961).
3. A. Cave, Ph.D. thesis, Univ. of Manchester, Inst. of Sci. and Technol. (1972) (unpublished).
4. R. Lohber, T. W. Eager, I. M. Puffer, and R. M. Rose, *Appl. Phys. Lett.* **22,** 69 (1972).
5. M. S. Lubell, *Cryogenics* **12,** 340 (1972).
6. T. W. Eager and R. M. Rose, *IEEE Trans. Magn.* **MAG-11,** 214 (1975).
7. D. L. Martin, M. G. Benz, C. A. Brach, and C. H. Rosner, *Cryogenics* **3,** 161 (1963).
8. E. Saur and J. P. Wurm, *Naturwissenschaften* **49,** 127 (1962).
9. J. J. Hanak, K. Strater, and R. W. Cullen, *RCA Rev.* **25,** 342 (1964).
10. R. A. Stauffer, *Conf. High Magn. Fields, Their Product, and Their Appl., Univ. of Oxford, July 10–12 1963* (unpublished).
11. M. G. Benz, D. L. Martin, and C. A. Brach, *Cryogenics* **3,** 248 (1965).
12. C. H. Rosner and M. G. Benz, *Proc. Int. Symp. Magn. Technol., Stanford Univ.* p. 597 (1965).
13. R. Hancox, *Phys. Lett.* **16,** 208 (1965).
14. R. Hancox, Culham Laboratory Rep. CLM, p. 121 (1966) (unpublished).
15. R. Hancox, *in Proc. Int. Conf. Low Temp. Phys., 10th, Moscow* **2B,** 45 (1966).
16. A. R. Kaufman and J. J. Pickett, *J. Appl. Phys.* **42,** 58 (1971).
17. K. Tachikawa and T. Tanaka, *Jpn. J. Appl. Phys.* **6,** 782 (1967).
18. K. Tachikawa, *Proc. Int. Cryogen. Eng. Conf., 3rd* p. 339. Iliffe Sci. and Tech. Publ., Berlin, 1970.
19. C. D. Henning, *IEEE Trans. Magn.* **MAG-13,** 15 (1977).
20. W. B. Sampson, M. Suenaga, and S. Kiss, *IEEE Trans. Magn.* **MAG-13,** (1), 287–289 (1977).
21. J. F. Bussière, *IEEE Trans. Magn.* **MAG-13,** (1), 131–137 (1977).
22. J. P. Charlesworth, I. Macphail, and P. E. Madsen, *J. Mater. Sci.* **5** (1970).
23. K. Tachikawa and Y. Iwasa, *Appl. Phys. Lett.* **16,** (6), 230 (1970).
24. J. B. Clark and F. N. Rhines, *Trans. Am. Soc. Met.* **51,** 199 (1959).
25. T. Luhman, O. Horigami, and D. Dew-Hughes, *Am. Chem. Soc. Appl. Polym. Symp. 29th,* 61–70 (1976). John Wiley and Sons, Inc.

26. D. Dew-Hughes and T. S. Luhman, *J. Mater. Sci.* **13**, pp. 1868–1876 (1978).
27. R. H. Hopkins, G. W. Roland, and R. Daniel, *Metall. Trans.* **8A**, 91 (1977).
28. E. M. Savitskii, Yu. V. Efimov, V. Ya Markiv, and O. I. Zrolinskii, *Izv. Akad. Nank SSSR Metall.* 199 (1976).
29. C. R. Hunt and A. Raman, *Z. Metallk.* **59**, 707 (1968).
30. J. D. Livingston, *J. Mater. Sci.* **12**, 1759 (1977).
31. J. D. Livingston, Private communication.
32. O. Horigami, T. Luhman, C. S. Pande, and M. Suenaga, *Appl. Phys. Lett.* **28** (12), 738 (1976).
33. T. S. Luhman and M. Suenaga, *Appl. Phys. Lett.* **129** (1), 61 (1976).
34. S. Ceresara, M. J. Ricci, N. Sacchett, and G. Sacerdoti, *IEEE Trans. Magn.* **MAG-11**, 263 (1975).
35. R. A. Sigsbee, *IEEE Trans. Magn.* **MAG-13**, 307 (1977).
36. A. B. Hallek, R. H. Hammond, T. H. Geballe, and R. B. Zubeck, *IEEE Trans. Magn.* **MAG-13**, 311 (1975).
37. I. Ahmad, W. J. Heffernan, and D. U. Gubser, *IEEE Trans. Magn.* **MAG-13** (1), 483 (1977).
38. A. Echarri and M. Spadoni, *Cryogenics* **11**, 274 (1977).
39. Y. Tanaka, K. Tachikawa, and K. Sumiyama, *Inst. Met.* **34**, 835 (1970).
40. A. Isao, T. Noguchi, Y. Uchida, and A. Kono, *J. Vacuum Sci. Technol.* **7**, 557 (1970).
41. F. J. Wonzala, W. J. Yang, B. P. Strauss, R. W. Boom, and J. Lawrence, *Proc. ICMC* **4**, 347 (1973).
42. J. J. Hanak, K. Strater, and G. W. Cullen, Preparation and properties of vapor-deposited niobium stannide, *RCA Rev.* **25**, No. 3 (1964).
43. J. Hanak, *in* "Metallurgy of Advanced Electronic Materials" (G. E. Brock, ed.), p. 161. Wiley (Interscience), New York, 1963.
44. L. J. Vieland and A. W. Wicklund, *Phys. Lett.* **49A**, 407 (1974).
45. H. Kawamura and K. Tachikawa, *Phys. Lett.* **50A**, 29 (1974).
46. A. I. Braginski and G. W. Roland, *Appl. Phys. Lett.* **25**, 762 (1974).
47. L. R. Newkirk, F. A. Valencia, A. L. Giorgi, E. G. Szklarz, and T. C. Wallace, *IEEE Trans. Magn.* **MAG-11**, 221 (1975).
48. A. I. Braginski, J. R. Gavaler, G. W. Roland, M. R. Daniel, M. A. Janocko, and A. T. Santhanam, *IEEE Trans. Magn.* **MAG-13**, No. 1, 300 (1977).
49. J. F. Bussière, M. Garber, and M. Suenaga, *J. Appl. Phys.* **45**, 4611 (1974).
50. E. Adam, P. Beischer, W. Marancik, and M. Yound, *IEEE Trans. Magn.* **MAG-13**, No. 1, 425 (1977).
51. Brookhaven National Laboratories Rep. No. 22202, Power Transmission Project, Progress Through Fiscal Year 1976, Fig. 2-2.
52. R. H. Hammond, *IEEE Trans. Magn.* **MAG-11** (2), 201 (1975).
53. R. E. Howard, *et al.*, Electrical properties of multilayered Nb_3Sn superconducting power line conductors, *Appl. Superconduct. Conf., Palo Alto* (1976) (unpublished).
54. R. H. Hammond, *et al.*, *J. Appl. Phys.* **40**, 2010 (1969).
55. S. D. Dahlgren, *IEEE Trans. Magn.* **MAG-11** (2) 217 (1975).
56. S. D. Dahlgren, *Metall. Trans. A* **7A**, 1375 (1976).
57. R. Wang and S. D. Dahlgren, *Metall. Trans. A* **8A**, 1763 (1977).
58. S. D. Dahlgren and D. M. Kroeger, *J. Appl. Phys.* **44**, 5595 (1973).
59. S. D. Dahlgren, *IEEE Trans. Magn.* **MAG-11**, 217 (1975).
60. S. D. Dahlgren, M. Suenaga, and T. S. Luhman, *J. Appl. Phys.* **45**, 5462 (1974).
61. J. R. Gavaler, *Phys. Today* 12 (1973).
62. See Applied Superconductivity Conferences, *IEEE Trans. Magn.* **MAG-13**, 1 (1977).

63. E. Gregory, W. G. Marancik, and F. T. Ormand, *IEEE Trans. Magn.* **MAG-11** (2), 295 (1975).
64. K. Tachikawa, Y. Yoshida, L. Rinderer, *J. Mat. Sci.* **7**, 1154 (1972).
65. M. Suenaga and W. B. Sampson, *Appl. Phys. Lett.* **20**, 443 (1972).
66. M. Suenaga, O. Horigami, and T. S. Luhman, *Appl. Phys. Lett.* **25** (10), 624 (1974).
67. M. Suenaga and M. Garber, *Science* **184**, 952 (1974).
68. C. C. Tsuei, *Science* **180**, 57 (1973); C. C. Tsuei, M. Suenaga, and W. B. Sampson, *Appl. Phys. Lett.* **25**, 318 (1974).
69. R. Roberge and J. L. Fihey, *J. Appl. Phys.* **48**, 1327 (1977); J. P. Harbison and J. Bevk, *J. Appl. Phys.* **48**, 5180 (1977); S. Foner, E. J. McNiff, Jr., B. B. Schwartz, R. Roberge, and J. L. Fihey, *Appl. Phys. Lett.* **31**, 853 (1977).
70. M. R. Pickus, K. Hemachalam, and B. N. P. Babu, *Mater. Sci. Eng.* **14**, 265 (1974); K. Hemachalam and M. R. Pickus, *IEEE Trans. Magn.* **MAG-13**, 466 (1977). See also H. L. Luo, *J. Mater. Sci.* **12**, 1841 (1977).
71. T. W. Lohberg, T. W. Eager, I. M. Puffer, and R. M. Rose, *Appl. Phys. Lett.* **22**, 69 (1973).
72. T. W. Eager and R. M. Rose, *IEEE Trans. Nucl. Sci.* **NS-20**, 742 (1973).
73. J. W. Hafstrom, *IEEE Trans. Magn.* **MAG-13** (1), 480 (1977).
74. R. Bruzzese, *et al.* Paper presented at the *Symp. Eng. Prob. of Fusion Res. 7th, Knoxville, Tennessee, October 25–28* (1977).
75. S. Ceresara, *et al., Proc. Int. Conf. Mag. Technol., 5th (MT-S), Rome* p. 685 (1977).
76. S. Ceresara, M. V. Ricci, N. Sacchetti, and G. Sacerdot, *IEEE Trans. Magn.* **MAG-11**, 263 (1975).
77. M. Suenaga, W. B. Sampson, and C. J. Klamut, *IEEE Trans. Magn.* **MAG-11**, 231 (1975).
78. T. Luhman and M. Suenaga, *Adv. Cryogen. Eng.* **22**, 356 (1977).
79. L. S. Weinman, R. A. Meussner, and D. G. Howe, *Solid State Commun.* **14**, 275 (1974).
80. D. G. Howe, T. L. Francailla, and D. U. Gubser, *IEEE Trans. Magn.* **MAG-13** (1), 815 (1977).
81. D. G. Howe and L. S. Weinman, *IEEE Trans. Magn.* **MAG-11** (2), 251 (1975).
82. H. H. Farral, G. H. Gilmer, and M. Suenaga, *J. Appl. Phys.* **45** (9), 4025 (1974).
83. H. Hillman, *et al., IEEE Trans. Magn.* **MAG-13** (1), 792 (1977).
84. M. Suenaga and W. B. Sampson, A Comparative Study, Magnet Tech., BNL (1970).
85. T. Luhman and M. Suenaga, *Symp. Stress Effects Supercond., Vail, Colorado* (1976) (unpublished).
86. J. F. Bussière, Private communication.
87. M. Suenaga, Private communication, BNL, Upton, New York.
88. R. M. Scanlan, W. A. Rietz, and E. F. Koch, *J. Appl. Phys.* **46** (5), 2244 (1975).
89. W. A. West and R. O. Rawlings, *J. Mater. Sci.* **12**, 1862 (1977).
90. E. Nembach and K. Tachikawa, *J. Less-Common Met.* **19**, 359 (1969).
91. O. Singh, A. E. Curzon, and C. C. Koch, *J. Phys. D* **9**, 611 (1976).
92. B. J. Shaw, *Appl. Phys. Lett.* **47**, 2143 (1976).
93. M. Suenaga, T. S. Luhman, and W. B. Sampson, *J. Appl. Phys.* **45** (9), 4049 (1974).
94. B. J. Shaw, *J. Appl. Phys.* **47** (5), 2143 (1976).
95. M. Suenaga and W. B. Sampson, *Appl. Phys. Lett.* **18** (12), 584 (1971).
96. J. E. Crow and M. Suenaga, *Proc. Appl. Supercond. Conf., Annapolis, Maryland* p. 472 IEEE, New York (1972).
97. E. J. Kramer, *J. Appl. Phys.* **44**, 1360 (1973).

98. T. Luhman, C. S. Pande, and D. Dew-Hughes, *J. Appl. Phys.* **47** (4), (1976).
99. C. S. Pande and M. Suenaga, *Appl. Phys. Lett.* **29,** 444 (1976).
100. A. W. West and R. D. Rawlings, *J. Mater. Sci.* **12,** 1862 (1977).
101. I. L. McDougall, *IEEE Trans. Magn.* **MAG-11** (5), 1467 (1975).
102. D. C. Larbalestier, J. E. Magraw, and M. N. Wilson, *IEEE Trans. Magn.* **MAG-13,** 462 (1977).
103. C. F. Old and J. P. Charlesworth, *Cryogenics* **16,** 469 (1976).
104. T. Luhman and M. Suenaga, *Appl. Phys. Lett.* **29,** 1 (1976).
105. T. Luhman and M. Suenaga, *IEEE Trans. Magn.* **MAG-13** (1), 800 (1977).
106. L. R. Testardi, *in* "Physical Acoustics" (W. Mason and R. N. Thurston, eds.), vol. 10, p. 194. Academic Press, New York, 1973.
107. J. W. Ekin, *Appl. Phys. Lett.* **29** (3), 276 (1976).
108. D. S. Easton and R. E. Schwall, *Appl. Phys. Lett.* **29** (5), 319 (1976).
109. M. Suenaga, T. S. Luhman, W. B. Sampson, and T. Onishi, Brookhaven National Laboratory Rep. No. 24102 (1977).
110. J. W. Ekin, *CEC/ICMC Conf., Univ. Colorado, Boulder, Colorado* Paper No. Ca-1 (1977).
111. T. S. Luhman, M. Suenaga, and D. O. Welch, *J. Appl. Phys.* (May 1979).
112. E. S. Fisher and S. H. Kim, *IEEE Trans. Magn.* **MAG-13** (1), 112 (1977).

Superconductors for Power Transmission

J. F. BUSSIÈRE

Accelerator Department
Brookhaven National Laboratory
Upton, New York

I. Introduction	267
II. Cable Designs and Superconductor Requirements	268
A. Introduction	268
B. Selection of Superconductor Materials	270
C. Conductor Configurations and Fabrication	271
III. Bulk Critical Currents of Type II Superconductors	278
A. Introduction	278
B. Experimental Results	279
IV. Surface Currents in Type II Superconductors	284
A. Theory	284
B. Experimental Observations	290
V. Theory of AC Losses in Type II Superconductors	296
A. Introduction	296
B. Bulk Losses	296
C. Effect of Surface Currents	298
D. AC Losses in the Meissner State	300
E. Extra Losses for Conductors in Cable Configurations	302
VI. AC Losses of Pure Niobium	303
A. Metallurgical Effects	303
B. Composite Conductors	304
C. Effect of Trapped Magnetic Flux	307
D. Temperature Dependence	308
VII. AC Losses of Nb_3Sn and Nb_3Ge	309
A. Metallurgical Effects	310
B. Temperature Dependence of Losses (Nb_3Sn, Nb_3Ge)	321
References	322

I. Introduction

At present, because of lower cost, long-distance high-power transmission consists mainly of high-voltage overhead (HVOH) lines. The highest-voltage line presently in operation is 765 kV with a single-circuit capacity of 2000 MVA, and work is proceeding on the next voltage level of 1200 kV, which will raise this capacity to 4000 MVA. Overhead lines, although economical in unpopulated areas, require increasingly large rights-of-way as their voltage, and hence power level, is increased. In heavily populated urban areas such rights-of-way are either too costly or simply unavailable, and utility companies in the future will be forced to underground transmission lines of increasingly large capacities over longer distances. This poses a number of problems. Currently available high-pressure paper–oil-filled cables cost approximately four times more than overhead lines. Furthermore they are rated at much lower power levels than HVOH lines. The highest installed rating at present is ~600 MVA for 345-kV cables, the next voltage level (not yet installed) being 550 kV with a power rating of 900 MVA. Hence, multiple circuits must be used to connect to HVOH lines. This increases the complexity, the width of trench, and hence the cost. Conventional cables are also plagued with substantial generation of reactive power because of their large capacitance. This limits the useful length of cable that can be put into service. Two improvements are presently under development: (1) the use of insulation with a lower dielectric constant, such as polyethylene or gas, and (2) the use of forced cooling to raise the current-carrying ability of the conductor [1, 2]. Superconducting cables may be regarded as the ultimate in force-cooled systems. Both ac and dc superconducting cables have higher power density than any other competing technology. For this reason they are economically attractive at high power levels (≥2000 MVA up to >10,000 MVA) since large power levels could be carried in a single trench [3, 4].

There are several projects in the United States [5, 6], Europe [7], and the Soviet Union [7a, 7b] to develop dc and ac superconducting transmission lines. This chapter will focus mainly on the properties of the superconductor to be used in a transmission line. However, to specify requirements for the superconductor, various cable designs will first be described in Section II. Low-field bulk critical currents for superconductors of interest to power transmission (Nb, Nb–Ti, Nb–Zr, Nb_3Sn, Nb_3Ge) will then be discussed in Section III. Surface currents, which strongly influence the loss behavior, will be the subject of Section IV. Loss theory and experimental results will then be treated in Sections V, VI, and VII.

II. Cable Designs and Superconductor Requirements

A. Introduction

The design of a superconducting transmission line can be divided into three interdependent parts: the cable itself; the refrigeration system, including the cryogenic enclosure, and the transmission line as a whole, including its relationship to existing networks and facilities. Hence, in order to eventually specify requirements for the superconductor itself, such as ac loss, critical current, and operating temperature, one must take into account the entire system.

Although superconductors can exhibit "zero resistance," superconducting transmission lines are not 100% efficient, even in the case of dc where essentially no loss occurs in the superconductor. For instance, both ac and dc systems suffer losses in the cryogenic enclosure. With ac cables additional losses also occur in the superconductor and dielectric, whereas with dc cables additional losses are encountered at the ends because of conversion to or from ac to dc. Conversion can be a dominant loss component for short dc cables [4]. Efficiencies of superconducting dc cables become comparable to overhead ac lines only for lengths of the order of 700 km, whereas comparable efficiency between overhead and underground superconducting ac cables ($\sim$99%) can be achieved with ac cables of short length ($\sim$30 km) [4].

Heat losses in the cryogenic enclosure, coupled with the efficiency of the refrigerator, determine the ultimate power loss of the line at a given temperature. In general an ac cable is designed so that the superconductor and dielectric losses are roughly equal to the cryostat heat leak. When the size of the cable and cryostat are taken into account, this leads to an acceptable loss of $\sim$10 μW/cm^2 for the inner conductor of the coaxial cable [8–10]. The inner conductor contributes most of the loss because it is exposed to a higher magnetic field than the outer conductor, and losses increase very rapidly with magnetic field.

The operating temperature of the transmission line is determined from the characteristics of both superconductor and dielectric, as well as from refrigeration efficiency. If the superconductor is one with a low T_c, such as Nb, the operating temperature will be limited by the rapid increase of losses and decrease of critical current density J_c with temperature. A niobium cable will probably operate between 4.5 and 5.5 K [11, 12]. If a higher-T_c superconductor such as Nb$_3$Sn or Nb$_3$Ge is used, losses will probably be acceptable up to $\sim$10 K or higher. In this case, however, the operating temperature may be limited by dielectric requirements since dielectric breakdown is a function of helium density and pressures in excess of 15 atm are required to maintain sufficient density above 10 K [13]. Hence, if helium is the coolant, ac cables incorporating a high-T_c superconductor will probably operate in a range of 6–9 K. For dc cables, loss and dielectric requirements are not as stringent and a temperature range of 11–14 K has been proposed [6, 14]. The use of a high-T_c superconductor allows both a higher operating temperature and a greater temperature range. This leads to a system that is more efficient overall than if pure niobium is used, even though losses may be lower in niobium [4, 9]. Similarly, the choice of surface current density σ (rms A/cm) and electric stress of the cable under normal operation is ideally determined by systems considerations, but may be limited by material constraints. Early considerations [8] dealt only with the materials constraints, which led to a choice of ~500 rms A/cm for the inner conductor because above this value losses of pure niobium were too high. Forsyth *et al.* [15], however, have shown that the value of ~500 rms A/cm represents a good compromise between too much inductive line drop and too much charging current. In recent systems studies [10, 16], σ values have tended to be slightly lower than the foregoing number under contingency conditions, whereas under normal operation much lower σ values are possible. This is due to the extra cables, connected in parallel, that are required for system security. A value $\sigma = 500$ rms A/cm corresponds to a peak surface magnetic field at the inner conductor of ~890 Oe.

During operation, the cable will occasionally be subjected to sudden rapid rises in current and/or voltage due to lightning strikes or to short circuits somewhere along the network. The occurrence of these surges or "faults" must be taken into account in the cable design. For conventional ac cables surges of up to ~10–20 times the normal rated current could occur and last for a few cycles before breakers open [17]. In very high-power cables, however, a somewhat lower fault capability may be adequate. Cables should not be permanently damaged by these faults and should recover rapidly from them. To cope with this problem conductors are made of composites that incorporate a good normal conductor, such as pure copper or aluminum, and sometimes a hard superconductor with high critical current density. The high critical current and high critical current density are desirable since they minimize losses during the fault and allow for a quick recovery.

B. Selection of Superconductor Materials

To meet the requirements listed above the superconductor should have as high a T_c as possible, exhibit low losses (≤ 10 μW/cm^2) for peak surface fields of ~900 Oe (for an ac cable), and have the largest possible J_c and I_c at the operating temperature, so that sizable faults can be handled in the superconductor. In addition, it should have acceptable mechanical properties such as adequate strength and bendability. Table I lists selected properties of a limited number of superconductors that are of interest for power transmission. Among niobium and the alloys Nb–Ti and Nb–Zr, the overwhelming choice for ac cables is pure Nb because of its low hysteretic losses. These low losses are associated with its exceptionally high

TABLE I

CHARACTERISTICS OF SUPERCONDUCTORS OF INTEREST FOR POWER
TRANSMISSION APPLICATION

Superconductor	T_c (K)	H_{c2} at 4.2 K (T)	J_c at 1 T, 4.2 K (A/cm²)	Remarks	Reference
Nb	9.2	0.28–0.47	0	Low loss, ductile	18
$Nb_{\sim 0.4}$–$Ti_{\sim 0.6}$	9.5	12	4×10^5	High loss, ductile	18,19
$Nb_{0.75}$–$Zr_{0.25}$	11	9	1.6×10^5	High loss, fair ductility	18, 19
Nb_3Sn	18	23	8.7×10^6	Low loss, forms readily, available commercially, brittle	18, 20
Nb_3Ge	22	35	2×10^6	Low loss, not commercially available, brittle	18, 21

value of H_{c1} [22]. Pure niobium, however, does have low T_c and H_{c2}, which make it undesirable for dc applications or for handling faults in ac cables. Nb–Ti and Nb–Zr, with high values of H_{c2}, higher J_c, and somewhat higher T_c, provide better fault-handling capacity and have been incorporated with Nb in some ac designs [23–25].

Among the high-T_c A15 compounds, Nb_3Sn is unique because it displays low ac losses, exhibits high J_c at low fields, and is commercially available. Because of its higher T_c, Nb_3Ge is also an interesting possibility for the future, provided it can be produced commercially with acceptable properties and at competitive cost. V_3Ga and V_3Si, although they could probably be made with low losses and high J_c, have lower T_c than Nb_3Sn and do not seem to offer any advantage over Nb_3Sn.

C. Conductor Configurations and Fabrication

1. INTRODUCTION

Most ac cable designs incorporate three coaxial cables—one for each phase—in a common cryostat (see Fig. 1). In the case of dc cables, a single coaxial cable can be placed in the cryostat [26] or two conductors can each be enclosed in separate cryostats [27]. The coaxial geometry is generally preferred because of compactness and lower eddy current losses associated with ripple current. The cryostat may either be flexible or rigid. Similarly, both flexible and rigid designs have been proposed for cables, although the present trend is toward flexible cables.

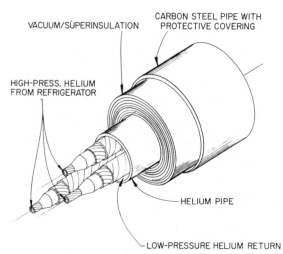

Fig. 1. Conceptual design of the flexible superconducting transmission line under development at Brookhaven National Laboratory. The three pairs of conductors and thermally insulated enclosure, or cryostat, are common elements of most superconducting designs.

2. Rigid Cables (Niobium Composites)

Rigid conductors are the simplest, both conceptually and electrically, and were the first to be developed. They consist of rigid pipes held concentric by solid spacers. Each pipe is made of a composite incorporating a superconductor (Nb) and a good normal metal (copper or aluminum) [23, 25]. The coolant (liquid helium) flows in the annular space between the conductors and also acts as the main dielectric. Thus, an advantage of this system is simplicity from the electrical point of view. Furthermore, ac losses will generally be lower in rigid cables than in flexible cables where current flows in a number of tapes and losses are increased by edge effects and eddy currents. The disadvantages, however, are the low dielectric strength of helium, the much larger number of field joints required, and the necessity for the accommodation of thermal contraction [6]. Rigid conductors have been developed extensively in England at the Central Electricity Research Laboratories (CERL), in Germany at Siemens, and in the United States at the Linde Division of Union Carbide; they are still being developed by the Krzhizhanovsky Power Engineering Institute in Moscow.

The simplest type of composite consists of concentric layers of copper and niobium as shown in Fig. 2. The niobium faces the annular space between inner and outer conductors and therefore under normal operation carries all the current. During faults, the current transfers to the

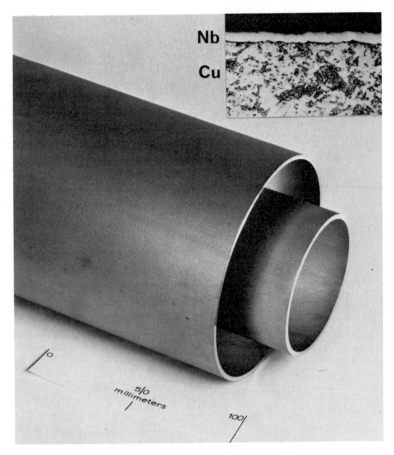

Fig. 2. Coaxial pair of niobium-clad copper tubes developed at the Central Electricity Research Laboratories, England. Outer diameters are 63.5 and 100 mm. Inset: etched cross section showing one of the thin 50-μm niobium surface layers on top of the copper. (From Graeme-Barber *et al.* [166].)

copper layer. Recovery from such faults has been demonstrated by Meyerhoff [17] for this type of composite. The copper also provides sufficient strength to withstand both the pressure of the helium coolant and the magnetic pressure during faults. A typical rigid cable such as that developed at Union Carbide is rated at 3400 MVA per circuit, and operates at 138 kV. The inner and outer conductors have diameters of 8 and 21.7 cm, respectively. The cryogenic enclosure has a diameter of ~69 cm [6]. Recently, composites have been made that include a hard type II superconductor such as Nb–Ti to increase the overall critical current and thereby improve fault-handling capacity [23]. Thermal contraction of rigid conductors is a problem that can be accommodated either by incorporating special mechanical constructions [28] or by designing a composite with low overall thermal contraction by including Invar in the composite [25].

Figure 3 illustrates two common fabrication techniques. The first (Fig. 3a) consists of coextrusion of the various components, followed by drawing. The second technique (Fig. 3b), better suited for yielding long lengths of conductor, consists of cladding an oxygen-free high-

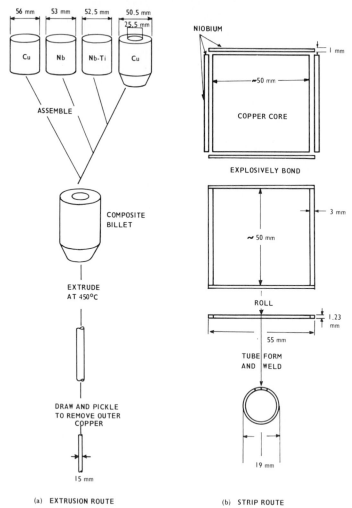

(a) EXTRUSION ROUTE   (b) STRIP ROUTE

Fig. 3. Production routes for tubular composites of copper, niobium, and niobium–titanium; (a) extrusion route; (b) strip route. (From Carter *et al.* [23].)

conductivity (OFHC) copper strip with niobium on its faces and edges, the edges being much thicker. The strip is then formed by passing it through a series of shaping rolls and is finally longitudinally seam welded. The thick Nb edges allow welding without the formation of any nonsuperconducting Nb–Cu alloys. Since the final stage in producing the composite strip before it is formed into a tube is one of rolling, the surface finish of tubes made this way can be very smooth. Another technique, which has been employed by Union Carbide, is to electroplate niobium from a fused salt bath using the process developed by Mellors and Senderoff [29]. This has been shown to produce tubes that exhibit acceptable losses [30]. Another technique that may be useful for joints but that yields high-loss niobium is plasma spraying [25].

3. FLEXIBLE CABLES

a. Introduction. Because of advantages such as longer fabrication lengths, higher dielectric strength, and accommodation of thermal contraction, most recent designs of superconducting cables favor flexible coaxial conductors with lapped plastic tape as dielectric. The basic design is shown in Fig. 4. Conductors (either tapes or wires) and dielectric tapes are helically wound on cylindrical forms. Flexibility results from butt gaps between individual conductor strands and dielectric tapes. Dielectric strength is a function of the density of supercritical helium impregnating the plastic dielectric, and is higher than for liquid helium in the pressure and temperature range of operation. Thermal contraction of the con-

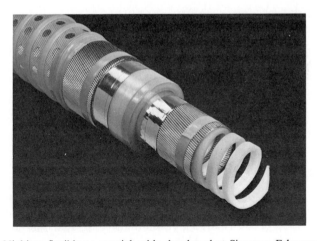

Fig. 4. Niobium flexible ac coaxial cable developed at Siemens, Erlangen, West Germany. (From Bogner and Penczynski [31].)

ductor is adjusted and matched to the contraction of the dielectric by choosing a suitable lay angle (or pitch length) [32]. This choice assures that the conductor–dielectric assembly remains radially tight upon cooldown, thereby preventing electromechanical vibrations and degradation of dielectric strength.

Because of the helical lay of conductors, flexible ac cables have problems not encountered in simple rigid coaxial cables. These problems arise because of the time-varying axial magnetic fields inside the inner conductor (hereafter "core") and annular space of the cable. Axial magnetic fields in the core can give rise to substantial eddy current losses in normal stabilizing metal or metallic supporting structures. Also, a net axial flux enclosed by the outer conductor causes both eddy current losses in those metal pipes enclosing the cable [33] and an appreciable voltage drop along the outer conductor, especially during faults [34]. This voltage would require substantial insulation of the outer conductor. Some of these loss mechanisms can be avoided by adjusting the pitch angles of inner and outer conductors so as to cancel either the net axial field in the core or the total net flux in the cable [33]. However, it is not possible to satisfy both these conditions simultaneously using a simple helical winding like that shown in Fig. 4. In addition, the lay angle, as discussed above, must satisfy mechanical requirements of thermal contraction, a requirement that results in an angle of ~20–30° for most plastic dielectrics. These problems have led to the development of two different flexible cable designs: single helical windings for Nb cables and double helices for Nb_3Sn ac cables, as will be discussed.

b. Niobium Flexible Cables. Niobium conductors can easily be fabricated so that a thin layer of niobium completely encloses a copper or aluminum stabilizer. Thus, eddy currents due to axial fields in the core can be avoided using niobium cables, even with a simple helical winding such as the one shown in Fig. 4, since the normal metal is shielded by the superconductor. The types of conductors that have been manufactured so far for niobium flexible cables are aluminum strips (7 mm × 3 mm) bonded on one side to niobium (7 mm × 50 μm) [34a], copper strips (10 mm × 1 mm) enclosed with a 5- to 50-μm-thick layer of niobium [35], and aluminum wires 2 mm in diameter clad with 100 μm of niobium [31, 36]. The last two were designed for cables of 4000 MW/275 kV and 2500 MW/120 kV by CERL and Siemens, respectively [7]. The problem of losses in pipes surrounding each cable can be avoided to a large extent by placing all three phases inside a common pipe so that the net axial flux vectorially cancels out [33]. This may impose cryogenic or mechanical restrictions, since some designs require a jacket around each cable to

withstand high helium pressures [2, 37]. It is also probably easier to manufacture and handle cables with individual jackets. A flexible cable design
that eliminates axial fields by incorporating double helices for both inner
and outer conductors has recently been described by Sutton and Ward
[38].

c. Nb$_3$Sn *Flexible Cables*.   With Nb$_3$Sn it is not possible to fully enclose the normal stabilizer since this would place the brittle compound far
away from the neutral axis and crack the Nb$_3$Sn during bending. The
Brookhaven group has therefore adopted a cable design that eliminates
most of the axial magnetic fields [34, 39] (see Fig. 5). Each coaxial "con-

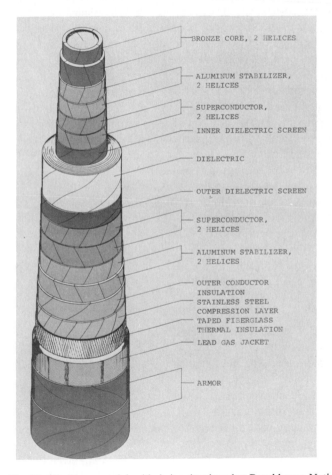

BRONZE CORE, 2 HELICES

ALUMINUM STABILIZER,
2 HELICES

SUPERCONDUCTOR,
2 HELICES

INNER DIELECTRIC SCREEN

DIELECTRIC

OUTER DIELECTRIC SCREEN

SUPERCONDUCTOR,
2 HELICES

ALUMINUM STABILIZER,
2 HELICES

OUTER CONDUCTOR
INSULATION
STAINLESS STEEL
COMPRESSION LAYER
TAPED FIBERGLASS
THERMAL INSULATION
LEAD GAS JACKET

ARMOR

Fig. 5.   Flexible Nb$_3$Sn ac coaxial cable being developed at Brookhaven National Laboratory. (From Forsyth *et al.* [13].)

ductor" consists of two sets of double helices—one of aluminum and one of Nb_3Sn tapes—wound with opposite helicity (Fig. 5).

Under normal operation current is shared approximately equally between the superconducting layers, thereby resulting in negligible residual axial fields [39]. During faults, current transfers to the aluminum layers, again sharing equally so that no net axial flux is present and voltage drop along the outer conductor is minimal.

The aluminum layers (Fig. 5) are approximately a skin-depth thick (~ 1 mm) and the superconducting layers are composites consisting of a thin Nb_3Sn tape, itself of composite structure, clad on one side with either stainless steel or copper and on the other with copper (see Fig. 6). The Nb_3Sn tape is itself a composite consisting of a substrate ($10-15$ μm thick) on which layers of Nb_3Sn are formed (~ 5 μm thick). The nature of the substrate depends on the method of fabrication. If the Nb_3Sn is formed by diffusion (either solid or liquid state), the substrate will be unreacted Nb (or Nb 1% Zr). If the Nb_3Sn is formed by physical or chemical vapor deposition, the substrate could be either Nb or some resistive material such as stainless steel or Hastelloy. The importance of the substrate arises because of parasitic losses due to currents crossing the substrate. Figure 7 shows schematically the patterns of current flow for the two superconducting layers of the inner conductor. On the top layer, current flows helically around each tape conductor and thus results in "substrate losses" if the Nb_3Sn does not fully enclose the substrate (as shown in Fig. 6) [40]. On the bottom layer, current flows along the tape provided no axial field is present in the core. The thickness of the copper cladding on each tape is insufficient to produce full stabilization but should be enough to prevent flux jumps. The thickness is limited by eddy current loss considerations. Current sharing between the superconducting and aluminum layers during faults has not yet been demonstrated.

III. Bulk Critical Currents of Type II Superconductors

A. Introduction

As discussed above, transmission line superconductors require a high critical current density J_c as well as low ac losses (in the case of ac cables). The loss reqirement is for peak surface magnetic fields of ~ 1 kOe or less whereas the high J_c is required for fields of up to ~ 10 kOe. This field corresponds to $\sim 2\times$ and $10\times$ faults, respectively, for dc and ac cables.

Low ac losses can be achieved in materials exhibiting either a very large bulk critical current density J_c or large surface currents, or a combi-

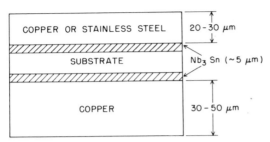

Fig. 6.   Schematic of a Nb₃Sn conductor for a flexible ac cable.

nation of both (see Section V). The critical current density (and its
dependence on magnetic field and temperature) is therefore a crucial
factor determining ac loss behavior. It is also important in determining the
behavior of the superconductor during fault conditions where the current
flows mainly in the bulk. Experimental data are presented below for the
field and temperature dependence of J_c of a few superconductors that are
of interest for power transmission applications.

B.   Experimental Results

1.   Nb, Nb–Ti, Nb$_{0.75}$–Zr$_{0.25}$

The self-, or low-field, J_c of niobium and the alloys Nb$_{0.75}$–Zr$_{0.25}$ and Nb
60 at. % Ti is generally well described by a simple linear dependence on
temperature [36, 41, 42]

$$J_c(T) = J_0(1 - T/T_\mu) \tag{1}$$

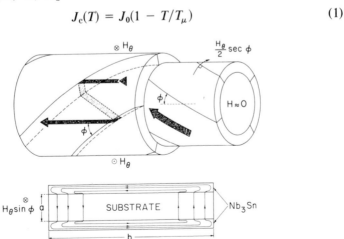

Fig. 7.   Schematic of current flow in the two superconducting layers of the inner coaxial
conductor (a) and across the substrate (b). This current flow is encountered in the ac cable
being developed at Brookhaven. (From Bussière *et al.* [40].)

where J_0 is a constant and T_μ is a temperature somewhat lower than T_c. This behavior is illustrated in Fig. 8 for a tube and a wire sample of cold-worked niobium prepared, respectively, for rigid and flexible cables [36]. The greater critical current density of the wire results from its greater degree of cold work. Equally good fits to Eq. (1) have also been obtained for Nb–Ti and Nb–Zr in self-fields [41, 42]. Table II lists J_0 and T_μ values for these alloys and for Nb. The temperature T_μ is usually within 0.5 K of T_c. The value J_0 depends on the particular material tested. Values listed in Table II correspond to the maximum observed in the references quoted. In the case of pure niobium, J_0 increases with cold work. For $Nb_{0.75}–Zr_{0.25}$ J_c can be optimized by anneals in the range of 550–800°C for times ranging between 168 and 1 h, respectively [42]. Similarly, annealing cold-worked Nb–Ti at 385°C for ~1 h was found to increase J_c by a factor of 6 [19].

2. Nb_3Sn and Nb_3Ge

Although most studies relating J_c to microstructure and/or metallurgical treatments of A15 compounds have concentrated on enhancement of J_c at high fields, $H > 50$ kOe, one could expect that the behavior at low fields ($H \approx 10$ kG) would be similar. High-field pinning in A15 materials

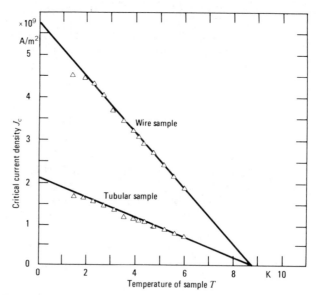

Fig. 8. Temperature dependence of the critical current density of wire and tubular samples of niobium. (From Penczynski *et al.* [36].)

TABLE II

CRITICAL CURRENT PARAMETERS FOR MATERIALS FOLLOWING
EQ. (1): (SELF-FIELD MEASUREMENTS)

Material	J_0 (A/cm²)	T_μ (K)	Remarks	Reference
Nb	5.8×10^5	8.8	Cold-worked wire	36
$Nb_{0.4}-Ti_{0.6}$	1.09×10^6	9.25	Commerical wire	41
$Nb_{0.75}-Zr_{0.25}$	2.48×10^6	10.4	Cold-drawn wire	42

can be increased by refining the grain size [43–46] or by introducing impurities, which in turn can affect the grain size. In cases where grain boundaries are the most active pinning centers, one often observes J_c to be a decreasing function of layer thickness [47, 48] and to be anisotropic with respect to magnetic field orientation [49–52]. This behavior is associated with columnar growth of the grains perpendicular to the substrate and can sometimes be avoided by adding impurities or producing precipitates, which either act as new pinning centers or modify the grain growth. Metallurgical treatments affecting J_c are described below for Nb_3Sn and Nb_3Ge produced by various processes.

Chemical vapor deposition (CVD) lends itself well to the addition of impurities and hence extensive studies have been made of the effect of impurities on J_c of Nb_3Sn and to a lesser extent of Nb_3Ge. In the case of Nb_3Sn, additions of oxygen in concentrations of up to 1000 ppm by weight were found to produce precipitates with ~50-Å particle size and to increase J_c by a factor of ~6 at 46 kOe [53]. Enstrom and Appert, however, find that CO_2, CO, and N_2 (but not O_2 or CH_4) are effective in increasing J_c of Nb_3Sn [54, 55]. The same authors have also studied the effects of a large number of other impurities on J_c [54, 55]. Ziegler *et al.* [56] found a linear increase of J_c with carbon concentrations of up to 2 at. % C with a slow decrease for larger concentrations. Carbon was also found to have no effect on the lattice parameter but it reduced T_c by a small amount. When carbon was present they found no correlation between grain size and J_c. However, they found a correlation between J_c and numerous defects within the grains when carbon was present. In the case of Nb_3Ge high critical current densities have been obtained in sputtered films [21, 57]. The high J_c in this case was attributed to the fine grain size (~1000 Å). For CVD material, however, large grains are usually obtained (several thousand angstroms to >1 μm in diameter) and pure A15 films have low J_c [21]. High critical current densities can be obtained by introducing other pinning centers, such as third-element impurities, or producing a fine dispersion of the tetragonal Nb_5Ge_3 phase [58]. The Nb_5Ge_3 phase particles have a mean diameter of ~600 and mean spacing of ~1600 Å [21]. Impurities such as CO_2 and C_2H_6 and N_2 have been found to lead to comparable enhancements of J_c [21, 58a]. The critical temperature of doped high-J_c samples, however, is depressed by 1–4 K. The most detrimental to T_c and the least effective impurity is CO_2.

In the case of liquid-diffused Nb_3Sn a common method of increasing J_c is to use internally oxidized Nb 1% Zr as substrate material. Internal oxidation produces ZrO_2 precipitates [59], which lead to smaller and equiaxed grains when compared to the columnar growth obtained with pure Nb. This results in a J_c that is independent of layer thickness and of field orientation [59]. Another parameter that affects the growth of the compound and J_c is the degree of cold work in the Nb or Nb–Zr starting material. Passotti et al. [60], using Nb–Zr base alloys with 0–7 at. % Zr, found that J_c increases both with the degree of cold work and with Zr content but that the former has a more noticeable effect (~30%). Additions of copper in the Sn bath can also produce significant improvements of J_c [61].

For bronze-processed Nb_3Sn, grain boundaries are the main pinning centers [45] and J_c can be strongly dependent on layer thickness [47, 48]. This dependence on thickness can be decreased or eliminated and high J_cs can be obtained for thick layers by adding impurities such as Zr and Sn in the Nb substrate and using high Sn concentrations in the bronze matrix [47, 48, 62, 63].

Another technique used to increase pinning, which has been extensively studied at Stanford University, is the production of multilayered structures using electron beam evaporation [64, 65]. Layers of Nb_3Sn (~1000 Å thick) are made to alternate with normal metal "barriers" (~50 Å thick). The barriers limit the growth of Nb_3Sn grains, which would otherwise grow in columnar fashion throughout the entire layer and result in a low J_c. A direct correlation has been found between the number of layers per unit thickness and the critical current density J_c at various temperatures [65].

Table III lists a number of J_c values in self-fields and at 1 T for Nb_3Sn and Nb_3Ge at various temperatures. Note that several techniques can lead to high J_cs and that the J_c of Nb_3Sn and that of Nb_3Ge are comparable at 4.2 K.

3. TEMPERATURE DEPENDENCE

The linear decrease of self-field J_c with temperature reported for Nb and the alloys Nb–Ti and Nb–Zr [Eq. (1)] is not usually observed for A15

TABLE III

Low-Field J_c of Nb_3Sn and Nb_3Ge

Material	Fabrication	T (K)	Self-field J_c (10^6 A/cm^2)	J_c at 1 T (10^6 A/cm^2)	Ref.
Nb_3Sn	Liquid diffusion with ZrO_2 precipitates	4.2	1.9	1.4	59
	Liquid diffusion (commercial)	4.2	2–8	—	66, 67
	Solid state diffusion	4.2	—	1–9	20
	Vapor deposited with 0.1 vol % CO_2	4.2	—	1.1	55
	Vapor deposited (commercial)	4.2	2.6	—	67
	Electron beam evaporation, multi-layered	4.2	8.5	—	65
		10	5	—	
	Magnetron sputtering	4.2	1.2–17		68
		7.7		1.2–2.7	
Nb_3Ge	CVD with Nb_5Ge_3 second-phase particles 5% second phase	4.2	4.4	—	69
		14	4.4	—	70
		14	1	—	70
		4.2	—	—	21
	CVD with Nb_5Ge_3 phase	4.2	1.5	1.5 at 5 kG	71
	With Nb N precipitates	—	4.2	1.8 at 5 kG	
	Sputtered film	4.2	2.8	—	72
		4.2	1	—	21

compounds such as Nb_3Sn and Nb_3Ge in the low- or self-field region. The temperature dependence is generally closer to parabolic and can be fitted to the more general equation

$$J_c = J_0[1 - (T/T_\mu)^2]^n \qquad (2)$$

where for Nb_3Sn [65–67, 73] $n = 1$ and T_μ is ~1 K lower than T_c and for Nb_3Ge [21, 71, 72] n is between 1 and 2 with $T = T_c$. The simple parabolic dependence was first reported for Nb_3Sn [66] and is illustrated in Fig. 9 for a commercial Nb_3Sn tape produced by liquid diffusion. Similar results were also observed for other diffused commercial tapes [67] and for multilayered Nb_3Sn prepared by electron beam evaporation [65]. For Nb_3Ge, the parabolic approximation gives a poorer fit to the data (see Fig. 9). Figure 10 shows data for Nb_3Ge prepared by CVD that are fitted to Eq. (2) with $n = 1$ and $n = 2$ [71].

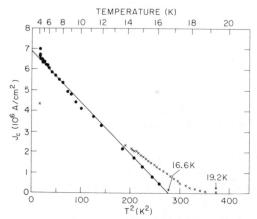

Fig. 9. Critical current density (X) of vapor-deposited Nb$_3$Ge (from Bartlett *et al.* [70]) and (●) of a commercial Nb$_3$-Sn tape made by liquid diffusion (from Bussière *et al.* [66]) as a function of temperature.

IV. Surface Currents in Type II Superconductors

A. Theory

As mentioned earlier and further discussed in Section V, "surface losses," which result when currents flow only in a thin layer at the surface of a superconductor, are usually much smaller than "bulk losses," which

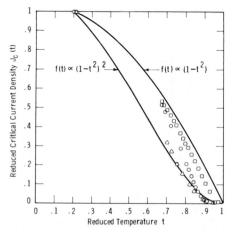

Fig. 10. Reduced critical current density versus reduced temperature for Nb$_3$Ge in self-field from transport measurements: O, sample 414 M10T; □, sample 377.2; △, sample 47. (From Daniel *et al.* [71].)

result from irreversible current flow in the bulk. Surface currents therefore play an important role in obtaining low losses. In type II superconductors, surface currents can arise from various physical mechanisms. These are (1) Meissner currents, (2) the surface sheath, (3) the surface barrier, and (4) surface pinning. Each of these mechanisms is described in this subsection. Experimental measurements are then described in Section IV,B and discussed in terms of these mechanisms.

1. MEISSNER CURRENTS

The magnitude of Meissner currents depends on the equilibrium B versus H curve, which is an intrinsic property of the material. Their maximum value is reached when the external magnetic field equals the lower critical field H_{c1} of the material. At this field the magnetic induction B inside the superconductor is equal to zero (assuming a reversible superconductor). As the external field is further increased, B increases more rapidly than H and shielding from Meissner currents decreases. For this reason, reversible type II superconductors cannot be operated in a transmission line for surface fields greater than H_{c1}. For high-κ superconductors the relation between B and H is given by [73a]:

$$B = H - H_{c1} + (\phi_0/8\pi\lambda^2)\{\ln[4\pi(H - H_{c1})\lambda^2/\phi_0] + 2 - 2\gamma\} \qquad (3)$$

where $H_{c1} \ll H \ll H_{c2}$ and $\gamma = 0.5772 \cdots$ is Euler's constant. Meissner currents are thermodynamically reversible and for a given external field H should be of equal magnitude and direction of flow regardless of whether the field has previously been increased from zero or decreased from a high value, or if the material has cooled through T_c in the presence of H. Current flow is not generally in agreement with Lenz's law. Low-κ superconductors can often be made with nearly reversible properties and H_{c1} determined with good accuracy; e.g., for pure niobium $H_{c1} \approx 1400$ Oe at 4.2 K [22]. Higher-κ materials, however, are usually difficult to prepare with reversible behavior. Irreversibility arises from interactions of vortices mostly with internal defects, but also with the surface. Surface interactions, for instance, play an important role in preventing flux expulsion after cooling through T_c in the presence of a magnetic field [74, 75]. For these reasons it is usually difficult to determine H_{c1} with any accuracy. This is the case for Nb_3Sn, for which H_{c1} values between ~100 and 1400 Oe have been reported [76, 77]. Similarly, large discrepancies between experimental and calculated values of H_{c1} were reported for V_3Ga [78] and for high-κ Nb–Ti alloys [79].

2. SURFACE SHEATH

The "surface sheath" or surface layer refers to a thin layer near the surface of a superconductor that remains superconducting while the rest of the bulk may be in the normal state. For type II superconductors in the presence of a magnetic field applied parallel to the surface, the surface layer will exist up to a field $H_{c3} = 1.695H_{c2}$ [80, 81]. For field angles other than parallel, H_{c3} decreases rapidly to H_{c2} for $\theta = 90°$. Above H_{c2} and for fields parallel to the surface, the thickness of the surface layer is of the order of the coherence length ξ [82]. Surface currents can be induced in the surface layer above H_{c2} and account for the magnetization tail observed in that field range. The current flow is in the direction given by Lenz's law, i.e., paramagnetic in decreasing fields and diamagnetic in increasing fields. The critical current of the surface layer was calculated on the assumption that at the critical current the free energy of the superconductor is equal to that in the normal state [83]. This gives reasonable agreement with experiment. The surface layer, characterized by a higher-order parameter in a thin layer near the surface, is also present below H_{c2}. In this case the bulk of the material is in the mixed state with the order parameter reaching a maximum near to, but not at, the surface [84]. As the field is decreased below H_{c3} and H_{c2}, the thickness of the surface layer increases rapidly. No discontinuity of the superconducting properties of the layer occur as the field is decreased through H_{c2} [82]. Since the layer is still present when the bulk of the material is in the mixed state, one can expect that surface currents similar to those observed above H_{c2} will also be present below H_{c2}. This mechanism has been proposed to explain hysteresis of magnetization curves of type II superconductors both above and below H_{c2} [83–86] as well as critical surface currents observed using transport currents above and below H_{c2} [87]. This interpretation is reinforced by the fact that plating the superconductor with a good normal metal such as copper dramatically reduces the magnitude of surface currents in low-κ alloys, especially above H_{c2} [87]. Plating with magnetic materials has an even more dramatic effect [86].

3. THE SURFACE BARRIER

The "surface barrier" refers to a maximum in the Gibbs free energy of a vortex as a function of its position relative to the surface of a superconductor. The position and magnitude of this energy maximum depend on the external magnetic field H and on the value of κ of the superconductor. For high-κ superconductors simple theory indicates that the barrier persists up to an external field $H_s \approx H_c$ where H_c is the thermodynamic critical field [88–91]. More recent two-dimensional treatments indicate that the ultimate barrier field is decreased to $H_s = 0.745H_c$ because of two-dimensional variations of the order parameter along the surface [92, 93]. Physically, the barrier arises because of two opposing interactions acting on a vortex at a distance x from the surface: a repulsive force from Meissner currents, which decreases as $e^{-x/\lambda}$, and an attractive image force which decreases as $e^{-2x/\lambda}$ where λ is the penetration depth [89]. For fields below H_s the attractive force is greater at short distances than the repulsive force and a barrier results. The net attractive force near the surface prevents flux entry until H_s is reached, at which point the attractive force is balanced by the repulsive force. Although early calculations of the barrier only considered the case where no flux is initially present in the superconductor ($B = 0$), more-recent calculations have shown that a barrier is still present when the bulk is in the mixed state [94, 95]. In this case the relation between B and the external field H_a depends on whether the external field is increasing (flux entry) or decreasing (flux exit) and a hysteretic magnetization curve is obtained even in the absence of bulk pinning. For a high-κ superconductor with no bulk pinning and $B \gg H_{c1}$, Clem [94] shows that in increasing fields

$$H_{en}^2 = B^2/\mu_0^2 + H_s^2 \qquad (4)$$

where H_{en} is the external field at the critical entry condition and $H_s \approx H_c$. In decreasing fields,

$$H_{ex} \approx B/\mu_0 + \phi_0/4\pi\lambda^2 \qquad (5)$$

where H_{ex} is the external field at the critical exit condition, ϕ_0 is the flux quantum, and λ is the penetration depth. Similar, although somewhat different results, were also obtained by Ternowskii and Shekhata [95]. According to this model, the external field differs significantly from B/μ_0 only in increasing fields, the difference being $\sim H_c$ when $B = 0$. Substantial surface currents are therefore present only in increasing fields where the Lorentz force is directed inside the material and the image force prevents flux entry. If a sample is cooled through T_c in the presence of a magnetic field, flux expulsion expected from the Meissner effect is limited because of a barrier to flux exit. In this case, the relationship between the external field H_a and B is not given by the intrinsic (reversible) $B(H_a)_{rev}$ relation but by the critical exit condition [Eq. (5)], where $H_{ex} = H_a$. Experimental evidence for this will be presented Section IV,B.

To our knowledge, the surface barrier is the only mechanism that leads to large asymmetries of critical surface currents with direction of current flow.

For discussions of experimental results to be presented in Section IV,B it is useful to define the following quantities for a given external field H_a and magnetic induction B near the surface.

(1) Barrier to flux entry: $\Delta H_{b_{en}}$ is, for a given B, the difference between the value of external field that causes flux entry $H_{en}(B)$ and the equilibrium value of the external field $H_{eq}(B)$:

$$\Delta H_{b_{en}}(B) = H_{en}(B) - H_{eq}(B) \tag{6}$$

(2) The barrier to flux exit $\Delta H_{b_{ex}}$ is similarly defined in terms of the exit field $H_{ex}(B)$:

$$\Delta H_{b_{ex}}(B) = H_{ex}(B) - H_{eq}(B) \tag{7}$$

The quantities $\Delta H_{b_{ex}}$ and $\Delta H_{b_{en}}$, however, cannot be determined experimentally unless the equilibrium curve $H_{eq}(B)$ is known. For this reason it is useful to define two "surface shielding" fields ΔH_{en} and ΔH_{ex} as follows [75].

$$\Delta H_{en}(B) = H_{en}(B) - B/\mu_0 \tag{8}$$

$$\Delta H_{ex}(B) = H_{ex}(B) - B/\mu_0 \tag{9}$$

These ΔH values are a measure of the effective shielding currents at the surface and include Meissner currents as well as actual "barrier currents."

4. SURFACE PINNING

Surface pinning is a rather loose term that has been used in the past to describe different mechanisms, including both enhanced bulk pinning near the surface and true interactions of vortices with or near the surface. Enhanced bulk pinning in the surface region often occurs in materials that have been cold worked by abrasion or machining, resulting in a surface region more heavily worked than the rest of the bulk. The term surface pinning should refer only to mechanisms where vortices interact with the surface itself. This usually occurs in a region of thickness approximately equal to a penetration depth λ at the surface. The surface barrier is one form of surface pinning; another can occur when quantized flux vortices or spots interrupt the surface of a sample and interact with surface irregularities or inhomogeneities as proposed by Hart and Swartz [96]. In this model, at a current below the intrinsic theoretical limit, given by either surface sheath or barrier models, the Lorentz force exceeds the pinning force, and flux moves across the surface. The critical surface current is defined in analogy to bulk pinning. Since this pinning mechanism is based on flux threading the surface, the model applies when the magnetic field has a component perpendicular to the surface. Perpendicular components are encountered in practice because of (1) intentional misalignment of the external field; (2) nonuniform current distributions, especially at the edges of a sample; and (3) rough surfaces. Surface pinning from fluxoids threading the surface arises because of variations in vortex energy with position. Such changes in energy, and hence pinning, have been obtained by modulation of the thickness of thin films [97, 98]. In this case, pinning is associated with changes in fluxoid energy that arise from changes in length between thin and thick regions of the film. Another technique, used by Autler, is that of coating the surface with a finely ordered array of magnetic particles [99]. In most materials, however, pinning is expected to be due to thickness variations arising from surface roughness or from inhomogeneities within a depth λ of the surface.

Inhomogeneities within a depth $\sim \lambda$ from the surface may be more effective than in the bulk because vortices are coupled over longer distances near the surface than in the bulk, forcing vortices to movie in unison [100, 101]. This difference in behavior near the surface is associated with changes in the electromagnetic characteristics of a vortex as it emerges near the surface region. Pearl [100, 101] has calculated that the electromagnetic region "spreads like a mushroom" at the surface. The current density at the metal–air interface follows a $1/r^2$ distribution characteristic of thin films, whereas deep inside the metal it falls off exponentially. The range of this transition is of the order of, λ away from the surface. As a consequence, Pearl [101] concludes that flux lines are loosely coupled inside the metal and experience a strong coulomblike repulsion at their ends [101]. Because of the long-range interaction, vortices do not assume a particular lattice order and the shear modulus vanishes at the superconductor surface. This long-range interaction results in a stiffening of the fluxoids in the surface layer and presumably forces the vortices to move in unison when subjected to a Lorentz force and enhanced pinning. According to Joiner and Kuhl [102], this mechanism is responsible for most of the critical current observed in their films with the magnetic field perpendicular to the surface. Thys et al. [103] have observed changes in pinning of Nb foils after plating the surface with very thin layers of copper. The results are interpreted as being due to a modification of the surface broadening of vortices because of proximity effects between the Cu and Nb.

The surface current mechanisms described above are directly analogous to bulk pinning where the current is determined by a balance between a Lorentz force and pinning force acting on vortices. Irreversible gradients and ac losses will result under ac conditions. Surface pinning may be useful to increase the overall critical current density, especially near edges of conductors where large perpendicular components of field are often encountered.

B. *Experimental Observations*

Experimental evidence for surface currents in type II superconductors comes from a variety of measurements, such as (1) critical transport currents; (2) magnetization in quasistatic, pulsed, or alternating magnetic fields; (3) microwave breakdown and resistance in the presence of magnetic fields; and (4) tunneling. Experimental observations of surface currents in the low-field region $H \ll H_{c2}$ of interest to power transmission are presented below. Of particular interest are the dependence of surface currents on magnetic field intensity and orientation, asymmetries with respect to direction of current flow, the effect of a normal metal on the surface, effects of surface roughness, and the dependence on temperature.

1. Dependence on Magnetic Field Intensity and Orientation; Asymmetries

As in the case of bulk currents, surface currents are observed to decrease rapidly with magnetic field intensity. Surface currents are much more sensitive to orientation of the magnetic field with respect to the surface, and often present a sharp maximum when the field is parallel to the surface. The maximum value of surface currents in any given material always occurs for the case where the magnetic induction B is zero in the bulk and the external field is parallel to the surface. In this case the critical surface current density $\sigma_c(A/m)$ is simply the field of first flux entry H_s. For a high-κ superconductor in a parallel magnetic field, surface barrier or superheating theory (see Section IV,A,3) predicts $H_s \approx H_c$ (the thermodynamic critical field). For low-κ materials H_s can even be greater than H_c. Although values close to H_c have never been reported for very-high-κ materials such as Nb_3Sn, flux entry fields in substantial excess of H_{c1} have been reported for various lower-κ alloys [104, 105] and pure niobium [106]. For Nb_3Sn, large microwave breakdown fields have recently been reported (89 mT at 4.2 K and 106 mT at 1.5 K) [107]. In comparison, the same authors report a maximum breakdown field of 159 mT for pure niobium at 1.5 K. Large flux entry fields have also been observed in Nb_3Sn by microwave resistance (1400 Oe at 4.2 K) [77] and tunneling (1800 Oe at 4.2 K) measurements [108]. Magnetization measurements in pulsed fields

gave a flux entry field $\Delta H_{en} \approx 800$ Oe at 4.2 K for a polished Nb_3Sn rod
[75]. As the external field was increased beyond 800 Oe, the value of ΔH_{en}
was observed to decrease, reaching a value of ~ 200 Oe for an external
field of 9.5 kOe (see Fig. 11). This field dependence was compared with
Clem's theory and gave reasonable agreement for $H_{en} > 6$ kOe [75]. Similar ΔH results have been reported for Pb–In and Nb–Zr alloys [109, 110],
with no attempt, however, to explain the field dependence of ΔH. The
failure of the theory, especially at low fields, is probably related to the fact
that the maximum superheating field H_s is never reached for $B = 0$.
Although the entry fields reported above are large for a high-κ superconductor such as Nb_3Sn, they are still considerably lower than $H_c (\sim 5000$ Oe)
and simple theory for the field dependence of ΔH is not expected to be applicable. It is expected that future materials with better surfaces will
exhibit even larger entry fields. Large differences occur for entry (or barrier) fields observed by different techniques. Microwave resistance and
tunneling measurements give much larger entry fields than microwave
breakdown and magnetization (or ac loss) measurements. Some of the difference is presumably due to variation in materials but some may be due
to more fundamental reasons yet to be explored.

As mentioned previously, surface currents are very sensitive to magnetic field orientation as well as to intensity. This orientation dependence
was first reported by Swartz and Hart [87] using transport currents in a
Pb–Tl alloy. Similar observations were then reported for other materials
[111, 112]. Figure 12 shows typical results, in this case for pure niobium at
4.2 K. Note that the critical surface current density J_F increases by

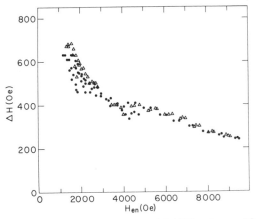

Fig. 11. Barrier to flux entry ΔH versus external field H_{en} observed for increasing fields
after cooling the sample in a magnetic field (●) or after decreasing the external field from
a high value (△). The sample is a Nb_3Sn cylinder at 4.2 K. (From Bussière [75].)

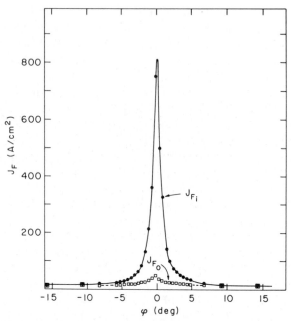

Fig. 12. Critical current density of an annealed niobium prism versus angle φ between magnetic field and face of prism at 2.6 kOe. When the Lorentz force is directed out of this face, J_{F_o} is measured; when it is directed into the face, J_{F_i} is measured. (From DasGupta and Kramer [111].)

approximately two orders of magnitude as the angle φ between external magnetic field and surface is decreased from 15 to 0°. Note also that two different values of surface critical current density are given for each field orientation. The values J_{F_i} and J_{F_o} correspond, respectively, to the cases where the Lorentz force is directed into and out of the surface. The large difference of critical current with direction of Lorentz force (and hence direction of current flow for a given field direction) was first reported by Swartz and Hart [87]. Such a large difference appears to be a common feature of surface currents below H_{c2} [111–115] and will be discussed further later. The angular dependence of the surface critical current density can be described over a limited range of angles by the following expression [114, 115]:

$$\sigma_c(\varphi) = a/\sin(\varphi - \alpha) \tag{10}$$

where σ_c is the surface critical current density, a and α are constants. The value of α is positive when the Lorentz force is directed into the surface ($\sigma_c = \sigma_{in}$) and negative when the Lorentz force is out of the surface ($\sigma_c = \sigma_{out}$). For pure niobium, Eq. (10) was found valid for σ_{in} (α positive) only for angles smaller than $\varphi_{crit} \approx 2\text{--}6°$ that are in turn greater than α ($\sim 1.5\text{--}3°$) [111]. This has the unfortunate consequence that Eq. (10) is not valid over the region where σ_{in} peaks sharply. Equation (10) predicts a maximum at $\varphi = \alpha$ and not at $\varphi = 0$, as is observed. Equation (10), however, is valid over the entire range of angles for σ_{out} (α negative).

In some cases the asymmetry of critical surface current, which in Fig. 12 gives a much greater value for σ_{in} than for σ_{out}, is reversed; i.e., σ_{out} is greater than σ_{in} [111, 112]. This behavior is expected only in low-κ materials in external fields close to H_{c1} and is due to an interplay between surface pinning, the surface barrier, and reversible Meissner currents [115]. In the case of the large-κ material Nb$_3$Sn, surface currents in increasing fields were found to be much greater ($\Delta H_{en} \lesssim 800$ Oe) than in decreasing fields ($\Delta H_{ex} < 10$ Oe) [75]. The asymmetry in this case is very large, with larger currents corresponding to the Lorentz force into the surface at all fields (see Figs. 11 and 12) [75].

2. EFFECT OF A NORMAL METAL AT THE SURFACE

Experimental observations show that thin normal metal layers at the surface of type II superconductors dramatically reduce surface currents above H_{c2} but also have a significant effect in fields less than H_{c2}. Barnes and Fink [86] report magnetization curves of Pb–Tl where electroplating the surface with Cr removed hysteresis above H_{c2}, and significantly decreased hysteresis for fields between H_{c1} and H_{c2}. Swartz and Hart [87] observe similar results using transport currents after plating a Pb–Tl alloy with copper. Figure 13 shows critical current measurements with and without copper plating as a function of external field for two different angles (2 and 30°) between field and surface. Note that for both angles and for fields far away from H_{c2} the critical current is decreased by almost a factor of 2 after electroplating. Both Barnes and Fink [86] and Swartz and Hart [87] conclude from their measurements that the surface sheath is responsible for surface currents in magnetic fields below but near H_{c2}, and that another mechanism becomes more important as the field is reduced away from H_{c2}. The second mechanism is probably associated with the surface barrier, which accounts for the asymmetry of critical currents observed below H_{c2}.

One may expect that the effectiveness of a normal metal at the surface of a superconductor will decrease as the value of κ is increased. This expectation is partly supported by measurements of Ullmaier and Gauster [109] that show no difference in ΔH for a Nb 25% Zr alloy after silver-plating the surface. However, good quality bond with niobium alloys is difficult to obtain by electroplating, and their negative observation may

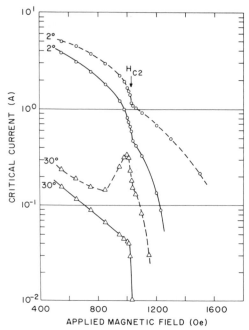

Fig. 13. Critical current of an annealed $Pb_{0.95}$–$Tl_{0.05}$ ribbon as a function of magnetic field at $\phi = 2°$ and $\phi = 30°$ before (———) and after (————) copper electroplating. (After Swartz and Hart [87].)

not be conclusive. Bussière and Kovachev [116] have reported an increase in ac loss due to thin layers of bronze on the surface of Nb_3Sn that may be attributed to a decrease of the barrier from a proximity effect. Finally, even in the case where the magnetic field has a significant component perpendicular to the surface, the critical current can be affected by the presence of a normal layer (see also Fig. 13) [102, 103]. In this case the proximity effect presumably affects the detailed shape of a vortex as it emerges from the superconductor. This has not been calculated theoretically.

3. Effect of Surface Roughness

The dependence of surface currents on surface roughness is of importance in optimizing materials for high surface currents and low ac losses. Samples exhibiting large entry fields [77, 104–108], or large asymmetries of critical surface currents [111–114] almost invariably have well-polished smooth surfaces. This is consistent with a barrier mechanism, which would be of reduced effectiveness for a rough surface, thus reducing the field of flux entry and the asymmetry of critical current with direction of Lorentz force. The critical surface current, as observed with transport currents, is often observed to *increase* as the surface is *roughened*. This is true whether the magnetic field has a component perpendicular to the surface or not [96, 102]. Introducing uniform bulk defects also causes an increase in the surface critical current [111]. When ac techniques are used, the resulting values of ΔH are usually higher for smooth surfaces [110, 117]. This is also the case, as mentioned above, for microwave breakdown measurements and measurements of the field of first penetration. A rough surface therefore reduces the barrier, leading to premature flux entry, but may also increase pinning in the surface region. When a significant component of magnetic field is perpendicular to the surface, roughness may significantly enhance pinning. Bulk defects can also stabilize the barrier, which is metastable, by preventing those fluxoids that have penetrated one end of the sample or sharp edges from propagating throughout the rest of the sample. This "reinforces" the barrier, which by itself would be insufficient except for a perfect surface with no edge effects. Thus, a combination of surface or bulk pinning as well as the barrier may be essential for observing surface barriers in type II superconductors.

4. TEMPERATURE DEPENDENCE

Data for the temperature dependence of surface currents are rather scarce. Some information has been obtained indirectly from ac loss measurements with temperature, either by fitting the data to a semiempirical model [66] or by assuming that the surface current is exceeded when a given loss value is reached [118]. The temperature dependence of the barrier of Nb_3Sn was also obtained from tunneling measurements [108]. In general, the dependence on temperature is described within experimental accuracy by

$$\sigma_c = \sigma_0 \left[1 - (T/T_\mu)^n \right] \tag{11}$$

For a commercial Nb_3Sn tape, Bussière *et al.* [66] found a good fit to the data assuming $n = 4$ and $T_\mu = T_c$. Thompson *et al.* [118] report $n = 2$ and T_μ lower than T_c for Nb_3Ge samples. Moore and Beasly [108] find for Nb_3Sn that the observed barrier decreases more slowly than parabolic ($n > 2$) at low temperatures, in qualitative agreement with the Bussière *et al.* data [66].

V. Theory of AC Losses in Type II Superconductors

A. Introduction

It was mentioned in Section II that one of the most stringent requirements for the superconductor of an ac transmission line is that it exhibits
low ac losses at power frequencies. Indeed, this requirement alone has for
many years narrowed the choice of superconductor to pure niobium.
More recently, however, it was shown that high-κ materials such as
Nb_3Sn and Nb_3Ge can also exhibit very low losses. The reasons for this
are associated with the very large surface and bulk current densities that
can flow in these materials (see previous sections). Presented below are
theoretical models of ac losses that take into account the presence of both
surface and bulk currents. First, the simple case where only bulk currents
are present is discussed. Surface currents are then taken into account for
the case where the total current is greater than the surface current.
Finally, theories of ac losses for surface fields below H_{c1} are presented.

B. Bulk Losses

The simplest model of ac losses is that originally proposed by London
[119] and Bean [120] in which it is assumed that currents flow uniformly in
the bulk of a superconductor characterized by a critical current density J_c
independent of magnetic field. For the case of a slab geometry with H parallel to the surface, this simple analysis leads to a power loss per unit surface area p increasing as the cube of the peak surface magnetic induction
B_p, inversely as J_c, and linearly with frequency f:

$$p = (2/3\,\mu_0^2)fB_p^3/J_c \quad (W/m^2) \tag{12}$$

The linear dependence with frequency arises because the magnetic profiles in the superconductor (shown schematically in Fig. 14a) are assumed
independent of frequency. This assumption is valid provided that the
depth of penetration of supercurrents is much less than the flux flow penetration depth [121]. This condition is valid for most type II superconductors below ~1 kHz. Because of their linear frequency dependence, ac
losses of type II superconductors are often termed "hysteretic."

Although it makes oversimplifying assumptions, the Bean–London
model provides considerable insight toward the understanding of ac
losses at relatively large amplitudes where surface currents can be

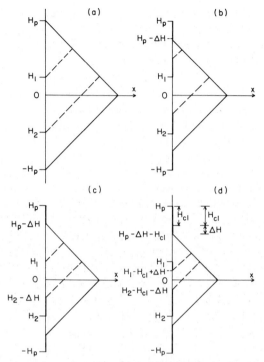

Fig. 14. Schematic of flux profiles for positive and negative peak field amplitudes ($\pm H_p$) and intermediate surface fields H_1 and H_2 obtained after decreasing the external field from $+H_p$. All figures assume a constant J_c independent of field; (a) is for no surface current, (b) assumes a "reversible ΔH," (c) assumes ΔH to be present in increasing fields only, and (d) assumes a reversible ΔH in conjunction with Meissner currents of magnitude H_{c1}.

neglected. The model can be made more realistic by introducing a field-dependent critical current density $J_c(B)$. Assuming

$$J_c = \alpha B^{-n} \tag{13}$$

where $0 \leq n \leq 1$ and α is a constant, leads to the following loss per unit area [122–124]:

$$p = 2fB_p^{3+n}I(n)/(2 + n)\mu_0^2 \tag{14}$$

where $I(n)$ takes values ~ 1, e.g., $I(0) = 0.667$, $I(\tfrac{1}{2}) = 0.868$, $I(1) = 1.025$ [122]. Equation (14) shows that the losses decrease with increasing critical current density, as in the Bean–London model, but increase more rapidly with magnetic field when J_c is a decreasing function of B.

C. Effect of Surface Currents

The foregoing type of calculation is valid provided that surface currents contribute only a negligible fraction to the total current. When the amplitude of the external field is comparable to H_{c1} or to H_s, the field of first entry of fluxoids, the foregoing models break down since surface currents are nearly lossless. A simple method of including surface currents in ac loss calculations is to assume that they are completely lossless; ie., that no loss occurs for surface fields below H_{c1} (or H_s). As the surface field is increased beyond H_s, losses are associated with the fraction of current flowing in the bulk. A small loss contribution also occurs from work done as vortices cross the surface current region.

As a first approximation, one can assume that lossless surface currents shield the bulk of the superconductor by an amount ΔH both in increasing and decreasing fields independent of field intensity. For peak surface magnetic fields $H_p < \Delta H$ no loss occurs. When H_p exceeds ΔH, losses are given by [125–127]

$$p = (f\mu_0/J_c)[\tfrac{2}{3}(H_p - \Delta H)^3 + 2(H_p - \Delta H)^2 \, \Delta H] \qquad (15)$$

The corresponding field profiles in the superconductor are shown in Fig. 14b. Figure 15 shows how different values of ΔH affect loss behavior, as calculated from Eq. (15). When $\Delta H = 0$, losses increase as the cube of the surface current magnetic field H_p.

When ΔH is different from zero, losses increase abruptly as H_p first exceeds ΔH; the slope then decreases asymptotically toward 3 as the surface field is further increased. For some of the parameters chosen in Fig. 15, e.g., $J_c = 5 \times 10^{10}$ A/m² and $\Delta H = 300$–400 rms A/cm, it may be seen that surface currents have a dramatic effect on losses in the range of current densities corresponding to an ac transmission line (400–500 rms A/cm). The model above reproduces many features of experimental data obtained for Nb_3Sn [65, 67, 117, 118] and other superconductors [125, 128]. It was also used to reproduce the temperature dependence of ac losses of Nb_3Sn by using the experimentally observed dependence of J_c on temperature and assuming that ΔH decreased as the fourth power of temperature [66]. The temperature dependence could not be reproduced adequately when the surface contribution was neglected. The model suffers, however, from a number of oversimplifying assumptions, such as considering J_c independent of magnetic field, neglecting Meissner currents, and not taking into account the asymmetries of surface currents discussed in Section IV. Taking into account the field dependence of J_c, while keeping ΔH independent of field, can be done fairly easily analyti-

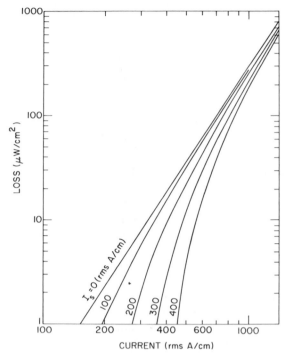

Fig. 15. Theoretical calculations of ac losses based on Eq. (15) for various values of surface currents I_s and a constant critical current density $J_c = 5 \times 10^{10}$ A/m²; $f = 60$ Hz.

cally for simple functions $J_c(B)$ [125, 127]. Treating the more general case where both J_c and ΔH are functions of field, however, is best done numerically. A simple but reasonably accurate method for taking into account both $J_c(B)$ and $\Delta H(B)$ is to calculate the loss with the appropriate $J_c(B)$, and the assumption that $\Delta H = \Delta H(H_p)$ throughout the entire cycle. Thus Eq. (15), for example, becomes

$$p = (f\mu_0/J_c)\{\tfrac{2}{3}[H_p - \Delta H(H_p)]^3 + 2[(H_p - \Delta H)^2(H_p)] \Delta H(H_p)\} \quad (16)$$

It was shown in Section IV that surface currents in type II superconductors such as Nb₃Sn are only present in increasing fields and that ΔH in decreasing fields is negligible compared to its value in increasing fields. Taking this into account and assuming J_c and ΔH to be field independent (see Fig. 14c for field profiles) leads to the loss equation

$$p = (f\mu_0/J_c)[\tfrac{2}{3}(H_p - \Delta H)^3 + \tfrac{4}{3}(H_p - \Delta H)^2(\Delta H)] \quad (17)$$

Equation (17) differs only slightly (in the second term) from Eq. (15), which assumes a "reversible" ΔH. This supports the notion that losses are more sensitive to the total amount of flux penetrating the sample at peak amplitude than to the details of flux motion during the cycle, as assumed in Eq. (16). The difference between Eqs. (15) and (17) is associated with losses due to flux crossing the surface current region [second term in Eq. (17)]. This loss is smaller for a nonreversible ΔH because flux from the bulk crosses a surface current during a smaller fraction of the cycle in this case. The validity of the models can be verified by comparing the theoretical voltage waveforms and hysteresis loops with experiment. The reversible ΔH model [Eq. (15)] predicts no change of flux as the external field is decreased from H_p to $H_p - 2\Delta H$, whereas the asymmetric ΔH model predicts the zero flux change only from H_p to $H_p - \Delta H$, in better agreement with experiment [75]. The asymmetric ΔH model also predicts a zero flux change as the external field is decreased through zero and increased by an amount ΔH in the other direction. This abrupt change in $\partial \varphi / \partial H$ as the external field is swept through zero is not usually observed [75] and should be the subject of more experimental and theoretical work if a completely realistic model of ac losses is to be developed.

To treat the presence of both reversible (H_{c1}) and irreversible (ΔH) currents, Dunn and Hlawiczka [127] proposed a "generalized critical-state model." In this model it is assumed that reversible currents have a constant magnitude equal to H_{c1} (unless $B = 0$) and that irreversible currents shield the bulk in increasing and decreasing fields by an equal amount ΔH_{ir}. Losses occur, therefore, only when the external field H_{ex} exceeds the value $H_{c1} + \Delta H_{ir}$. For $H > H_{c1} + \Delta H_{ir}$ losses are given by

$$p = \frac{5 \times 10^{-7} b}{12 \pi^2 J_c} (H_m')^2 (1.5 H_{c1} + 3\Delta H + H_m') \quad \text{W/cm}^2 \qquad (18)$$

where $H_m' = H_m - \Delta H - H_{c1}$; J_c is in amperes per square centimeter and H in oersteds.

The foregoing model has been applied successfully to pure niobium [36, 129] and to a number of commercial superconductors [128]. Because of its greater complexity and also because the voltage waveforms predicted by the model are not usually observed in high-κ superconductors, the previous models in which all currents are grouped into ΔH are often preferred for high-κ superconductors.

D.   AC Losses in the Meissner State

When the external magnetic field is lower than H_{c1}, fluxoids do not penetrate the bulk of the superconductor and one could expect in principle to observe no hysteresis loss. Losses are observed, however, even for amplitudes much smaller than H_{c1}, with the magnitude varying considerably from sample to sample (see Melville [122] and Wipf [130]). Buchhold [131] has suggested that the losses are due to enhancement of the field at surface irregularities and subsequent flux penetration. Melville [122, 132] has calculated losses resulting from rough surfaces, making the assumption that the field enhancement at peaks in the surface is infinite. This results in fluxons which enter peaks in the surface at very low fields but that do not penetrate the bulk unless the external field exceeds H_{c1}. This leads to the following equation for flux gradients in the region of surface roughness [132]:

$$\frac{dB}{dx} = \frac{-(dA/dx)(\mu_0 H_{rev} - B) \pm \mu_0 A J_{cs}}{1 - A(1 - \mu_0 dH_{rev}/dB)} \qquad (19)$$

where x is the distance inside the superconductor measured from the highest peak in the surface, $A(x)$ is the proportion of superconducting material at a distance x from the surface and thus describes the surface profile, J_{cs} is the critical current density in the region of roughness, and the other symbols have their usual meaning. Unfortunately, Eq. (19) is not readily integrated analytically for situations of practical interest. Using numerical integration, Melville [132] finds good agreement with experiment. He also finds that losses increase linearly with the depth of surface roughness and are strongly dependent on H_{c1} (and hence on purity of the surface material), but depend only to a limited extent on pinning. The pinning, however, has to be relatively large, $J_c \approx 3 \times 10^8$–$3 \times 10^{10}$ A/m^2, to explain the hysteresis loops. The loss behavior is approximately described by the expression [132]

$$p \approx 2^{w-1} f \mu_0 Q s H_{c1}^{-w} H_{c2}^{-\epsilon} H_p^{w+2+\epsilon} \qquad (20)$$

Here s corresponds approximately to the depth of surface roughness. The constants Q and ϵ depend in a complicated manner on surface profile and on the pinning force in the surface layer ($-0.2 < \epsilon < 1.6$ and $0.01 < Q < 0.3$). Large values of Q correspond to small values of w, where w describes the type of surface profile. Small values of w correspond to sharp-edged peaks, large values of w to rounded peaks of the surface. For example, $w = 1$ corresponds to a triangular profile and $w = 2$ to parabolic or sinusoidal profiles. The value of ϵ is usually small when compared to $w + 2$, hence the field dependence of the loss is chiefly determined by w, i.e., by the shape of the surface. The losses vary rapidly with H_{c1}. Hence, decreasing the purity of the superconductor, for instance by oxidizing niobium, may be more important than pinning in determining the loss.

DeSorbo [133] found that $\mu_0 H_{c1}$ decreases from ~ 0.14 T for pure Nb to ~ 0.058 T and 0.02 T, respectively, for 0.7 at. % and 2.6 at. % oxygen. Since oxygen tends to concentrate at the surface, it may have an important affect on ac losses even for much smaller concentrations.

E. Extra Losses for Conductors in Cable Configurations

In flexible cable designs the current is carried in a number of parallel conductors (tapes or wires) wound helically, with small gaps between them to accommodate flexing. It was already mentioned in Section II that such windings lead to parasitic losses that are mainly associated with axial magnetic fields. Enhanced losses will also arise because of edge effects and currents circulating around the conductors, especially in the case of Nb_3Sn cables. These two effects are discussed below.

1. EDGE EFFECTS

One consequence of dividing the conductor into strips or wires is an enhancement of the magnetic field in the vicinity of the gap between conductors. This results in increased losses as compared to simple tubular conductors. The problem has been analyzed for the case of wires and tapes [134–136]. In each case losses per unit area p are assumed to vary as

$$p = \alpha H_p{}^n \qquad (21)$$

where H_p is the peak surface field and $3 \le n \le 10$. For the case of wires of much smaller diameter than the cable diameter that are touching each other, the increase of loss compared to a single tubular conductor is 27% for $n = 3$, 56% for $n = 4$, and a factor of 2.51 for $n = 6$. Assuming a constant current in the cable, these numbers increase as the spacing between wires is increased [134].

For tape conductors the loss enhancement is generally less and depends not only on the spacing between strips but also on the radius of curvature of the strips at their edges. It is concluded by Ward et al., however, that the loss enhancement from edges can be offset by a drop in current density of only a few percent [135, 136].

2. CURRENTS CIRCULATING AROUND THE
CONDUCTORS

As discussed in Section II and shown schematically in Fig. 7, currents in flexible cables flow helically along the conductors. The helical flow can be resolved into a component of current along the conductor and a component circulating around the conductor (not contributing to the net transport current). The magnetic field H_{11} along the tape is given by [39]

$$H_{11} = H_\theta \sin \varphi \qquad (22)$$

where H_θ is the azimuthal magnetic field at the outer surface of the inner coaxial conductor (typically 400–500 rms A/m) and φ is the pitch angle [39]. Assuming H_θ values between 400 and 500 rms A/cm and pitch angles between 30 and 45° gives rise to axial fields H_{11} in the range of ~200–350 rms A/cm for the inner coaxial conductor. H_{11} is, therefore, relatively large and can produce significant losses when the superconductor consists of a "sandwich" structure such as Nb_3Sn tapes with Nb_3Sn layers on each side of a niobium or other substrate. In this case the axial field H_{11} will induce currents to flow across the substrate as shown schematically at the bottom of Fig. 7.

The losses can be significant when the substrate is near its critical temperature or in the normal state. Both of these cases have been described by Garber *et al.* [39]. For Nb_3Sn made by solid or liquid diffusion the substrate is either pure Nb or a Nb–Zr alloy and hence will be operated in the superconducting state. In this case losses are reduced by increasing both the J_c and T_c of the substrate [39]. This can best be done by adding a few percent Zr to the Nb substrate [40]. A Nb–2% Zr substrate should allow operation of the cable up to ~9 K. The use of pure Nb or oxidized Nb–1% Zr substrates will limit the maximum temperature to ~8.5 K [40].

VI. AC Losses of Pure Niobium

A. Metallurgical Effects

Early measurements on pure niobium indicated that ac losses occurred for surface fields well below the first critical field H_{c1} [131, 137]. It was postulated that these losses arose because imperfections such as protrusions or voids in the surface caused premature flux penetration and hysteresis in the surface region [131]. This hypothesis was supported by the observation that the losses vary linearly with frequency up to the kilohertz range [131, 138], and increase very rapidly with peak surface field,

$$p = cH_p{}^n \qquad (23)$$

where p is the loss per unit area, H_p the peak surface field, c a constant, and $2.5 \leq n \leq 6$. This ruled out the possibility of eddy currents in normal regions of the surface as the loss mechanism [131].

Later investigations essentially confirmed the findings of Buchhold and showed that the losses were also very dependent on the characteristics of the particular Nb sample. For peak surface fields of ~ 10 mT, losses measured in a large number of samples can differ by up to three orders of magnitude [122, 130]. This wide variation is associated with metallurgical differences, variations in surface roughness, and also with edge effects, which sometimes dominate the measured loss [139]. The effect of metallurgical treatments such as cold work, heat treatment, and mechanical and electrochemical polishing are considered later.

For field amplitudes below H_{c1} the losses are found to be very sensitive to surface roughness. Rocher and Septfonds [138] were able to decrease the loss of their Nb sample by a factor of 20–50 by electropolishing. Other workers have also found a qualitative correlation between loss and surface roughness as measured with a Talysurf profilometer [129, 140]. For a large number of niobium samples Brankin and Rhodes [140] found that the losses increase approximately linearly with the average height of the surface asperities. This is in agreement with their semiempirical model and with Melville's model presented in Section V.

Mechanical polishing usually results in the smoothest surfaces and the lowest losses [129, 140, 141]. Surface roughness, however, is not the only parameter determining ac loss since losses can differ considerably for a given average roughness reading.

Brankin and Rhodes [140] report that losses of an initially annealed polycrystalline Nb sample are drastically reduced by chemical polishing, although this treatment did not change the average roughness reading (0.3 μm) (see Fig. 16). Mechanical polishing produces even lower losses, although the roughness reading decreases only slightly (to 0.22 μm) (see Fig. 16). Similar results were obtained by Al-Huseini [142], although the roughness was not measured. The very low losses resulting from mechanical polishing are probably associated with increased pinning by surface defects introduced during the polishing. Changes of losses introduced by chemical polishing may also be due to hydrogen pickup, as postulated by Brankin and Rhodes, or to the removal of a surface layer that is more contaminated with impurities than the bulk.

In general, a combination of cold work and mechanical polishing leads to the lowest ac losses. This is understandable since mechanical polishing produces very smooth surfaces and cold work increases pinning, which in turn reduces losses (see Section V).

B. Composite Conductors

Composite conductors for power transmission have been fabricated using several techniques, such as coextrusion of niobium and copper or

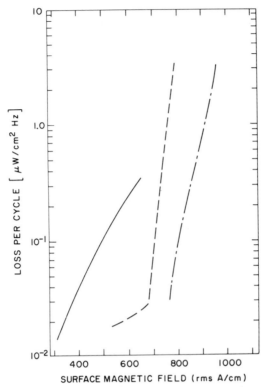

Fig. 16.   Loss per cycle of polycrystalline niobium samples in the as-annealed condition (————) and after chemical (– – –) and mechanical (—·—) polishing. (After Brankin and Rhodes [140].)

niobium and aluminum, co-rolling, plasma spraying and electroplating (see Section II,C). For practical and economic reasons the conductors are not usually treated after manufacture to reduce ac losses. Except for plasma spraying [25], all of the foregoing techniques can, however, lead to losses that are acceptable, although much higher than for polished Nb. Figure 17 compares losses at 4.2 K for a number of prototype tubular (Nos. 1–5) and flexible (Nos. 6 and 7) conductors developed for power transmission cables at Union Carbide (UC) (Nos. 1–3), Central Electricity Research Laboratories (CERL) (Nos. 4–6), and Siemens (No. 7). The loss measurements of the UC conductors (Nos. 1–3) were made on 6-m lengths of 1-cm- and 3-cm-diameter tubes used as inner conductors of a coaxial pair shorted at one end [30]. Tubes No. 1 and No. 3 were roll formed from 0.063-cm-thick annealed niobium sheet and longitudinally seam welded. Tube No. 2 was electroplated using a process developed by Mellors and Senderoff [29]. The CERL conductors [35] are all composite

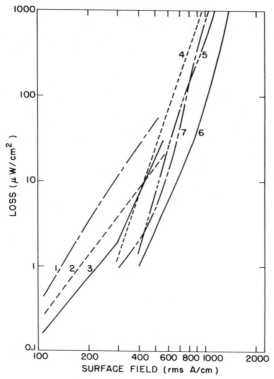

Fig. 17. Comparison of losses at 4.2 K for a number of prototype tubular (Nos. 1–5) and flexible (Nos. 6 and 7) niobium conductors developed for power transmission cables at Union Carbide (Nos. 1–3) [30], Central Electricity Research Laboratories (Nos. 4–6) [35], and Siemens (No. 7) [143]. Details of conductor designs are given in the text.

conductors consisting of a thin layer of high-purity niobium on a substrate of high-conductivity copper and with an intermediate layer of Nb 44 wt. % Ti in the case of sample No. 4. Both tubes No. 4 and No. 5 were processed by extrusion and have an external diameter of 15 mm. Other tubes processed by press-forming and seam-welding composite strips gave losses very similar to tube No. 5. Because of the particular manufacturing processes, the surface structure is anisotropic and losses depend on the direction of current flow (axial or circumferential). The data presented in Fig. 17 are for axial currents only, since this type of flow is encountered in tubular axial conductors. The CERL flexible conductor (No. 6) was a copper strip 17.6 mm wide by 1 mm thick coated with 20 μm of niobium. The strip was wound into a solenoid, which closely simulated the conditions in a flexible cable [35]. The data for the Siemens flexible conductor (No. 7 in Fig. 17 [143]) are for a cable model consisting of ~3-mm-diam wires of high-purity aluminum coated with a thin layer of niobium. The wires are manufactured by coextruding a Nb tube filled with high-purity aluminum. It is somewhat surprising at first to note that rigid tubular conductors exhibit higher losses than the simulated flexible cable losses, which include edge effects. This is probably due to the greater ease of fabrication and quality control for smaller conductors, which ultimately leads to better niobium for the flexible conductors.

C. *Effect of Trapped Magnetic Flux*

Early measurements of ac losses in high-purity lead [137] showed that the losses increased by an order of magnitude when the samples were cooled in the presence of external magnetic fields not much greater than the earth's field. They also showed that when the critical field of their Nb specimens was exceeded, losses increased considerably. Rocher and Septfonds [138] report that cooling in a field normal to the sample axis produces a large increase of loss, whereas trapping a magnetic field along the axis has little influence. In general, the presence of trapped flux causes losses to appear at a much lower surface field, thereby reducing the high power law dependence of the loss on field and substituting for it an approximately cubic dependence [144, 145]. This is illustrated in Fig. 18, for pure niobium. Male has shown that applying a small component of

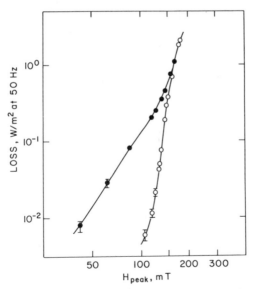

Fig. 18.   Ac losses of pure niobium with (●) and without (○) trapped magnetic flux. (After Male [144].)

transverse magnetic field has an effect similar to trapping flux along the sample and concludes that trapping longitudinal fields produces small transverse components that are responsible for the increased loss. Melville [132] has shown that this mechanism leads to the observed cubic dependence of loss on surface field.

Although trapped flux can cause very large increases of ac losses it is not clear at present if this would have a significant effect on the operation of a transmission line since the trapping of flux in actual transmission line configurations, e.g., during faults, has not been demonstrated.

D. Temperature Dependence

The temperature dependence of losses for pure Nb has been investigated by a number of workers for short samples [36, 146, 147] and for a large tubular composite conductor [148]. Below H_{c1}, the temperature dependence of losses is determined by the dependence of H_{c1} and H_{c2} on temperature. Introducing a parabolic dependence for H_{c1}

$$H_{c1}(T) = H_{c1}(0)[1 - (T/T_c)^2] \tag{24}$$

and a modified parabolic dependence for H_{c2} [36]

$$H_{c2}(T) = H_{c2}(0)[1 - (T/T_c)]^2 \tag{25}$$

into Eq. (20), we obtain the following dependence of losses on temperature.

$$p(T) = p(0)H_m^{w+2+\epsilon}[1 - (T/T_c)^2]^{-w-2\epsilon} \tag{26}$$

where $p(0)H_m^{w+2+\epsilon}$ is the loss value at zero temperature.

Penczynski et al. [36] find good agreement with Eq. (26) assuming $w = 2.23$, $\epsilon = 0.17$, giving $w + 2 + \epsilon \approx 4.4$.

For amplitudes greater than H_{c1} one gets reasonable agreement with theory by introducing temperature-dependent values of J_c, ΔH, and H_{c1} in the Dunn and Hlawiczka model [Eq. (18)] [36].

Penczynski et al. [36] find that the losses of their Nb samples remain below 10 μw/cm^2 provided the surface field does not exceed H_{c1}. For a surface field of 0.1 T this condition is satisfied provided that the temperature does not exceed 6 K. At 360 A/cm Grigsby and Rogers find that losses remain below 10 μw/cm^2 up to 7 K [147]. Measurements on the CERL tubular conductor are presented in Fig. 19. The measurements were made in a coaxial arrangement of Nb–Cu composite conductors cooled with helium gas between 4.4 and 8.0 K using transport currents [148]. Note that at 500 rms A/cm the losses increase by an order of magni-

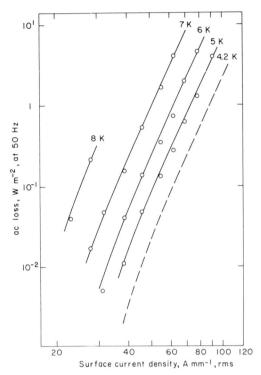

Fig. 19. Temperature dependence of losses for a large tubular Nb–Cu composite. (After Baylis *et al.* [148].)

tude as the temperature is raised from 4.2 to 6 K. At 500 rms A/cm losses remain acceptable (<10 μw/cm^2) up to 5 K.

It therefore appears that Nb transmission lines are limited by losses to a maximum temperature below ~6 K.

VII. AC Losses of Nb$_3$Sn and Nb$_3$Ge

It has been shown earlier that ac losses of Nb are not an intrinsic material property and can vary considerably, even by orders of magnitude, depending on metallurgical treatment. Nb$_3$Sn and Nb$_3$Ge are both subject to this rule and, depending on the mode of preparation, can manifest very different loss characteristics. Nb$_3$Sn, a relatively stable compound, forms readily and can be fabricated with high T_c (~18 K) by several different processes, such as reacting niobium with liquid tin (heretofore referred to as liquid diffusion), chemical vapor deposition (CVD), electron beam

(EB) evaporation, sputtering, and solid state diffusion from a bronze matrix. At present two of these techniques, liquid and solid state diffusion, are commercially used by Intermagnetics General Corp. (IGC) and Airco Industries, respectively, to produce Nb_3Sn tapes for power transmission applications [149, 150]. Chemically vapor-deposited tapes initially pioneered by RCA and then taken up by Canada Superconductor and Cryogenics Co. (CSCC) are no longer available. For future applications, however, any of the foregoing processes could probably be developed for commercial-scale production.

On the other hand, Nb_3Ge is highly unstable and leads to high T_c only when formed by nonequilibrium processes such as CVD, EB evaporation, and sputtering. The ac loss behavior of the two compounds can be optimized by metallurgical treatments described in Section VII,A. The temperature dependence of losses is discussed in Section VII,B.

A.   Metallurgical Effects

1. Nb_3Sn

Figure 20 shows 60 Hz losses at 4.2 K as a function of surface current density σ (rms A/cm) for various bare Nb_3Sn tapes. None of the tapes was given any special treatment after reaction. Two of the samples (IGC-4 and KB-15 [117]) are commercial liquid-diffused tapes, two are commercial CVD tapes (RCA-2119 and CSCC-21 [117]), and BT 40E is a tape prepared at Brookhaven by solid state diffusion [20, 151, 152]. Losses are seen to vary by up to two orders of magnitude at 500 rms A/cm and approximately one order of magnitude at 1000 rms A/cm. This behavior can be understood on the basis of two major factors affecting the losses: bulk pinning and surface condition, which, respectively, affect the bulk critical current density and surface currents. The high losses shown in Fig. 20 for samples RCA-2119 and IGC-4, for instance, can be largely accounted for by the presence of a rough and porous outer surface. This can be observed from optical as well as scanning electron micrographs [117]. Figure 21 shows six scanning electron micrographs [117] of samples exhibiting high (Fig. 21a,b), intermediate (Fig. 21b,e), and very low (Fig. 21f) losses. Micrographs (b), (c), and (d) (Fig. 21) correspond to samples KB-15, RCA-2119B, and CSCC-21, whose loss characteristics are shown in Fig. 20. The surface structure of sample IGC-4 (loss given in Fig. 20) is similar to that of IGC-1, whose surface is shown in Fig. 21a. In general, the qualitative relationship between surface roughness and ac loss is good. However, because the porosity or roughness may be as-

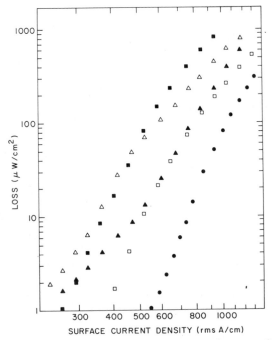

Fig. 20. Losses at 60 Hz at 4.2 K for various bare Nb_3Sn tapes: ●, BT 40E, solid state diffused; ■, IGC-4, and □, KB-15, liquid diffused; and ▲, CSCC-21, and △, RCA-2119, vapor deposited. (From Bussière [151].)

sociated with the presence of second-phase material or can extend deep into the material, the correlation remains a qualitative one.

The loss behavior of Nb_3Sn depends on the method of fabrication. In the case of liquid-diffused material rough and porous surfaces usually result when thick layers are grown and/or when free tin is left on the surface after reaction. This is the case with commercial tapes made by IGC for high-field magnet applications (see Fig. 22). In this instance the free tin left on the surface is desirable because it facilitates the lamination by soldering of the Nb_3Sn to copper and/or stainless steel. The irregular growth of the compound is due to a solution–dissolution process at the liquid-tin–Nb_3Sn interface [153]. This irregular growth, and consequently the loss characteristics, can be considerably improved either by limiting the amount of free Sn on the surface during reaction and growing a thin layer or by treating after reaction. The first approach was used by Kawecki-Berylco Industries (sample KB-15) and their tapes exhibit lower losses than the IGC tapes (see Fig. 20). Treatment after reaction can consist of either removing the porous layer by polishing [154] or converting tin-rich

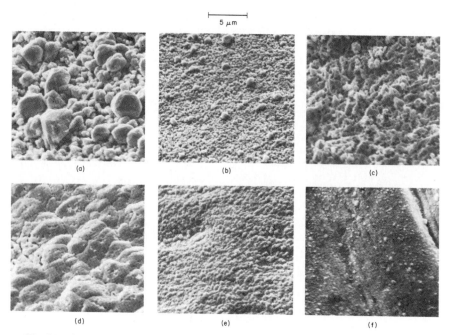

Fig. 21. Scanning electron micrographs of a number of Nb$_3$Sn tapes: sample (a) IGC-1, (b) KB-15, (c) RCA-2119B, (d) CSCC-21, (e) BT-19, and (f) A-17. (From Bussière *et al.* [117].)

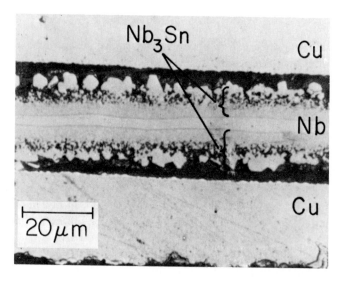

Fig. 22. Optical micrograph of a commercial tape made by liquid diffusion. (From Garber *et al.* [67].)

phases to Nb_3Sn by additional heat treatment [151]. Figure 23 shows losses of a commercial liquid-diffused tape both in the as-reacted (as-received) condition and after a short heat treatment. Prior to heat treating the tape, free tin was removed from the surface; the tape was then heated in vacuum at 950°C for 30 min. This treatment was sufficient to convert most of the second-phase material to Nb_3Sn and produced a significant reduction of loss. Even more dramatic reductions of loss have been obtained by chemical etch, electropolishing, and mechanical polishing [154]. Very thick layers (up to 90 μm) of Nb_3Sn were grown on the surface of Nb 1% Zr. A typical cross section taken after reaction is shown in Fig. 24a. Note the presence of a porous outer layer that extends as deep as 10–15 μm from the surface. Low losses resulting after chemical etch or polishing are attributed mainly to the removal of this porous outer layer (see Fig. 24b). The lowest losses were obtained by mechanical polishing with relatively coarse paper (3/0) and followed by annealing for $\sim \frac{1}{2}$ h at 850°C. Mechanical polishing with 3/0 paper without the annealing treatment actually caused an increase of ac loss at low current densities (see Fig. 25). This increase is attributed to damage or strain introduced in a surface layer approximately 1 μm thick at the surface [154]. The subsequent heat

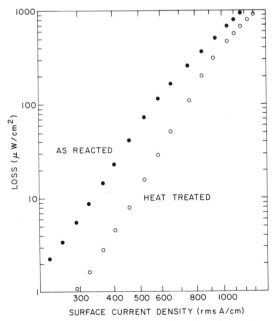

Fig. 23. Reduction of loss of a commercial liquid-diffused tape (bare IGC-5) by heat treatment in vacuum; 60 Hz, 4.2 K. (From Bussière [151].)

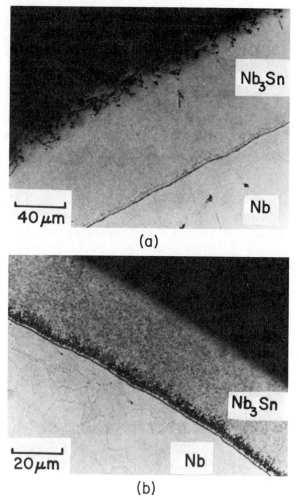

Fig. 24. Optical micrographs of Nb₃Sn made by liquid diffusion (a) as reacted and (b) after 5 h of mechanical polishing with 0.3-μm alumina powder, followed by electropolishing. (From Bussière and Suenaga [154].)

treatment removes this damage and results in increased pinning at the surface. Although this technique leads to the lowest losses, excellent results can also be obtained by electropolishing and chemical etch, which give similar results. The effect of electropolishing is shown in Fig. 25. The chemical etching technique lends itself to commercial application and is presently used by IGC to reduce ac losses of their Nb₃Sn tapes [149]. These polishing techniques have also been used successfully to reduce ac losses of Nb₃Ge, as will be discussed later.

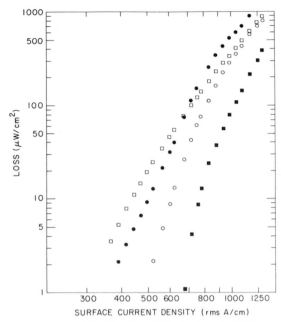

Fig. 25. Reduction of loss of liquid-diffused Nb₃Sn (sample IG3) by polishing: ●, as reacted; ○, electropolished; □, 15 min mechanical polish (3/0 paper); ■, annealed. (From Bussière [151].)

In the case of solid state diffused Nb_3Sn the surface layer is usually relatively smooth and dense as compared to liquid-diffused material, and low losses can result without the need for further surface treatment [20, 152, 155]. Loss results are shown in Fig. 20, (sample BT 40E) and a typical cross section is shown in Fig. 26. Note that losses at 500 rms A/cm are less than 1 $\mu w/cm^2$ (Fig. 20) and that the interface between the bronze and Nb_3Sn is very smooth and well defined (Fig. 26) as compared to that of the liquid-diffused tape of Fig. 22. Optimizing loss and critical current characteristics of solid state diffused tapes is similar to the problem of improving the critical current of multifilamentary wires [63]. The parameters available are reaction time and temperature, concentration of Sn in the Cu–Sn matrix, and third-element additions to either the Nb or the Cu–Sn matrix (or both). Best results (low losses) are usually obtained with a high concentration of Sn in the Cu–Sn matrix, without exceeding the solubility limit of Sn in Cu [152]. This usually means, for the temperature range used (650–800°C), an optimum composition of ~13 wt. % Sn in the copper. The high Sn concentration causes a higher growth rate of the Nb_3Sn layer and results in a smaller grain size, which in turn leads to high J_c values. The growth rate is also increased with temperature in the range of interest (~650–800°C); there is, however, a compromise between layer growth

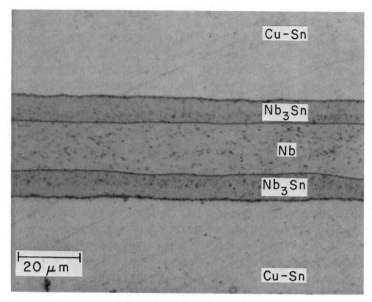

Fig. 26. Optical micrograph of a typical tape made by solid state diffusion.

and grain growth. Grains are usually larger and J_c somewhat smaller for Nb_3Sn reacted at 800°C than for Nb_3Sn growth at 700°C. Reaction at a low temperature, however, leads to somewhat lower T_c values and higher losses at current levels of the order of 500 rms A/cm [152]. An optimum temperature would therefore be close to 750°C. This gives reasonable growth kinetics, low losses, and good critical currents [20].

Most loss data reported in the literature for solid state diffused Nb_3Sn is for material with the bronze matrix removed [117, 152, 155]. Recent measurements have shown that the compressive stress produced by contraction of the matrix upon cooldown causes significant reduction of T_c (~1 K) [156, 157] and an increase of ac loss [20, 116]. For transmission line applications, however, this is not a problem since the bronze matrix will be removed and the tape laminated to copper and stainless steel (see Section II).

The use of electron beam evaporation to produce Nb_3Sn was pioneered by workers at Gulf Atomic Co., who produced some of the first Nb_3Sn with losses below 20 $\mu w/cm^2$ at 500 rms A/cm [158, 159]. The samples produced at Gulf Atomic and more recently at Stanford University [64, 65, 73, 160, 161] consist of alternate layers of Nb_3Sn and other materials, such as Nb, Sn, Y, and Nb 5% Sn, which act as barriers to fluxoid motion. The fabrication technique and electrical properties have been reviewed by Hammond [64] and Howard *et al.* [65], respectively. Figure 27 shows a schematic of the multilayered structure that is deposited on a Hastelloy B substrate together with a SEM micrograph of the cross section of an actual sample. The flux pinning barriers of nonsuperconducting material range in thickness between 30 and 2500 Å. The Nb_3Sn layers vary between 400 Å and 1.4 μm in thickness. The barrier layers act not only as flux-pinning sites but also as renucleation centers, which disrupt the tendency toward columnar growth and limit the Nb_3Sn grain size [65]. The growth morphology of the Nb_3Sn depends on the substrate temperature T_s during deposition. For unlayered Nb_3Sn the highest J_c values were obtained for $T_s < 600°C$. The lowest losses, however, were obtained in the range $600 < T_s < 720°C$ [65]. When high temperatures are used, the compound interacts with the Hastelloy and forms a diffusion layer. A substrate temperature ~690°C seems to be a good compromise between desirable growth morphology and excessive diffusion [65].

Typical loss results are shown in Fig. 28 for a temperature of 10 K. Note that losses at 500 rms A/cm are ~20 μw/cm². The experimental data are fitted to a theoretical model similar to Eq. (15) but assuming a field-dependent J_c

$$J_c = J^*/(1 + B/B_0) \tag{27}$$

and a constant $\Delta H = 200$ Oe. The values of J^* and B_0 are given in the fig-

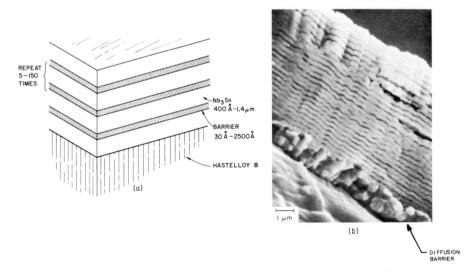

Fig. 27. (a) Schematic of layered superconducting composite (total effective superconductor thickness 3–7 μm, total evaporated metal thickness 7–10 μm) together with (b) a SEM micrograph showing the cross section of such a sample fabricated by electron beam evaporation. (From Howard *et al.* [65].)

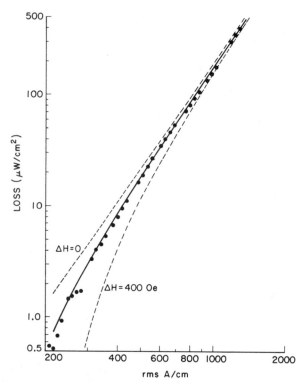

Fig. 28. Alternating-current loss of Nb_3Sn made by electron beam evaporation measured at 10 K and 50 Hz; $J^* = 7.6 \times 10^6 \, A/cm^2$, $B_0 = 7800$ G. The solid line is based on a critical state model using the measured $J_c(B)$ and a surface barrier ($\Delta H = 200$ Oe). Dashed curves show the effect of choosing different barrier fields. (From Howard et al. [65].)

ure. The value of J_c was determined from flux penetration measurements. Also shown in Fig. 28 are calculations for $\Delta H = 0$ and $\Delta H = 400$ Oe. It is seen that a relatively small barrier is required to fit the data and that the low losses are mostly due to a high value of J_c.

2. Nb_3Ge

Although Nb_3Ge has been produced with high J_c and T_c by sputtering and CVD, at present ac loss characteristics have been investigated only for CVD-produced material [21, 118, 162–164].

Braginski et al. [21] report losses in the range of 30–100 $\mu w/cm^2$ at 500 rms A/cm and 4.2 K for Nb_3Ge deposited on one side of a Hastelloy substrate. In their measurements, however, Braginski et al. induce currents both on the free Nb_3Ge surface and along the Hastelloy–Nb_3Ge interface and the losses include hysteretic and eddy current losses from the Hastelloy. Similar results were also obtained by Kim *et al.* [164] for Nb_3Ge deposited on copper substrates. Again, the high losses were attributed to the copper–Nb_3Ge interface, in which a diffusion layer is also present.

Thompson *et al.* [162, 163] have measured losses of Nb_3Ge deposited on the outside of Nb or Nb 1% Zr tubes containing various amounts of Nb_5Ge_3 second-phase material. They have also studied the effects of mechanical and chemical polishing. Figure 29 shows typical loss results at 50 Hz and 4.0 K for Nb_3Ge deposited at 900°C (sample V444) and 835°C (sample V469). Prior to any surface treatment, the samples deposited at 900°C and 835°C have losses at 500 rms A/cm of 50 and ~10 $\mu w/cm^2$, respectively. After mechanical polishing that is followed by a chemical etch to remove the strained region of the surface, the losses of the sample formed at 835°C (V469) were reduced to less than 1 $\mu w/cm^2$ at 500 rms A/cm. The loss behavior was also studied for a number of samples containing Nb_5Ge_3 second-phase material [162, 163]. For a range of deposi-

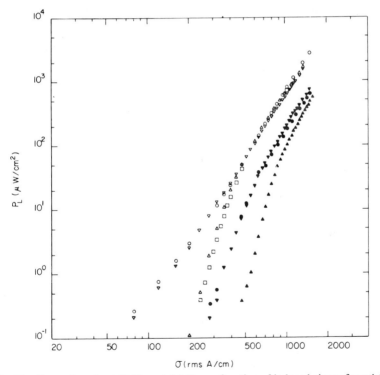

Fig. 29. Power loss P_L at 50 Hz and 4.0 K as a function of induced circumferential current density σ for two vapor-deposited Nb_3Ge samples: sample V444; $\triangledown$, as deposited; $\bigcirc$, light mechanical polish; $\triangle$, chemically etched; $\square$, annealed; sample V469; $\blacktriangledown$, as deposited; $\bullet$, mechanically polished; $\blacktriangle$, chemically etched. Losses of both samples are reduced after surface polishing. (From Thompson *et al.* [162].)

TABLE IV

Alternating-Current Losses of Nb_3Sn and Nb_3Ge above 4.2 K

Material	Description	Current density (rms A/cm)	Loss ($\mu W/cm^2$)	Temperature (K)	Ref.
Nb_3Sn	Multilayered cylinders	500	40	14	158
	made by electron	500	40	11.5	159
	beam evaporation	500	8	9.0	159
		500	10	10	160
			25	12	
		500	4.5–120	8	73
		500	20	10	65
	Commercial tape (liquid diffusion)	470	10	11	66, 151
	Bronze-processed tape	500	11	12.0	165
Nb_3Ge	Vapor-deposited	500	12.7	12	118
	cylinders	500	19	12	163

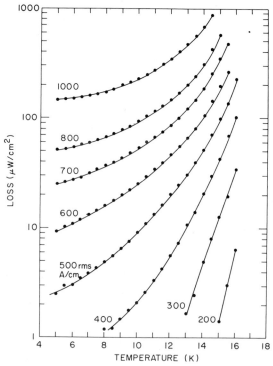

Fig. 30. Losses (60 Hz) of bronze-processed (750°C, 94 h) (Nb 1% Zr)$_3$Sn as a function of temperature. (From Bussière and Kovachev [165].)

tion temperatures between 790 and 835°C the concentration of second phase was varied between 0 and 40 wt. %. For deposition temperatures greater than 810°C, Nb_5Ge_3 precipitates were found to enhance J_c and reduce losses to a suitable level (losses of ~20 $\mu w/cm^2$ at 500 rms Å/cm up to 12 K). For deposition temperatures less than 870°C mechanisms other than second-phase precipitates appear to dominate the flux-pinning process and ac losses. Optimal values of Nb_3Ge_3 second phase seem to lie between ~5 and 10% for $T_d > 810$°C.

B.   Temperature Dependence of Losses (Nb_3Sn, Nb_3Ge)

Loss measurements above 4.2 K have been reported by a number of authors for Nb_3Sn [65, 66, 73, 151, 158–160, 165] and Nb_3Ge [118, 163]. Loss values at 500 rms A/cm for these two compounds are summarized in Table IV. The dependence of loss on temperature can generally be described by Eq. (15) as shown by Bussière *et al.* for a commercial Nb_3Sn tape [66], Howard *et al.* for multilayered Nb_3Sn [73], and Thompson *et al.* for CVD Nb_3Ge [118]. Figure 30 shows losses of bronze-processed Nb_3Sn at various current densities in the temperature range of 5–15 K. Note that losses remain below 10 $\mu w/cm^2$ up to ~10.5 K and ~20 $\mu w/cm^2$ at 12 K. Figure 31 shows results obtained for as-deposited and

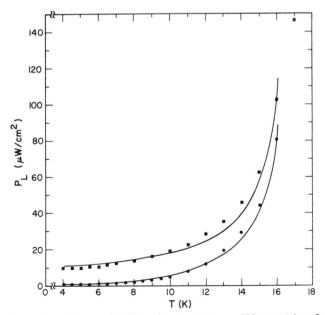

Fig. 31. Power loss P_L as a function of temperature at 500 rms A/cm for a Nb_3Ge sample (No. V469) before (■) and after (●) surface treatment ($\sigma = 500$ rms A/cm). (From Thompson *et al.* [118].)

polished and etched Nb_3Ge at 500 rms A/cm. For the polished sample losses at 12 K are ~ 13 $\mu w/cm^2$. The solid lines are computer fits using Eq. (15). The temperature dependence of J_c and $\Delta\sigma$ were determined independently.

References

1. G. Bogner, Transmission of electrical energy by superconducting cables, *in* "Superconducting Machines and Devices, Large Scale Applications" (S. Foner and B. Schwartz eds.), Nato Advanced Study Institutes Series B, Physics, Vol. 1, Chapter 7. Plenum, New York, 1974.
2. E. B. Forsyth, "Energy," Vol. 1, pp. 111–122. Pergamon, Oxford, 1976.
3. B. C. Belanger, *Adv. Cryogen. Eng.* **20,** 1 (1975).
4. N. P. Laguna, B. C. Belanger, and A. S. Clorfeine, *IEEE Trans. Magn.* **MAG-13,** 161 (1977).
5. B. C. Belanger, *Cryogenics* **15,** 88 (1975).
6. A. S. Clorfeine, B. C. Belanger, and N. P. Laguna, *IEEE Trans. Magn.* **MAG-13,** 915 (1976).
7. G. Bogner, *Cryogenics* **15,** 79 (1975).
7a. G. I. Meschanov, D. I. Belyi, P. I. Dolgosheev, I. B. Peshkov, and G. G. Svalov, *IEEE Trans. Magn.* **MAG-13,** 154 (1977).
7b. I. M. Bortnik, V. L. Karapazuk, V. V. Lavrova, S. I. Lurie, Y. V. Petrovsky, and L. M. Fisher, *IEEE Trans. Magn.* **MAG-13,** 188 (1977).
8. M. T. Taylor, *Bull. Int. Inst. Refrig. Comm.* **1,** 119 (1969).
9. E. B. Forsyth *et al., Proc. Appl. Supercond. Conf., Annapolis, Maryland* Publ. No. 72CH0682-5-TABSC, p. 202. IEEE, New York, 1972.
10. E. B. Forsyth, G. A. Mulligan, J. W. Beck, and J. A. Williams, *IEEE Trans. Power Apparatus Syst.* **PAS-94,** 161 (1975).
11. J. A. Baylis, *Philos. Trans. R. Soc. London* **A275,** 205 (1973).
12. R. W. Meyerhoff, *Adv. Cryogen. Eng.* **21,** 60 (1975).
13. E. B. Forsyth, A. J. McNerney, A. C. Muller, and S. J. Rigby, *IEEE Trans Power Apparatus Syst.* **PAS-97,** 734 (1978).
14. J. W. Dean and H. L. Laquer, *Underground Transmiss. Distribut. Conf., September 27-October 1* IEEE Conf. Record No. 76CH1119-7-PWR, p. 417 (1976).
15. E. B. Forsyth *et al., IEEE Trans. Power Apparatus Syst.* **PAS-92,** 494 (1973).
16. E. B. Forsyth, J. R. Stewart, and J. A. Williams, *Underground Transmiss. Distribut. Conf., September 27–October 1* IEEE Conf. Record No. 76CM1119-7-PWR, p. 446 (1976).
17. R. W. Meyerhoff, *IEEE Trans. Power Apparatus Syst.* **PAS-94,** 1734 (1975).
18. B. W. Roberts, *J. Phys. Chem. Ref. Data* **5,** 581–821 (1976).
19. R. G. Hampshire, *J. Phys. D* **7,** 1847 (1974).
20. M. Suenaga, C. Klamut, and J. F. Bussière, *IEEE Trans. Magn.* **MAG-13,** 436 (1977).
21. A. I. Braginski, J. R. Gavaler, G. W. Roland, M. R. Daniel, M. A. Janocko, and A. T. Santhanam, *IEEE Trans. Magn.* **MAG-13,** 300 (1977).
22. D. K. Finnemore, T. F. Stromberg, and C. A. Swenson, *Phys. Rev.* **149,** 231 (1966).
23. C. N. Carter, J. C. Male, and C. Graeme-Barber, *Cryogenics* **14,** 332 (1974).
24. C. N. Carter and J. Sutton, *Cryogenics* **15,** 599 (1975).

25. W. T. Beall, *IEEE Trans. Magn.* **MAG-11**, 381 (1975).
26. H. L. Laquer, J. W. Dean, and P. Chowdhuri, *IEEE Trans. Magn.* **MAG-13**, 182 (1977).
27. E. Bochenek, H. Franke, and R. Wimmershoff, *IEEE Trans. Magn.* **MAG-11**, 366 (1975).
28. J. A. Baylis, *Proc. Appl. Superconduct. Conf., Annapolis, Maryland* Publ. No. 72CH0682-5-TABSC, p. 182. IEEE, New York, 1972.
29. G. W. Mellors and S. Senderoff, *J. Electrochem. Soc.* **112**, 266 (1965).
30. R. W. Meyerhoff and W. T. Beall, Jr., *J. Appl. Phys.* **42**, 147 (1971).
31. G. Bogner and P. Penczynski, *Cryogenics* **16**, 355 (1976).
32. H. Heumann, *Mitt. Kabelwerke* **4**, 1 (1972).
33. J. Sutton, *Cryogenics* **15**, 541 (1971).
34. G. H. Morgan and E. B. Forsyth, *Adv. Cryogen. Eng.* **22**, 434 (1977).
34a. P. Dubois, I. Eyraud, and E. Carbonell, *Proc. Appl. Supercond. Conf., Annapolis, Maryland* Publ. No. 72CH0682-5-TABSC, p. 173. IEEE, New York, 1972.
35. C. N. Carter and J. Sutton, *Cryogenics* **15**, 599 (1975).
36. P. Penczynski, H. Hentzelt, and G. Eger, *Cryogenics* **14**, 503 (1974).
37. G. A. Morgan and J. E. Jensen, *Cryogenics* **17**, 259 (1977).
38. J. Sutton and D. A. Ward, *Cryogenics* **17**, 495 (1977).
39. M. Garber, J. Bussière, and G. Morgan, *Proc. Magn. Magn. Mater. Conf.*, (J. J. Becker and G. H. Lander, eds.), AIP Conf. Proc. **34**, p. 84 (1976).
40. J. Bussière, V. Kovachev, C. Klamut and M. Suenaga, *Adv. Cryogen. Eng.* **24**, 449 (1979).
41. R. G. Hampshire, J. Sutton, and M. T. Taylor, *Inst. Int. Froid Comm.* **1**, 251. (1969), Annexe 1969-1.
42. I. Milne and D. A. Ward, *Cryogenics* **12**, 176 (1972).
43. J. J. Hanak and R. E. Enstrom, *Proc. Int. Conf. Low Temp. Phys. 10th, Moscow* **IIB**, 10 (1967).
44. R. E. Enstrom, J. J. Hanak, J. R. Appert, and K. Strater, *J. Electrochem. Soc.* **119**, 743 (1972).
45. R. M. Scanlan, W. A. Fietz, and E. F. Koch, *J. Appl. Phys.* **46**, 2244 (1975).
46. B. J. Shaw, *Appl. Phys. Lett.* **47**, 2143 (1976).
47. M. Suenaga, T. S. Luhman, and W. B. Sampson, *J. Appl. Phys.* **45**, 4049 (1974).
48. T. Luhman and M. Suenaga, *Adv. Cryogen. Eng.* **22**, 356 (1977).
49. F. F. Dettmann and F. K. Lange, *Sov. Phys. JETP* **20**, 42 (1965).
50. S. Takacs and M. Jergel, *Czech. J. Phys.* **B23**, 636 (1973).
51. M. Jergel, S. Takacs and V. Cernusko, *Phys. Status Solidi* **33a**, 85 (1976).
52. H. F. Braun, E. N. Haeussler and E. J. Saur, *IEEE Trans. Magn.* **MAG-13**, 327 (1977).
53. P. B. Hart, C. Hill, R. Ogden and C. W. Wilkins, *J. Phys. D* **2**, 521 (1969).
54. R. E. Enstrom and J. R. Appert, *J. Appl. Phys.* **43**, 1915 (1972).
55. R. E. Enstrom and J. R. Appert, *J. Appl. Phys.* **45**, 421 (1974).
56. G. Ziegler, B. Blos, H. Diepers, and K. Wohlleben, *Z. Angew. Phys.* **31**, 184 (1971).
57. J. R. Gavaler, M. A. Janocko, A. I. Braginski, and G. W. Roland, *IEEE Trans. Magn.* **MAG-11**, 192 (1975).
58. A. I. Braginski, M. R. Daniel, and G. W. Roland, *Proc. Magn. Magn. Mater. Conf.* AIP Conf. Proc. No. 34, p. 78 (1976).
58a. A. I. Braginski, G. W. Roland, M. R. Daniel, A. T. Santhanam, and K. W. Guardipee, *J. Appl. Phys.* **49**, 736 (1976).
59. M. G. Benz, *Trans. Metall. Soc. AIME* **242**, 1069 (1968).
60. G. Pasotti, M. V. Ricci, N. Sacchetti, G. Sacerdoti, and M. Spadoni, *Proc. Int. Conf.*

Magn. Technol. 5th, Rome, Italy, (N. Sacchetti, M. Spadoni, and S. Stipcich, **MT-5,** 679 (1975).
61. J. S. Caslaw, *Cryogenics* **11,** 57 (1971).
62. M. Suenaga, O. Horigami, and T. S. Luhman, *Appl. Phys. Lett.* **25,** 624 (1974).
63. M. Suenaga, W. B. Sampson, and C. J. Klamut, *IEEE Trans. Magn.* **MAG-11,** 231 (1975).
64. R. H. Hammond, *IEEE Trans. Magn.* **MAG-11,** 201 (1975).
65. R. E. Howard *et al., IEEE Trans. Magn.* **MAG-13,** 137 (1977).
66. J. F. Bussière, M. Garber, and S. Shen, *Appl. Phys. Lett.* **25,** 756 (1974).
67. M. Garber, S. Shen, J. Bussière, and G. Morgan, *IEEE Trans. Magn.* **MAG-11,** 373 (1975).
68. R. T. Kampwirth, J. W. Hafstrom, and C. T. Wu, *IEEE Trans. Magn.* **MAG-13,** 315 (1977).
69. R. V. Carlson, R. J. Bartlett, L. R. Newkirk, and F. A. Valencia, *IEEE Trans. Magn.* **MAG-13,** 648 (1977).
70. R. J. Bartlett, H. L. Laquer, and R. D. Taylor, *IEEE Trans. Magn.* **MAG-11,** 405 (1975).
71. M. R. Daniel, A. I. Braginski, G. W. Roland, J. R. Gavaler, R. J. Bartlett, and L. R. Newkirk, *J. Appl. Phys.* **48,** 1293 (1977).
72. L. R. Testardi, *Solid State Commun.* **17,** 871 (1975).
73. R. E. Howard, C. N. King, R. H. Norton, R. B. Zubeck, T. W. Barbee, and R. H. Hammond, *Adv. Cryogen. Eng.* **22,** 332 (1977).
73a. A. L. Fetter and P. C. Hohenberg, Theory of type II superconductors, *in* "Superconductivity" (R. D. Parks, ed.), Vol. II, p. 817. Dekker, New York, 1969.
74. J. Lowell and J. B. Sousa, *Solid State Commun.* **5,** 829 (1967).
75. J. F. Bussière, *Phys. Lett.* **58A,** 343 (1976).
76. R. Hecht, *RCA Rev.* **25,** 453 (1964).
77. R. Shaw, B. Rosenblum, and F. Bridges, *IEEE Trans. Magn.* **MAG-13,** 811 (1977).
78. D. L. Decker and H. L. Laquer, *J. Appl. Phys.* **40,** 2817 (1969).
79. M. S. Lubell and R. H. Kernohan, *J. Phys. Chem. Solids* **32,** 1531 (1971).
80. D. Saint-James and P. G. de Gennes, *Phys. Lett.* **7,** 306 (1963).
81. J. G. Park, *Adv. Phys.* **18,** 103 (1969).
82. H. J. Fink and R. D. Kessinger, *Phys. Rev.* **140,** A1937 (1965).
83. H. J. Fink and C. J. Barnes, *Phys. Rev. Lett.* **15,** 792 (1965).
84. H. J. Fink, *Phys. Rev. Lett.* **14,** 309 (1965).
85. J. G. Park, *Phys. Rev. Lett.* **15,** 352 (1965).
86. L. J. Barnes and H. J. Fink, *Phys. Lett.* **20,** 583 (1966).
87. P. S. Swartz and H. R. Hart, Jr., *Phys. Rev.* **137,** A818 (1965).
88. V. L. Ginzburg, *Sov. Phys.-JETP* **34,** 78 (1958).
89. C. P. Bean and J. D. Livingston, *Phys. Rev. Lett.* **12,** 14 (1965).
90. P. G. DeGennes, *Solid State Commun.* **3,** 127 (1965).
91. Orsay group on superconductivity, "Quantum Fluids" (D. F. Brewer, ed.). North-Holland Publ., Amsterdam, 1966.
92. L. Kramer, *Z. Phys.* **259,** 333 (1973).
93. H. J. Fink and A. G. Presson, *Phys. Rev.* **182,** 498 (1969).
94. J. R. Clem, "Low Temperature Physics-LT13" (K. D. Timmerhaus, W. J. O'Sullivan, and E. F. Hammel, eds.), Vol. 3, p. 102. Plenum, New York, 1974.
95. F. F. Ternowskii and L. N. Shekhata, *Sov. Phys.-JETP* **35,** 1202 (1972).
96. H. R. Hart, Jr. and P. S. Swartz, *Phys. Rev.* **156,** 403 (1967).
97. D. D. Morrison and R. M. Rose, *Phys. Rev. Lett.* **25,** 356 (1970).

98. O. Daldini, P. Martinoli, J. L. Olsen, and G. Berner, *Phys. Rev. Lett.* **32**, 218 (1974).
99. S. H. Autler, *J. Low-Temp. Phys.* **9**, 241 (1972).
100. J. Pearl, *Appl. Phys. Lett.* **5**, 65 (1964).
101. J. Pearl, *J. Appl. Phys.* **37**, 4139 (1966).
102. W. C. H. Joiner and G. E. Kuhl, *Phys. Rev.* **163**, 362 (1967).
103. W. Thys, A. Bouwen, E. Deprez, Y. Bruynseraede, and L. Van Gerven,*J. Low Temp. Phys.* **6**, 257 (1972).
104. A. S. Joseph and W. J. Tomasch, *Phys. Rev. Lett.* **12**, 219 (1964).
105. R. W. DeBlois and W. DeSorbo, *Phys. Rev. Lett.* **12**, 499 (1964).
106. J. C. Renard and Y. A. Rocher, *Phys. Lett.* **24A**, 509 (1967).
107. B. Hillenbrand, H. Martens, H. Pfister, K. Schnitzke, and Y. Uzel, *IEEE Trans. Magn.* **MAG-13**, 491 (1977).
108. D. F. Moore and M. R. Beasley, *Appl. Phys. Lett.* **30**, 494 (1977).
109. H. A. Ullmaier and W. F. Gauster, *J. Appl. Phys.* **37**, 4519 (1966).
110. H. A. Ullmaier, *Phys. Status Solidi* **17**, 631 (1966).
111. A. Das Gupta and E. J. Kramer, *Philos. Mag.* **26**, 779 (1972).
112. J. Lowell, *J. Phys. C* **2**, 372 (1969).
113. P. S. Swartz and H. R. Hart, Jr., *Phys. Rev.* **156**, 412 (1967).
114. E. J. Kramer and A. DasGupta, *Philos. Mag.* **26**, 769 (1972).
115. P. H. Melville and A. DasGupta, *Philos. Mag.* **28**, 275 (1974).
116. J. F. Bussière and V. Kovachev, *J. Appl. Phys.* **49**, 2526 (1978).
117. J. F. Bussière, M. Garber and M. Suenaga, *IEEE Trans. Magn.* **MAG-11**, 324 (1975).
118. J. D. Thompson, M. P. Maley and L. R. Newkirk, *IEEE Trans. Magn.* **MAG-13**, 429 (1977).
119. H. London, *Phys. Lett.* **6**, 162 (1963).
120. C. P. Bean, *Rev. Mod. Phys.* **36**, 31 (1964).
121. A. Campbell and J. Evetts, *Adv. Phys.* **21**, 199 (1972).
122. P. H. Melville, *Adv. Phys.* **21**, 647 (1972).
123. I. M. Green and P. Hlawiczka, *Proc. IEEE* **114**, 1329 (1967).
124. F. Irie and K. Yamafuji, *Phys. Lett.* **24A**, 30 (1967).
125. G. Fournet and A. Mailfert, *J. Phys. Radium* **31**, 357 (1970).
126. S. T. Sekula and J. H. Barrett, *Appl. Phys. Lett.* **17**, 204 (1970).
127. W. I. Dunn and P. Hlawiczka, *J. Phys. D* **1**, 1469 (1968).
128. M. J. Chant, M. R. Halse and H. O. Lorch, *Proc. IEEE* **117**, 1441 (1970).
129. R. M. Easson and P. Hlawiczka, *J. Phys. D* **1**, 1477 (1968).
130. S. L. Wipf, *Proc. Summer Study on Supercond. Devices and Accel.* Brookhaven National Laboratory Rep. No. 50155 (C-55), 511 (1968).
131. T. A. Buchhold, *Cryogenics* **3**, 141 (1963).
132. P. H. Melville, *J. Phys. C* **4**, 2833 (1971).
133. W. DeSorbo, *Phys. Rev.* **132**, 107 (1963).
134. D. A. Ward and R. J. Bray, *Cryogenics* **15**, 21 (1975).
135. D. A. Ward, W. S. Kyte, and B. J. Maddock, Central Electricity Generating Board Rep. No. RD/L/N128/74 (August 1974).
136. D. A. Ward, B. J. Maddock, and W. S. Kyte, Central Electricity Generating Board Rep. RD/L/N114/75 (October 1975).
137. T. A. Buchhold and P. J. Molenda, *Cryogenics* **2**, 344 (1962).
138. Y. A. Rocher and J. Septfonds, *Cryogenics* **7**, 96 (1967).
139. R. Grigsby and R. J. Slaughter, *J. Phys. D* **3**, 898 (1970).
140. P. R. Brankin and R. G. Rhodes,*Proc. Int. Cryogen. Eng. Conf., 4th, Eindhoven, May* p. 136. IPC Science and Technology Press, Guilford, England, 1972.

141. D. R. Salmon and J. A. Caterall, *J. Phys. D* **6**, 211 (1973).
142. F. A. Al-Huseini, *Proc. Int. Cryogen. Eng. Conf., 6th, Grenoble* (K. Mendelssohn, ed.), p. 438. IPC Science and Technology Press, Guilford, England, 1976.
143. G. Bogner, P. Penczynski, and F. Schmidt, *IEEE Trans. Magn.* **MAG-13**, 400 (1977).
144. J. C. Male, *Cryogenics* **10**, 381 (1970).
145. P. R. Brankin and R. G. Rhodes, *Proc. Int. Conf. Low Temp. Phys. 12th, Kyoto, September* (E. Kanda, ed.), Vol. LT-12, p. 389. Academic Press, New York, 1970.
146. E. C. Rogers, B. E. Redmonds, and G. D. Chan, *Cryogenics* **9**, 431 (1969).
147. R. Grigsby and E. C. Rogers, *Cryogenics* **13**, 100 (1973).
148. J. A. Baylis, K. G. Lewis, J. C. Male, and J. A. Noe, *Cryogenics* **14**, 553 (1974).
149. P. H. Brisbin, W. D. Markiewicz, R. E. Wilcox, and C. H. Rosner, *IEEE Trans. Magn.* **MAG-13**, 421 (1977).
150. E. Adam, P. Beischer, W. Marancik, and M. Young, *IEEE Trans. Magn.* **MAG-13**, 425 (1977).
151. J. F. Bussière, *IEEE Trans. Magn.* **MAG-13**, 131 (1977).
152. M. Suenaga, J. F. Bussière, and M. Garber, *Adv. Cryogen. Eng.* **22**, 326 (1977).
153. C. Old and I. Macphail, *J. Mater. Sci.* **4**, 202 (1969).
154. J. F. Bussière and M. Suenaga, *J. Appl. Phys.* **47**, 707 (1976).
155. M. Suenaga and M. Garber, *Science* **184**, 952 (1974).
156. T. S. Luhman and M. Suenaga, *Appl. Phys. Lett.* **29**, 61 (1976).
157. T. S. Luhman and M. Suenaga, *IEEE Trans. Magn.* **MAG-13**, 800 (1977).
158. C. H. Meyer, Jr., D. P. Snowden, and S. A. Sterling, *Rev. Sci. Instrum.* **42**, 1584 (1971).
159. D. P. Snowden, C. H. Meyer, Jr., and S. A. Sterling, *J. Appl. Phys.* **45**, 2693 (1974).
160. R. E. Schwall, R. E. Howard, and R. B. Zubeck, *IEEE Trans. Magn.* **MAG-11**, 397 (1975).
161. R. B. Zubeck, *et al., Thin Solid Films* **40**, 1 (1977).
162. J. D. Thompson, M. P. Maley, L. R. Newkirk, and F. A. Valencia, *Phys. Lett.* **57A**, 351 (1976).
163. J. D. Thompson, M. P. Maley, and L. R. Newkirk, *Appl. Phys. Lett.* **30**, 190 (1977).
164. K. S. Kim, Y. B. Kim, J. W. Savage, and L. R. Newkirk, *IEEE Trans. Magn.* **MAG-13**, 433 (1977).
165. J. Bussière and V. T. Kovachev, *J. Appl. Phys.* (in press).
166. Graeme-Barber *et al., Cryogenics* **12**, 317 (1972).

Metallurgy of Niobium Surfaces

M. STRONGIN, C. VARMAZIS,† and A. JOSHI‡

Physics Department
Brookhaven National Laboratory
Upton, New York

I. Introduction 327
II. Impurities at Nb Surfaces 330
   A. High-Vacuum Results 330
   B. Exposed and Anodized Surfaces 336
III. Superconducting Properties 339
   A. Sample Preparation for Superconductivity Measurements 339
   B. The Penetration Depth and the Surface Critical Field 340
   C. rf Superconductivity and Magnetic Field Breakdown 345
   References 346

I. Introduction

This chapter is a review of how the properties of Nb surfaces affect certain fundamental superconducting parameters, including the superconducting penetration depth. The importance of this problem is highlighted by the necessity of measuring fundamental properties of superconductors that are surface sensitive. Applications to rf superconducting cavities are stressed. Since the emphasis of the chapter is on the fundamental properties of superconductors, it is well to introduce this subject by discussing some of the problems encountered when superconductors are used for rf devices. The attempt to understand these problems led to much of the work discussed here.

In many ways the development of superconducting devices for rf purposes has paralleled the development of superconductivity for high-field

† Present address: Department of Physics University of Crete, Iraklion, Crete, Greece.
‡ Present address: Physical Electronics Industries Inc., Edina, Minnesota.

magnets. Just as it was recognized that zero-resistance materials could be used for magnets, it was also noted that the very small surface resistance of superconductors could be used for low-loss (high-Q) rf devices. In the case of magnet development, the materials with the highest critical temperatures and the highest critical fields, such as Nb_3Sn, are still not routinely used in the construction of large devices, and in an analogous way neither has the promise of superconductors for rf devices been entirely realized. Although superconducting accelerators, which use Nb rf cavities, are in operation, the highest rf field levels that should be achievable have not been attained. Some of the reasons for this are the subject of this chapter.

A further analogy between high-field magnets and rf devices can be found in the question of stability. When the first large magnets were built for bubble chambers, the necessary stability against flux jumping was obtained by putting the superconducting filaments in a high-conductivity copper matrix. Hence, if there was flux jumping or some kind of instability in the high-normal-state-resistivity superconductor, the current could be instantaneously shunted into the copper matrix. Furthermore, the matrix could also provide high thermal conductivity to rapidly reduce the temperature in the normal region of the superconductor. In a similar way it appears that at high-rf field levels a normal region that forms can propagate and drive the sample normal. Because the rf field only penetrates a penetration depth λ into the sample, the surface region is crucial in this breakdown process. For a good Nb sample λ can be of the order of 350 Å and hence the importance of the surface of the sample is evident. It is also clear that if a normal region appears, both high electrical conductivity and high thermal conductivity in the surface region will be beneficial for restricting the growth of the normal region and keeping the temperature rise in the region to a minimum.

It was recognized many years ago [1] that niobium would be a superior material for rf devices. This was due to several reasons. First, it had the highest T_c of any easily obtainable pure metal. Second, H_{c1} of Nb was higher than H_c of Pb and thus it appeared reasonable that higher peak rf fields could be achieved. At present, the superheating field provides a more reasonable estimate of the ultimate breakdown field, and this will be mentioned later. Finally, Nb is a better structural material than Pb and construction and machining of devices would be easier.

The initial results [1] using niobium that had been treated by high-temperature annealing in an ultrahigh vacuum were extremely promising. However, it was ultimately found that although low rf losses can be achieved, the breakdown field was still much lower than theoretically possible, and this has limited and delayed the use of niobium in proposed devices.

To provide a brief introduction to the main emphasis of the chapter, the following discussion seeks to provide some sense of what is occurring when the rf field impinges on the sample surface, and why this case is somewhat more complex than the low-frequency case. If one takes a two-fluid point of view, it is clear that in the dc case the superfluid essentially short-circuits the normal fluid and zero resistance is achieved. In the rf case the superelectrons cannot completely shunt the normal electrons because of phase differences, i.e., the superelectrons appear inductive, and a simple analysis on the basis of the London equations indicates the presence of ohmic losses. These losses depend on the amount of normal fluid; according to the London theory (see the introductory chapter by Dew-Hughes, this volume) the normal fluid density decreases at t^4 ($t = T/T_i$). Hence it can be seen that high T_c and low temperatures are necessary to reduce ohmic losses.

It is now evident that the details of surface preparation must be crucial. For example, if the surface region is high in impurities, then T_c might be lower and thereby give higher losses. Also, regions with high oxygen concentration might be regions where breakdown occurs. In any case it should be noted that the region at the surface is the key to understanding rf superconductivity. As a means of understanding this region, measurements of oxygen segregation at surfaces in addition to measurements of the superconducting penetration depth in small samples will be discussed.

The penetration depth can be measured by placing the sample in a coil and then measuring the change in inductance as the field penetration into the sample changes with temperatures. In measurements made at Brookhaven National Laboratory [2], the inductance was part of the resonant circuit of a sensitive tunnel-diode oscillator and changes in penetration were detected by the changes in the resonant frequency of the oscillator. With this technique changes of a few angstroms could be detected. Initial investigations of Nb surfaces showed values of λ that were much larger than expected, given the high quality of the bulk sample, which could be determined from the residual resistance ratio. This immediately implied that the surface was different from the bulk. For example, a shorter surface mean free path l might be expected to change λ through the formula

$$\lambda = \lambda_0 \sqrt{\xi_0/l}$$

where ξ_0 is the coherence length. One problem is whether l is actually changed over a region as large as 500–1000 Å due to impurities or whether something else is happening. The discussion to follow gives some answers to this question, although a complete picture of what happens at the actual Nb surface has not been obtained. The answer to this kind of question is also related to other surface-sensitive measurements, such as tunneling [3] and low-frequency ac losses [4].

One emphasis of this discussion is on how the special properties of even the best surfaces to date affect some values of the fundamental properties of superconductors. A description of the segregation of oxygen and other impurities to the niobium surface is given first. This is followed by a discussion of how changes in the surface regime affect measurements of some basic properties of superconductors, such as the penetration depth and the surface critical field. Of continuing importance are the changes in the properties of the most carefully prepared surfaces. Finally, a brief discussion of rf cavity problems and how nonideal surfaces affect these devices is given. Some recent promising work on using Nb_3Sn surfaces for rf cavities is also mentioned.

II. Impurities at Nb Surfaces

A. High-Vacuum Results

In recent years the availability of techniques such as Auger spectroscopy, secondary ion spectroscopy, and low-energy electron diffraction (LEED) have led to the possibility of convenient methods for the study of surface impurities. One of the simplest experiments is to study the properties of a clean surface of Nb, and to determine whether an ideal surface is possible due to the high reactivity to oxygen. It has been recognized for some time by investigators trying to obtain clean Nb for rf cavities [1] that high-temperature degassing is a crucial step in the achievement of high-Q rf devices.

When Nb is heated in ultrahigh vacuum, oxygen leaves the Nb as NbO_2 in the relatively low-temperature regime and as NbO at higher temperatures. Figure 1 shows how the relative proportions of NbO_2, NbO, and Nb change as the temperature is raised from 1500 to near 1800°C. As the temperature is raised, the vapor pressures of Nb and the oxides are raised; at the same time the relative amount of oxygen on the surface is reduced, as is shown by the oxygen Auger traces. Note that at 2×10^{-6} Torr, which was the ambient pressure for the data shown in Fig. 1, there is a steady state bulk concentration that is calculated to be about 2.3% at 1500°C, and a corresponding steady state amount of oxygen on the surface of about 55%. Both amounts decrease as the temperature is raised or the pressure is lowered. At temperatures near to ~1700°C some 6.5% of the surface is covered with oxygen and the bulk concentration is about 0.26%. The decreasing amounts of oxygen on the surface as T is increased are shown by the oxygen Auger traces shown in Fig. 2.

The various parts of Fig. 1 give the fluxes of Nb, NbO, and NbO_2 as

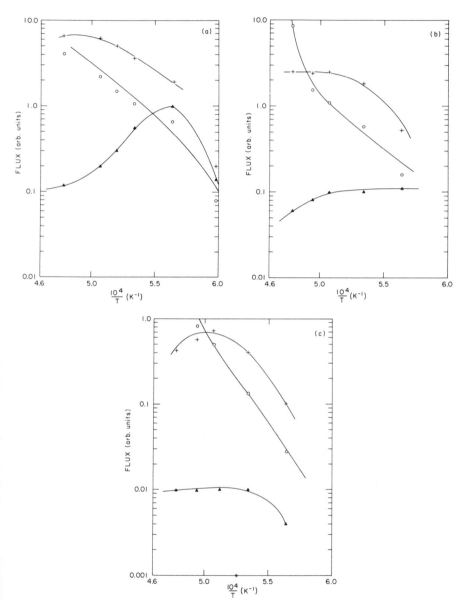

Fig. 1. Flux of Nb, NbO, and NbO$_2$ as a function of temperature for three different pressures (a) 2×10^{-6} Torr, (b) 8×10^{-7} Torr, (c) 2×10^{-7} Torr; ▲, NbO$_2$; +, NbO; O, Nb. At high pressure the flux of NbO$_2$ is dominant near 1400°C, whereas at lower pressure the NbO flux dominates. At high temperatures NbO and Nb dominate at all pressures.

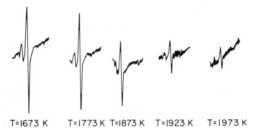

Fig. 2.   Oxygen Auger line for different temperatures, showing the decreased amount of oxygen on the surface as the temperature is raised. We estimate that at about 1773 K the surface is about half covered with oxygen.

functions of temperature for different ambient pressures. The absolute flux has not been calibrated and the figure is meant to indicate relative fluxes. Even though the oxide vapor pressure is increasing, the flux is expected either to fall off or to increase less rapidly, since the amount on the surface is rapidly decreasing as is the bulk concentration. This is shown clearly in Fig. 1. Also shown is that at the highest pressures and lowest temperatures shown here, NbO_2 removal becomes important. The total flux of oxides coming off the surface actually determines the rate of oxygen removal since little comes off as free oxygen. The diffusion times of oxygen in Nb are fairly rapid compared to the time needed to reduce the bulk concentration of oxygen by a factor of 10 due to the flux of oxides from the surface, i.e., the rate-limiting step is the evaporation of the oxides rather than diffusion for samples on the order of ~0.1 cm thick. An initial summary of some of the problems to be discussed in the following sections is given by Strongin et al. [5]. Rovner et al. [6] have given the NbO flux as $Z = K_3 C$ where C is the bulk concentration, Z is the flux of atoms/cm² sec, and $\log K_3 = 10.98 - 3.5 \times 10^4/T$. For NbO_2 removal they give $Z = K_4 C^2$ where $\log K_4 = 9.64 - 3.6 \times 10^4/T$.

For small C and high temperature, K_3 is the rate-limiting constant. The rate equation for oxygen removal is given as

$$dC/dt = -K_3 C/l$$

where $2l$ is the sample thickness. Then

$$t = l \ln(C_0/C)/K_3.$$

Table I gives some characteristic times $t(K_3)$ for $l = 0.1$ cm and $C_0/C = 10$ that give some indication of the difficulty of oxygen removal from Nb. Some values of the diffusion constant for oxygen in Nb and the time to reduce the oxygen concentration at the center of a sample 0.2 cm thick to 10% of its original value are also given. The steady state amount

TABLE I

Sᴏᴍᴇ Cʜᴀʀᴀᴄᴛᴇʀɪsᴛɪᴄ Tɪᴍᴇs ꜰᴏʀ Oxʏɢᴇɴ Rᴇᴍᴏᴠᴀʟ

T (°C)	$t(K_3)^a$	D (cm²/sec)	$t = l^2/D$ (sec)^b
1800	1.9×10^5	3.2×10^{-5}	313
1900	3.02×10^4	4.34×10^{-5}	230
2000	6.1×10^3	5.7×10^{-5}	175

^{a} $t(K_3)$ is the time to attain $C_0/C = 10$ in the case where diffusion is fast and K_3, the rate constant for NbO removal, is the limiting factor.

^{b} $t = l^2/D$ is the condition for $C \approx 0.11C_0$ for a slab of thickness $2l$. Hence, diffusion is relatively fast compared to NbO removal. Calculations are for $2l = 0.2$ cm.

of oxygen in the bulk for a given temperature can be calculated from log $C_0 = \log P_{02} - 3.35 + 16,700/T$ [7] and some characteristic values are given in Table I. For a given amount of oxygen, after degassing at high temperatures, a crucial question concerns the amount that remains on the surface and that which comes to the surface during cooling. This problem can be studied by conventional techniques and by Auger spectroscopy [8–10]. Such measurements indicate that during cooling oxygen migrates from the bulk to the surface region due to the high heat of formation of the surface oxide. Figure 3 shows the fraction of occupied surface sites as the temperature is changed for different bulk oxygen concentrations. The sample was kept in an ultrahigh vacuum environment during cooling from elevated temperatures so that the oxygen, measured by the height of the oxygen Auger line, came to the surface from the bulk. The data can be described by the equation

$$\theta_2/(1 - \theta_2) = K_2 C$$

where C is the solute concentration in atomic percent, θ_2 is the fraction of occupied sites, and $K_2 = K_0 \exp(\Delta H/RT)$. K_0 is about 2×10^{-4} at. $\%^{-1}$ and ΔH is about 20–30 kcal/mole [11] and changes with concentration. Hence it can be seen that even when the bulk concentration is only about 0.1%, upon cooling from 1800 to 800°C the surface concentration can change from 2 to 90 at. %. It is clear that this poses a problem when trying to prepare samples for surface-sensitive measurements.

Besides the problem of oxygen occurring at the Nb surface, carbon also plays a role in the quality of Nb surfaces. While the removal of oxygen, nitrogen, and hydrogen can be accomplished by high-temperature degassing in ultrahigh vacuum, carbon must be reacted with oxygen before it can be removed. Both carbon and oxygen restrict grain growth in high-purity niobium, and to get large crystallites in niobium containing

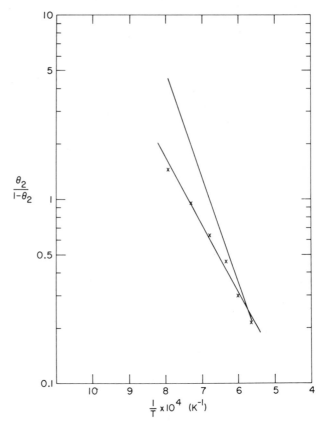

Fig. 3. $\theta_2/(1 - \theta_2)$ versus $1/T$ for oxygen on niobium surfaces, where θ_2 is the fraction of filled sites, and $\theta_2/(1 - \theta_2) = K_2 C$, where $K_2 = K_0 \exp(\Delta H/RT)$ and $K_0 \approx 3 \times 10^{-4}$ at. $\%^{-1}$ and $\Delta H = 27,000$ cal/mol. The solid line is computed from the given values of K_0 and ΔH, which are taken from Pasternak and Evans [8]. Bulk oxygen concentration was 0.37 at. %.

carbon, oxygen must be added, so that carbon is removed by CO evaporation. Excess oxygen can then be removed by heating at high temperatures in ultrahigh vacuum. Carbon also segregates to the surface, as does oxygen. In Fig. 4, θ_2 is plotted as a function of temperature for an alloy of 0.5 at. % C in Nb. As in the case of oxygen, the ratio of the surface concentration to the bulk concentration is on the order of 10^4. In this case $K_0 \approx 10^{-3}$ at. % and $\Delta H \approx 15$ kcal/g atom. These values are close to those given earlier for oxygen in niobium. It is conjectured that the segregation of carbon to grain boundaries is a phenomenon similar to surface segregation. This would explain why the removal of carbon is necessary for grain growth. In these experiments it was found that a resistivity ratio greater

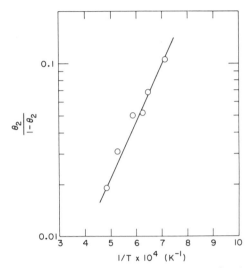

Fig. 4. $\theta_2(1 - \theta_2)$ versus $1/T$ for 0.5 at. % C in Nb. θ_2 is the fraction of filled sites; the partition coefficient K_2 for this sample is given by $K_2 = K_0 \exp(\Delta H/RT)$ where $K_0 \approx 10^{-3}$ at. $\%^{-1}$ and $\Delta H \approx 15$ K cal/g atom.

than 40 could not be obtained before carbon was removed. When carbon was removed by treatment with oxygen, as mentioned previously the resistance ratio increased to 1000 along with an increase in the grain size.

Besides the quantity of oxygen occurring on the surface, two other questions arise. One involves the crystal structure of the oxide, and the other involves the depth of the oxygen-rich region. The question of structure has been investigated in some detail and it appears that after high-temperature heating the oxygen that segregates to the surface is in the form of NbO [8]. For greater amounts of oxygen and lower temperatures, NbO_2 becomes important [6]. Bulk NbO itself is a superconducting semimetal with a T_c near 1 K [12], so its presence on the surface can be expected to affect the properties of superconductors, assuming that the NbO structure seen in this surface layer is equivalent to bulk NbO.

The extent of segregation has been measured by monitoring the oxygen Auger line during sputtering of the surface atoms [9, 10]. Figure 5 shows the results of this work. It is estimated that the region over which the oxygen goes from its surface concentration back to the bulk concentration is of the order of 10 Å. It should be emphasized that these samples were heated in different ambient pressures to get different steady state amounts of bulk oxygen. They were then cooled in ultrahigh vacuum and the oxygen segregation was measured in the high-vacuum environment. The steady state amount can be checked by a comparison with the increase in

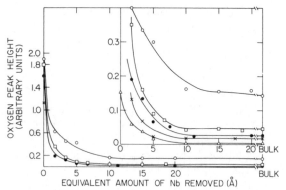

Fig. 5. Oxygen peak height as a function of the thickness of material removed from the surface by ion sputtering: O,

resistivity, and these values are consistent with the values in Table II. Before leaving this question of impurities on the surface of clean Nb, it is briefly mentioned that segregation also occurs in Nb–Zr alloys [13]. In Fig. 6 it can be seen that when a 1% Nb–Zr alloy is heated above about 800°C, Zr begins to segregate to the surface. About 30% Zr is estimated to appear eventually on the surface. The sputtering profile shows a falloff region of about 5 Å for the Zr.

B. Exposed and Anodized Surfaces

In tunneling and penetration depth measurements on clean Nb, surface layers of the order of 30 Å deep have been estimated. In contrast to the case previously discussed, where segregation occurs from the bulk to the surface in an ultrahigh-vacuum environment, it appears that when Nb is removed from the vacuum system, or when it is deliberately oxidized, a

TABLE II

CONCENTRATION OF OXYGEN IN Nb

T (°C)	log C		
	P 10^{-6} Torr	P 10^{-8} Torr	P 10^{-9} Torr
1600	−0.43	−2.43	−3.43
1700	−0.89	−2.89	−3.89
1800	−1.29	−2.29	−4.29
1900	−1.66	−3.66	−4.66
2000	−2	−4	−5

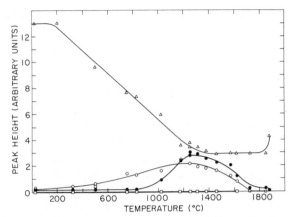

Fig. 6. Peak heights of Nb ($\triangle$), Zr ($\bullet$), O ($\bigcirc$), and N ($\square$) as a function of temperature while the sample is held at the specific temperature indicated on the abscissa.

heavily doped oxide or oxygen-rich region with a broader distribution of oxygen into the bulk than that shown in Fig. 5 is formed. From inductive measurements it is possible to see this region go normal with a T_c near 6 or 7 K, and this is discussed later. This region can also influence λ very close to the transition temperature, and this will be discussed below [2]. The situation for niobium surfaces that are actually used for rf cavities appears to be much worse than the foregoing and there can be layers of successively lower oxides and inclusions, as discussed by Schwartz and Halbritter [14]. In the work of Lindau and Spicer [15] NbO appears as the first stage. At higher amounts of oxygen the layer transforms into a mixture of NbO and NbO_2 and finally it is covered by Nb_2O_5. In some sense this follows the stability of the oxides of Nb at different temperatures of oxidation. It is also expected that on the surfaces of samples that are not single crystals, phenomena similar to segregation and oxidation on surfaces will occur and hence the oxygen distribution can be expected to be nonuniform. These regions may well be regions where rf breakdown is initiated. This problem of oxygen and other impurities segregating to grain boundaries can be studied with modern Auger apparatus that has spatial resolution of less than 1 μm. This does not yet appear to have been done on Nb.

Anodized niobium has been shown to be a suitable surface for rf superconductivity and this has led to studies of the oxygen distribution at the interface between Nb_2O_5 and Nb. Initial Auger studies at Brookhaven National laboratory showed that this interface region extended from about 50 to 75 Å; recent studies by Hahn and Halama [17] show a somewhat larger region. This is in contrast to the case of segregated oxygen on the surface, where the interface region is smaller. Studies by Gray [18] using ion scattering spectroscopy (ISS) showed an interface region that was similar qualitatively to that of the work of Lindau and Spicer [15], i.e., Nb_2O_5 forms the surface, followed by NO_2, and finally NbO. Lindau and Spicer describe the NbO layer as thin; this is interpreted by Gray to mean no more than a few atomic layers in thickness, whereas he believes the NbO layer in his work to be ~ 80 Å thick. Recent work by Hahn and Halama [17] did not show the separate interface as seen by Gray [18], and there is some possibility that their anodization process is somewhat different. This is shown in Fig. 7.

That nonideal regions and surface oxides including NbO [17] would have a detrimental effect on the properties of rf devices was suggested some time ago [19]. However, it is only in the past few years that the complexity of the surfaces used for rf cavities has become appreciated [5]. It seems likely that oxygen-rich regions or other surface imperfections serve as nucleation sites for the normal state, and because of the low thermal conductivity in such regions, due to the presence of impurities, a "runaway" condition is promoted [5]. It is also mentioned that thermal faceting has been observed on Nb surfaces heated in poor vacuum [20]. It appears that oxygen hitting the surface forms oxides preferentially along certain crystallographic directions and the resultant high-oxide vapor pressure leads to etching of the surface along these directions. Figure 8 shows an example of this effect.

In their recent work Schwarz and Halbritter [14] suggest that some of the problems with impure surfaces can be caused by a layer hundreds of angstroms thick that is composed of oxygen domains. This kind of model can certainly explain various nonideal results and this will be discussed briefly in the section on superconducting properties. Their samples,

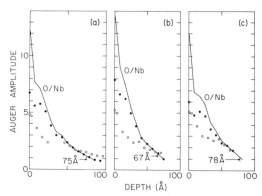

Fig. 7. Composition depth profile from ultrahigh vacuum furnace degassed Nb: ●, O; ○, C. (We thank Dr. H. Hahn for this figure.)

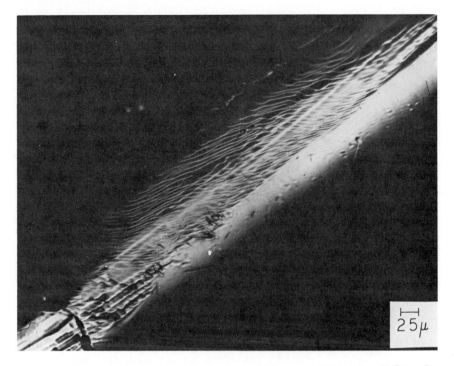

Fig. 8. Faceted Nb surface. The sample was heated to over 1900°C at 2×10^{-6} Torr for about 5 h.

which were larger, were treated for shorter times than those used for the measurements to be discussed, and this presumably leads to a nonideal surface layer with large deviations from the BCS theory. They discuss some of the unsolved questions, such as the possible mechanism to account for oxygen pentration at room temperature, which can occur up to depths of 200 Å and in concentrations of 2–5 at. %, whereas simple oxygen diffusion takes place at less than 1 Å/day. Furthermore they conjecture that stress fields can also affect the growth of oxygen-rich regions.

III. Superconducting Properties

A. *Sample Preparation for Superconductivity Measurements*

Some discussion of how clean Nb surfaces are generally obtained for laboratory measurements is worthwhile. Samples of MRC Marz-grade Nb wire, about 0.060 in. in diameter, were electropolished with a mixture of

HF and H_2SO_4 at 0°C. The samples were then heated to 1800°C in 10^{-7} Torr of oxygen for 5–10 h to remove carbon and to promote grain growth. This was followed by a high-vacuum treatment in which the samples are heated for 24 h at 10^{-8}–10^{-9} Torr. As a result of this treatment, the residual bulk ratio increased from 40 to 1000. Electron and optical microscopy indicated a relatively smooth surface with a few large grains in the region enclosed by the sample coil. The samples were removed from the vacuum and placed in the cryogenic apparatus within a few hours. The rate-limiting process in cleaning the niobium was the removal of Nb oxides, as discussed previously. The diffusion times calculated for adequate NbO removal, shown in Table I, mandate heating times on the order of a day at 1800°C for a sample with a diameter of about 0.061 in.

It should be mentioned that Hopkins and Finnemore [21] have used elaborate outgassing procedures in which a sample is heated at 2000°C for 24 h, rapidly cooled, and then sealed in a glass capillary.

B.   The Penetration Depth and the Surface Critical field

The effect of a nonideal surface on the properties of superconductors can be described in a phenomenological way by generalizing the usual Ginzberg–Landau boundary condition for an ideal surface, i.e., that $d\psi/dx = 0$, to a more general result where $1/b = (1/\psi)\, d\psi/dx$. It will be seen that the ratio of the coherence length to the extrapolation length, $\xi(0)/b$, parametrizes the surface. For an ideal surface $b \to \infty$, and $\xi(0)/b \to 0$. For a fairly good surface [2, 21] $\xi(0)/b \approx 0.02$, as will be seen from the discussion to follow. For extremely dirty surfaces many other effects are possible, such as increased surface area and/or segregation of impurities in a large surface region, which can lead to the impurity effects mentioned previously. This section is concerned with rather delicate changes in the properties due to a layer on the surface of a lower-T_c material that is about 30 Å thick. This nonideal layer can be seen directly in tunneling measurements [3] and inductive measurements [2, 14], and it is undoubtedly responsible for problems with measuring fundamental quantities on Nb surfaces, such as the temperature dependence of the penetration depth and the surface critical field H_{c3} [21].

The basic effect of this layer, at least on clean Nb, is to lower the Ginzburg–Landau (GL) order parameter in the surface region. Since λ goes inversely as the number of superfluid electrons ψ^2, it is easy to see physically that λ must increase as ψ^2 decreases at the surface. The GL equations can be solved in this case with the boundary conditions given above [2, 22], and the increased λ can be represented as λ_{eff}. In the limit as $\xi/b \to \infty$ deGennes and Matricon [22] have obtained the results that as $\kappa \to 0$.

$$\lambda_{\text{eff}}/\lambda = 1.76\kappa^{-1/2}$$

when $\kappa = 1/\sqrt{2}$, $\lambda_{\text{eff}}/\lambda \approx 2.3$, and when $\kappa \gg 1$, $\lambda_{\text{eff}}/\lambda = 1$. Table III gives the variation of $\lambda_{\text{eff}}/\lambda$ with $\xi(T)/b$ for two κ values. For further details of the solutions see Varmazis and Strongin [2].

Figure 9 shows the frequency of the sample inductor, which is proportional to λ, plotted against Z_{BCS}. Near T_c, $\lambda_{\text{BCS}}(t) = \lambda_L(0)Z_{\text{BCS}}$ and $\lambda \to \infty$ as $T \to T_c$. Note that Z_{BCS} increases very rapidly near T_c, being $= \sqrt{2}(1 + t^4)^{-1/2}$. The crucial point is that λ versus Z_{BCS} should be linear if

TABLE III

VARIATION OF $\lambda_{\text{eff}}/\lambda$ WITH $\xi(T)/b$ FOR TWO κ VALUES

$\kappa = 0.4$		$\kappa = 0.8$	
$\zeta(T)/b$	$\lambda_{\text{eff}}/\lambda$	$\xi(T)/b$	$\lambda_{\text{eff}}/\lambda$
0.01	1.006	0.01	1.005
0.02	1.011	0.02	1.009
0.03	1.017	0.03	1.014
0.04	1.022	0.04	1.018
0.05	1.028	0.05	1.023
0.06	1.033	0.06	1.027
0.07	1.039	0.07	1.032
0.08	1.045	0.08	1.036
0.09	1.050	0.09	1.041
0.1	1.056	0.1	1.045
0.12	1.067	0.12	1.054
0.16	1.090	0.16	1.072
0.2	1.113	0.2	1.090
0.25	1.142	0.25	1.113
0.3	1.171	0.3	1.135
0.4	1.229	0.4	1.179
0.6	1.344	0.6	1.264
0.8	1.454	0.8	1.342
1	1.559	1	1.413
2.01	1.966	2.01	1.667
3.01	2.21	3.01	1.806
4.01	2.362	4.01	1.880
5.01	2.464	5.01	1.943
6.01	2.536	6.01	1.981
7.01	2.59	7.01	2.009
8.01	2.631	8.01	2.03
9.01	2.664	9.01	2.046
9.96	2.689	9.96	2.059

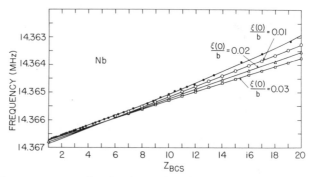

Fig. 9. Frequency versus Z_{BCS} for clean NB with residual resistance ratio greater than 1000. Curvature is removed for $\xi(0)/b \approx 0.02$. $\Delta\lambda$ is proportional to the change in frequency.

the sample is ideal and the BCS theory applies. A look at the data for large Z, where $T/T_c \approx 0.99$ or greater, starts showing curvature. This curvature is interpreted as arising from deviations from the BCS theory due to the nonideal surface, which can be represented by the ratio λ_{eff}/λ. This figure shows how the curvature is removed close to T_c for $\xi(0)/b \approx 0.02$. It should be noted that when $Z_{BCS} = 15.8$, $T/T_c = 0.998$, which for the case of Nb is only ~ 0.02 K below T_c (9.2 K). Physically it can be seen that very close to T_c, where superconductivity is weak, any nonideal effects become important. There is some evidence that the nonideal surface can also serve as sites for the nucleation of the normal–superconducting phase transition [23].

The nonideal surface also affects measurements of the nucleation field for surface superconductivity H_{c3}, and this problem has been discussed by Hopkins and Finnemore [21] and others. Figure 10 shows a plot of the ratio of H_{c3}/H_{c2}. One finds that as T_c is approached the ratio of $H_{c3}/H_{c2} \to 1$, which can be understood on the basis of a nonideal surface and corrections due to $\xi(0)/b$ if it is assumed that about 30 Å of a lower-T_c material is on the surface. The existence of this layer is consistent with the metallurgical work mentioned previously, and furthermore, in Fig. 11 [2] it is shown that this layer yields a diamagnetic behavior in the susceptibility when the transition at 6–7 K is reached. Note that 6 K corresponds to $Z_{BCS} \approx 1.2$ and 4 K corresponds to $Z_{BCS} \approx 1.1$. The net change below $Z \approx 1.1$ corresponds to a layer ~ 30 Å thick. With surface treatments at higher oxygen pressures, much thicker layers [24] can be seen (Fig. 12), along with other kinds of structures. Hahn and Halama [17] have assumed the existence of a two-phase region of NbO and Nb solid solution where the T_c has been found to be about 6–7 K. This kind of region could well explain the extended drop shown in Fig. 12 and, to a lesser extent, on the

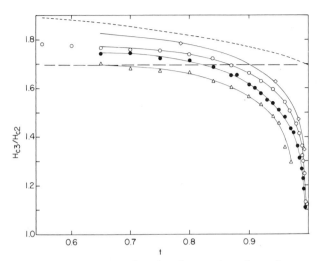

Fig. 10. Temperature dependence of H_{c3}/H_{c2} for a variety of samples: – – –, clean limit theory: ——, dirty limit theory; O, pure Nb; ●, Nb 5000 ppm Ta; △, Nb 10,000 ppm Ta; ◇, anodized pure Nb. We thank Prof. D. K. Finnemore for this figure.

cleaner surface in Fig. 11. Any solution of oxygen in niobium is expected to have a significant effect on T_c, since the depression rate is of the order of about 1 K/at. %. Even a small NbO layer can affect T_c, since the T_c of NbO is only 1.4 K [12]. It should be mentioned, however, that it is not clear that a monolayer of NbO has the same T_c and electronic properties as does the bulk material.

Schwarz and Halbritter [14] have observed large diamgnetic changes at low temperatures in their inductive measurement, and, as mentioned pre-

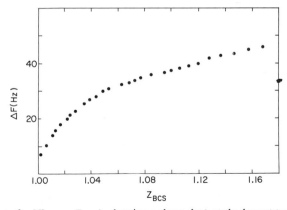

Fig. 11. Data for Nb near $Z = 1$, showing a sharp drop at the lowest temperatures due to diamagnetism of a lower-T_c surface layer.

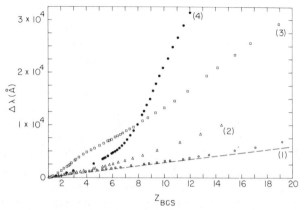

Fig. 12. Effect of oxygen on λ; curve 1, clean Nb; Curve 2, RRR 170 heated close to 1800°C at 8×10^{-7} Torr; Curve 3, RRR 85 heated over 1900°C at 2×10^{-6} Torr; curve 4, RRR 8 heated close to 1800°C at 2×10^{-6} Torr.

viously, they attribute this to oxygen inclusions. This is reasonable for dirty surfaces and may explain some of the data in Fig. 11. It must be emphasized that their samples, which have a diameter of 1 cm and are treated for 2 h at 1850°C at about 10^{-7} Torr, do not have clean surfaces, which would require heating for many days at 10^{-8} Torr for their diameter. They quote a bulk mean free path (mfp) of about 1000 Å, whereas the best Nb samples heated at 10^{-8}–10^{-9} Torr for about 24 h show resistance ratios greater than 1000. This implies that the bulk mean free path is greater than 30,000 Å; Hopkins and Finnemore have measured a mfp of 12,000 Å. Of course, the Schwarz–Halbritter treatment probably approximates more closely that which is obtained when Nb rf cavities are degassed, and hence their speculations about oxygen domain formation and rf breakdown are relevant. However, one must be careful to emphasize that surfaces can be obtained with which relatively small deviations from fairly ideal properties can be achieved, as shown in Fig. 10, by the H_{c3}/H_{c2} measurements of Hopkins and Finnemore, and by Nb tunneling measurements of Shen [3]. There is no obvious need to postulate Nb inclusions in these cases. It should be mentioned that in this regard Schwarz and Halbritter are extremely misleading, since we would assume from their paper that all Nb surfaces have extremely nonideal properties such as the ones that they measure. This is not true, and their poor surfaces are due to the particular UHV treatment that they used, as discussed above. We emphasize that either type of surface can be obtained.

Ta behaves in a way similar to Nb with respect to deviations in penetration depth measurements. Furthermore, Auger measurements show similar oxygen profiles and it appears that problems with surfaces where the order parameter is depressed occur generally in materials, such as Nb and Ta, with very reactive surfaces. Good surfaces can be obtained more readily in Ta, since it can be heated to significantly higher temperatures, making the degassing more effective.

C. rf Superconductivity and Magnetic Field Breakdown

One problem with using Nb for rf applications is that the field at which the ac field enters the superconductor is significantly below the ultimate field at which flux entry occurs, the superheating field. In type I superconductors the resistance to the entry of ac fields has persisted to fields higher than H_{c0}; in fact, flux exclusion has been achieved in Sn, Pb, and various Sn–In and In–Bi alloys up to fields very close to the superheating (SH) field [25]. The breakdown field in Nb is a different problem, and even though smooth surfaces have been obtained, the field for flux entry is typically only about 500 Oe, whereas $H_{c1} \approx 1200$ Oe and $H_{c0} \approx 2000$ Oe. Furthermore, the superheating field should be significantly higher than H_{c0}. Even in the best cases, breakdown fields are only ~ 1600 Oe, which are still well below H_{c0} or H_{SH}. It is emphasized that fields above H_{c1} are achievable and this field is not a limit on the flux entry field of type II superconductors [26]. In fact in Nb_3Sn cavities fields far above H_{c1} can be obtained [27].

It has been suspected for many years that the problem with Nb cavities involved the complex interactions of oxygen at the Nb surface [19] and this led to much of the work discussed here. In recent work Halbritter [28] discusses the effect of oxygen domains or islands of Nb suboxides as places that grow with cold working and electron impact and can be the cause of the weak spots that in turn cause breakdown. It is unfortunate that the size and stability of rf structures do not in practice allow surfaces to be obtained that approach the ideal features achievable in small samples.

Some optimism with regard to rf cavities is warranted from recent work that has investigated rf superconductivity in Nb structures coated with Nb_3Sn [27, 29]. This possibility of using Nb_3Sn for cavities has been under consideration for some time [30] and early work on cavities showed that, by reacting Sn with a Nb cavity to form Nb_3Sn on the inner surface, results could be obtained that were essentially comparable to results with niobium cavities. In the recent work by Hillenbrand and Martens [27], properties comparable to the best niobium cavities are obtained and there appears to be great promise for practical applications. An additional advantage is that, because of the higher T_c of Nb_3Sn, the operating temperature can be 4.2 K, He bath temperature, rather than near 1.5 K, as is necessary with niobium cavities. The stability of Nb_3Sn to oxides and the like also makes this kind of material attractive for practical use. A problem that may arise is the much poorer thermal conductivity of Nb_3Sn as compared to pure niobium, which implies that the layers must be as thin as possible to avoid thermal breakdown when normal regions form.

The details of compound formation at surfaces is an interesting problem in itself that can only be briefly mentioned here. There have been some LEED studies of the formation of Sn compounds [31], and also of Ge compounds, on Nb. After the formation of a crystalline structure in the first monolayer or so, disorder rapidly sets in and the Nb_3Sn is probably composed of many randomly oriented crystallites. The first crystalline layer is a surface structure that is not Nb_3Sn [32]. With the success obtained by Hillenbrand and Martens [27], it is hoped that in the near future more study will be devoted to the formation of surface materials and to how they can be improved.

ACKNOWLEDGMENTS

We are grateful to Dr. H. Hahn for many discussions during the course of this work, and also for the use of Figure 7. We also thank Professor D. K. Finnemore for the use of Figure 10. We are especially grateful to Dr. T. Luhman, without whose offer of tennis lessons we never would have finished this review.

References

1. See J. P. Turneaure and I. Weismann, *J. Appl. Phys.* **39**, 4417 (1968); J. P. Turneaure and Nguyen Tuong Viet, *Appl. Phys. Lett.* **16**, 333 (1970).
2. C. Varmazis and M. Strongin, *Phys. Rev. B* **10**, 1885 (1974).
3. L. Y. L. Shen, "Superconductivity in d and f-band Metals" (D. H. Douglass ed.), AIP Conf. Proc. No. American Institute Physics, 1972.
4. R. M. Easson and P. Hlawiczka, *J. Phys. D* **1**, 1477 (1968).
5. M. Strongin, H. H. Farrell, H. J. Halama, O. F. Kammerer, C. Varmazis, and J. M. Dickey, *Part. Accel.* **3**, 209 (1972).
6. L. H. Rovner, A. Drowart, F. Degreve, and J. Dowart, Tech. Rep. No. AFML-TR-68-200, Air Force Mater. Laboratory, Wright-Patterson Air Force Base, Dayton, Ohio.
7. E. Fromm and H. Jehn, *Vacuum* **19**, 191 (1969).
8. See Strongin *et al.* [5], also H. H. Farrell, H. S. Isaacs, and M. Strongin, *Surf. Sci.* **38**, 31 (1974).
9. A. Joshi and M. Strongin, *Scripta Met* **8**, 413 (1974).
10. S. Hofmann, G. Blank, and H. Schultz, *Z. Metallk.* **67**, 189 (1976).
11. R. A. Pasternak and B. Evans, *J. Electron Soc.* **114**, 452 (1967).
12. J. K. Hulm, C. K. Jones, R. A. Hein, and J. W. Gibson, *J. Low-Temp. Phys.* **7**, 291 (1972).
13. A. Joshi, M. N. Varma, and M. Strongin, *Metall. Trans.* **5**, (1974).

14. W. Schwarz and J. Halbritter, *J. Appl. Phys.* **48**, 4618 (1977).
15. I. Lindau and W. E. Spicer, *J. Appl. Phys.* **45**, 3720 (1974).
16. H. Martens, H. Diepers, and R. K. Sun, *Phys. Lett.* **34A**, 439 (1977).
17. H. Hahn and H. J. Halama, *J. Appl. Phys.* **47**, 4629 (1976).
18. K. E. Gray, *Appl. Phys. Lett.* **27**, 462 (1975).
19. See, for example, J. M. Dickey, H. H. Farrell, O. F. Kammerer, and M. Strongin, *Phys. Lett.* **32A**, 483 (1970); and P. Kniesel *et al., Proc. Int. Conf. High Energy Accel., 8th, Geneva* p. 275. CERN, Geneva, 1971.
20. C. Varmazis, A. Joshi, T. Luhman, and M. Strongin, *Appl. Phys. Lett.* **24**, 394 (1974).
21. J. R. Hopkins and D. K. Finnemore, *Phys. Rev. B* **9**, 108 (1974).
22. P. G. deGennes and J. Matricon, *Solid State Commun.* **3**, 51 (1965).
23. C. Varmazis, Y. Imry, and M. Strongin, *Phys. Rev. B* **13**, 2880 (1976).
24. C. Varmazis, *et al. IEEE Trans. Magn.* **MAG-11**, 423 (1975).
25. T. Yogi, G. J. Dick, and J. E. Mercereau, *Phys. Rev. Lett.* **39**, 826 (1977).
26. See, for example, K. Schnitzke, H. Martens, B. Hillenbrand, and J. H. Diepers, *Phys. Lett.* **45A**, 241 (1973); S. Giordano, H. Hahn, H. J. Halama, T. S. Luhman, and W. Bauer, *IEEE Trans. Magn.* **MAG-11**, 437 (1975); and J. A. Yasaitis and R. M. Rose, *ibid.* **MAG-11**, 434 (1975).
27. B. Hillenbrand and H. Martens, *J. Appl. Phys.* **47**, 4151 (1976).
28. J. Halbritter, Kernforschungzentrum Karlsruhle Institut für Experimentelle Kernphysik Internal Rep. No. 76-197-LIN (October 1976).
29. H. Hahn, *Herbstschule über Anwendung der Supraleitung in Elektrotechnek und Hochenergiephysik, Titisee, Germany, 1972.*
30. J. M. Dickey, M. Strongin, and O. F. Kammerer, *J. Appl. Phys.* **42**, 5808 (1971).
31. A. G. Jackson and M. P. Hooker *in* "Structure and Chemistry of Solid Surfaces" (G. Somorjai, ed.), p. 71. Wiley, New York, 1969.
32. J. A. Strozier, D. L. Miller, O. F. Kammerer, and M. Strongin, *J. Appl. Phys.* **47**, 1611 (1976).

Irradiation Effects in
Superconducting Materials

A. R. SWEEDLER† and C. L. SNEAD, JR.

Department of Energy and Environment
Brookhaven National Laboratory
Upton, New York

and

D. E. COX

Physics Department
Brookhaven National Laboratory
Upton, New York

I. Introduction 349
II. Elements 350
   A. Effect on T_c 350
   B. Changes in H_{c2} 353
   C. Critical Currents 355
III. Body-Centered-Cubic Alloys 359
IV. Non-A15 Compounds 370
   A. Laves Phases 370
   B. Chevrel Phases 370
   C. $NbSe_2$ 371
V. A15 Compounds 372
   A. Introduction 372
   B. Transition Temperatures 373
   C. Critical Fields and Currents 409
   References 422

I. Introduction

The effect of irradiation on the properties of superconductors has become an active area of research during the past several years. The large-scale use of superconducting materials in high-energy accelerators and fusion reactors has prompted the need to gain a deeper understanding of the effect of various radiation fields on the critical superconducting properties such as the transition temperature T_c, the upper critical field

† Present address: Department of Physics, California State University, Fullerton, California

H_{c2}, and the critical-current density J_c. The study of radiation effects has also proved to be a valuable tool in gaining a better understanding of the parameters and mechanisms of superconductivity in technologically important materials such as the A15 compounds. This is primarily due to the fact that it is possible to introduce controlled amounts of defects and to study their various effects. The defects can be annihilated and reversion to the unirradiated state can be achieved by annealing. This type of controlled, reversible procedure has added greatly to the understanding of the role played by defects in affecting the superconducting properties.

This chapter considers the effects of various irradiation fields on several superconducting and normal-state properties of elements, alloys, and compounds. The effects of high-energy neutrons (usually reported as either $E > 0.1$ or $E > 1$ MeV), energetic charged ions, protons, and electrons on T_c, H_{c2}, J_c, the lattice parameter a_0, and the degree of order are discussed. Detailed isothermal and isochronal annealing experiments on neutron-irradiated compounds are also described. Emphasis will be placed on the A15 compounds since these materials currently possess the most desirable combination of superconducting properties for technological applications and offer a constant challenge to any comprehensive theory of superconductivity.

II. Elements

A. *Effects on* T_c

Most of the recent work on irradiation effects in superconductors has focused on alloys and compounds since these materials offer the greatest promise for practical applications. However, elements, particularly Nb, provide an opportunity to study various flux-pinning models since it is possible to introduce defects in a controlled fashion by irradiation. Some of the work on the effects on the T_c's of the elements that have been studied is discussed. Subsequent sections deal with effects on H_{c2} and J_c. An excellent review concerning some of the early work on irradiation effects has been given by Cullen [1].

In early irradiation experiments, decreases in T_c were observed as a function of fluence, and these decreases were related to increases in the normal-state resistivity ρ_n. Rinderer and Schmid [2] irradiated samples of tin with 5.3-MeV alpha particles at helium temperature and observed decreases in T_c of ~5% and an irradiation-induced resistance increase of a factor of 4 over the initial value. Beyond that point, however, continued irradiation produced increasing resistivity toward saturation, but T_c stopped decreasing and actually increased slightly. The expression

$$-\Delta T_c = aR + bR \ln R \qquad (1)$$

where $R = \rho_n/(\rho_{300} - \rho_n)$ where ρ_{300} is the room temperature value of the resistivity and a and b are adjustable parameters, gives a qualitative fit in the region of decreasing T_c but fails in the higher-fluence region when T_c begins to increase with increasing ρ_n. Equation (1) was obtained empirically by Seraphin et al. [3], and theoretically established from BCS theory by Markowitz and Kadanoff [4]. Rinderer and Schmid also noted that the T_c decreases associated with resistivity increases due to irradiation-induced defects were a factor of 50 larger than those associated with equivalent resistivity increases brought about by alloying. Complete recovery of the induced property changes was obtained following a 24-h anneal at room temperature, demonstrating that the Frenkel defects produced by the irradiation were responsible for the effects seen.

Coffey et al. [5] irradiated several type-II materials at low temperature with 15-MeV deuterons and measured the changes in several properties, including ρ_n, as a function of fluence and subsequent annealing. They were able to obtain a qualitative fit to their data for Nb with Eq. (1). The fluence levels for this Nb irradiation were not as high as those of Rinderer and Schmid for Sn, and the leveling off of T_c at a lower saturation level is not clearly in evidence in the Coffey et al. work. The maximum decrease of T_c in Nb observed by Coffey et al. was ~1%.

Berndt and Sernetz [6] irradiated Nb at 4.6 K with fast neutrons to a $\Delta\rho_n$ of 1.2 $\mu\Omega$ cm and observed a decrease in T_c of ~1.5%. This is in reasonable agreement with the results of Coffey et al., whose irradiation-induced decrease in T_c of 0.1 K was accompanied by an identical increase in ρ_n. A striking observation made both by Coffey et al. and by Berndt and Sernetz was that when the Nb was annealed and some of the radiation-induced changes recovered, there was a hysteresis in the ΔT_c versus $\Delta\rho_n$ curve. The remaining resistivity after annealing was less "effective" in decreasing T_c than the low-temperature as-irradiated resistivity values. This implies that the suppression of T_c is not uniquely correlated with the induced resistivity, but it is strongly dependent upon the type or configuration of the defect giving rise to the increasing resistivity. Although Rinderer and Schmid did not report low-temperature annealing results on T_c following their irradiations, their results on alloy-induced resistivity changes bear out this dependence of T_c suppression on defect configuration.

One explanation of this behavior of T_c and ρ_n is for a BCS superconductor with an anisotropic energy gap. The expression for the critical temperature is

$$T_c = 1.14(W_\Delta/k) \exp[-1/N(0)V_{\text{eff}}], \qquad (2)$$

where W_Δ is the Debye frequency, k is Boltzmann's constant, $N(0)$ is the electronic density of states, and V_{eff} is the effective electron–electron interaction. In the absence of impurities V_{eff} can be written $V_{eff} = V(1 + \langle a^2 \rangle)$, where the factor $1 + \langle a^2 \rangle$ represents the anistropy of the interaction. Thus, substituting for V_{eff} gives

$$T_c = 1.14(W_\Delta/k) \exp\{-1/[N(0)V(1 + \langle a^2 \rangle)]\} \qquad (3)$$

As the mean collision time for scattering decreases (ρ_n increases) due to increasing impurity or defect concentrations, $\langle a^2 \rangle$ decreases. When $\langle a^2 \rangle$ becomes much less than 1, Eq. (3) reduces to

$$T_c = 1.14(W_\Delta/k) \exp\{-1/[VN(0)]\} \qquad (4)$$

The maximum relative change in T_c resulting from complete destruction of the anisotropy is then given by the change between Eq. (3) and Eq. (4); i.e.,

$$\Delta T_c/T_{cmax} = [1/VN(0)]\langle a^2 \rangle. \qquad (5)$$

With typical values for these quantities, a value of $\sim 4\%$ is obtained for the maximum relative change in T_c from the anisotropy effect alone.

This maximum T_c change is in good agreement with the maximum decrease seen by Rinderer and Schmid of $\sim 5\%$ in Sn. All of the decrease in T_c in irradiated Sn could well be attributed to the reduction in the anisotropy, and no other contributing mechanism need be involved. Note from Eq. (5) that if the electronic density of states $N(0)$ decreases, a decrease in T_c also results. In Sn, for the fluence levels investigated and for the type of defects produced by the irradiation, $N(0)$ was apparently unchanged. The changes in T_c for Nb reported by Coffey et al. were only 1%, so it can be concluded that the saturation value of ΔT_c due to complete reduction of the anistropy effects has not been reached. At this point, the resistivity is roughly half the saturation value ($\Delta\rho_n$ was ~ 1.2 $\mu\Omega$ cm whereas the radiation-induced saturation resistivity is ~ 2.6 $\mu\Omega$ cm in Nb).

Klaumünzer et al. [7] irradiated Nb at <20 K with 25-MeV oxygen ions and investigated the changes in T_c and ρ_n. Excellent agreement with Eq. (1) was noted in this work, and also in comparison with the results of Coffey et al. [5] and Berndt and Sernetz [6]. A nonlinear relation between ΔT_c and ρ_n was found upon annealing the irradiated samples, in agreement with Rinderer and Schmid [2], and the value obtained for $\langle a^2 \rangle$ was in good agreement with the theoretical value [4]. A summary of the available data for Nb is given in Fig. 1. Also shown are fitted theoretical curves based on the anisotropy model as discussed above. Some qualitative conclusions can be drawn concerning the depression of T_c for Nb from Fig. 1: (a) the functional dependence of ΔT_c is in good agreement with the anisotropy model; (b) ΔT_c is not a unique function of ρ_n as in the case of chemical impurities,

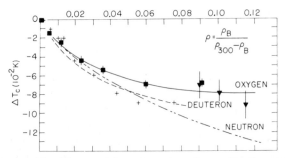

Fig. 1. Decrease ΔT_c of the transition temperature T_c as a function of the irradiation-induced electrical resistivity for Nb irradiated under the different conditions. (After Klaumünzer *et al.* [7].)

but depends on the particular defect; and (c) the $\langle a^2 \rangle$ value of 0.005 is in good agreement with values for pure Nb obtained from nonirradiation-induced defects.

B. Changes in H_{c2}

Just as decreases in T_c are related to increases in the normal-state resistivity ρ_n, changes in the upper critical field H_{c2} can be related to resistivity changes. The Ginzburg–Landau parameter κ, introduced in the introductory chapter by Dew-Hughes, this volume, is related to the normal-state resistivity by the relation

$$\kappa = \kappa_0 + A\gamma^{1/2}\rho_n \tag{6}$$

This expression is derived from the Ginzburg–Landau–Abrikosov–Gorkov (GLAG) theory [8], where κ_0 is the "pure" material value, A is a constant, and γ is the electronic specific-heat coefficient. This linear dependence of κ on ρ_n is valid for $T \approx T_c$ and for a spherical Fermi surface with only s-wave scattering and relaxation time τ. Eilenberger [9] extended the analysis to temperatures below T_c and allowed p-wave scattering, which introduced a new transport relaxation time $\sim \tau_{tr}$. The Ginzburg–Landau parameter κ is replaced by two temperature-dependent parameters $\kappa_1(t)$ and $\kappa_2$hr (t), where κ_1 is defined by the relation

$$H_{c2}(t) = \sqrt{2}H_c(t)\kappa_1(t) \tag{7}$$

where t is the reduced temperature T/T_c. Both $\kappa_1(t)$ and $\kappa_2(t)$ are strong functions of τ_{tr}/τ, which reflects the non-s-wave scattering character. In this approximation Eq. (6) becomes

$$\kappa_1(t) = \kappa_{01}(t) + A\gamma^{1/2}\rho_n\text{hr } g(t, \tau_{tr}/\tau) \tag{8}$$

where κ_{01} relates to the "pure" conductor and g is a function dependent upon the ratio τ_{tr}/τ. Although this function cannot be expressed in analytic form, it can be qualitatively understood as a "weighting" factor modifying the effect of the resistivity ρ_n on $\kappa_1(t)$. $g(t, \tau_{tr}/\tau)$ has its maximum value for pure s-wave scattering ($\tau_{tr}/\tau = 1$) and its minimum when p-wave scattering dominates ($\tau_{tr}/\tau = \sqrt{2}$).

In the preceding section it was noted that the changes observed in T_c were not uniquely correlated with ρ_n, and the conclusion was reached that the nature of the scattering (either s or p wave) due to the defects present played a role. This would lead to a modification of Eq. (1) to include a similar weighting factor for either or both of the arbitrary constants to account for defects with different scattering characteristics.

With respect to the relationship between κ_1 and H_{c2} described by Eq. (7), H_{c2} is only weakly dependent [10] on ρ_n, so that the linear dependence of κ on ρ_n is reflected in a linear dependence of H_{c2} on ρ_n. Experiments where measurements are made near T_c tend to confirm both the linearity and the predicted magnitude of the increase in H_{c2} with $\Delta\rho_n$. Most measurements, however, are made near liquid-helium temperature, which in many cases is well below T_c, in which case Eq. (7) must be used with the temperature-dependent $\kappa_1(t)$ previously defined.

Ullmaier et al. [11] irradiated niobium samples at 4.6 K with 3-MeV electrons to several fluences. They measured the upper critical field H_{c2} as a function of ρ_n for the as-irradiated samples at low temperature, and also following isochronal annealing of their samples. These results are shown in Fig. 2.

During the irradiation, the linear increase of H_{c2} with $\Delta\rho_n$ is quite ap-

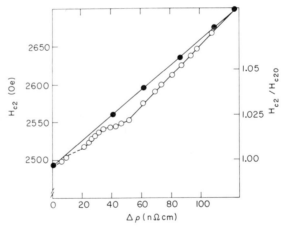

Fig. 2. Experimental dependence of H_{c2} on the resistivity increase $\Delta\rho$ ($\rho_0 = 4.99$ n Ω cm). The closed circles are values for different electron doses. After irradiation with a maximum dose of 3.88×10^{18} e/cm^2 the sample was subjected to an isochronal annealing program, $T_0 = 4.6$ K (open circles).

parent. The value of the slope of the production curve $\Delta H_{c2}/\Delta\rho_n$ is 1.68 G/$\mu\Omega$ cm. The total change in resistivity, ~ 130 nΩ cm, is well below the saturation value for niobium, so that no overlap effects are taking place. The primary knock-on energy is sufficiently low for 3-MeV electron irradiation that essentially only single Frenkel pairs are created. Thus the $\Delta\rho_n$ observed is simply due to a linear buildup of isolated interstitials and vacancies.

This homogeneous character of identical defects is altered, however, upon warming the samples. As the temperature is raised through isochronal annealing, interstitial-vacancy annihilation, interstitial-cluster formation, and interstitial trapping at impurities all occur. This changing character of the defects is evidenced in Fig. 2 in a departure from linear behavior of the H_{c2} versus $\Delta\rho_n$ curve, with the resistivity change after the anneals being less "effective" in producing an increase in H_{c2} than an identical change in an as-irradiated sample. This change in effectiveness of the radiation-induced resistivity is best seen in Fig. 3, taken from Ullmaier et al. [11], where the ratio $\Delta H_{c2}/\Delta\rho_n$ is plotted against annealing temperature along with the resistivity change $\Delta\rho_n$ versus temperature. A decrease in $\Delta H_{c2}/\Delta\rho_n$ with progressive annealing is seen, being most precipitous at the stage-I free-migration temperature [12] and turning around at the end of that stage. Again, there is a change in the nature of the behavior of the $\Delta H_{c2}/\Delta\rho_n$ curve at ~ 220 K, the temperature where vacancy migration is probably taking place [12]. The changes in the nature of the $\Delta H_{c2}/\Delta\rho_n$ curve are obviously closely connected with the defect configurations. This is quite analogous to the nonlinear relationship between ΔT_c and resistivity discussed in Section II,A.

Ullmaier et al [11] point out that their results are in qualitative agreement with an interpretation in which Eq. (7) is used with $\kappa_1(t)$ given by Eq. (8). As previously mentioned, the value for the function $g(t, \tau_{tr}/\tau)$ has its maximum value for pure s-wave scattering (in this case corresponding to only isolated interstitials and vacancies present) and this value decreases as p-wave-scattering contributions come into play (when interstitial clusters and/or interstitial-impurity complexes develop). These results show that just as with the T_c changes, the changes in H_{c2} associated with resistivity are not simply related to the magnitude of the resistivity changes, but depend on the value of the ratio τ_{tr}/τ. This ratio reflects the *nature* of the defects as evidenced by their electron-scattering characteristics.

C. Critical Currents

The behavior of type-II superconductors is governed by the "mixed state" in which magnetic flux penetrates the conductor in the form of

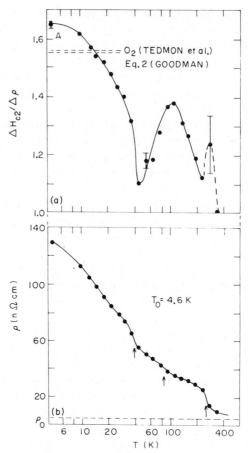

Fig. 3. The ratio $\Delta H_{c2}/\Delta\rho_n$ and $\Delta\rho_n$ plotted versus annealing temperature.

quantized flux or vortices. To a large degree the electrical properties of the superconductor are determined by the motion of these flux lines through the conductor, as described in the introductory chapter by Dew-Hughes, this volume. A simple explanation is that the more difficult it is to cause a flux line to move under a driving force (the Lorentz force due to magnetic fields present), the less the hysteretic energy loss due to the flux-line motion and the "harder" the superconductor. The metallurgical means for pinning flux lines and thereby optimizing the critical properties of type-II superconductors have been treated in detail in the introductory chapter by Dew-Hughes, as well as in the chapter by McInturff and the chapter on the physical metallurgy of A15 compounds by Dew-Hughes, this volume. When defects are introduced into the material by irradiation, the critical properties can be altered. This alteration comes about chiefly through two means—(a) the creation or destruction of pinning sites, and (b) the alteration of the effectiveness of existing pinning sites through changes of mechanical or electrical properties of the material. Either of these can be manifested in changes in the volume pinning force F_p, and in many cases both mechanisms may be operative, and the problem becomes one of sorting out their relative contributions to the changes observed.

In elemental superconductors, many investigators have shown that large pinning effects result from irradiation [13]. A few examples that demonstrate the kinds of effects that arise from various types of irradiations are considered here. To study the effects of irradiation-produced interstitials and vacancies on flux pinning, electron irradiation at low temperature is most desirable. Isolated Frenkel pairs uniformly distributed throughout the sample can be achieved in this way owing to the fact that the energy transfer to the primary knock-on atom is only a few times the threshold energy for Frenkel-pair production, thereby eliminating clusters of point defects from direct consideration. To act as a strong flux-line pinner, the size of the defect must be of the order of the coherence length ξ of the material (i.e., ~ 350 Å for niobium at 4.2 K). For uniformly distributed point defects no pinning effects are expected from low-temperature electron irradiations.

Ullmaier *et al.* [14] irradiated niobium at less than 9 K with 3-MeV electrons and monitored the changes in pinning produced using magnetization measurements. A small increase in pinning was observed, which was qualitatively interpreted as due to the interactions of the flux lines not with the individual defects, but with fluctuations in the *density distribution* of the Frenkel pairs. Efforts to achieve quantitative agreement with a calculated volume pinning force resulted in order-of-magnitude estimates. Surface interactions and also a variation of resistivity over these fluctuation regions that would produce a fluctuation in κ [refer to Eq. (6)] could also contribute to the effects found.

The importance of defect cluster size in flux-line pinning was nicely shown in the work of Berndt *et al.* [15], in which magnetization measurements were made following reactor-neutron irradiation at 4.6 K. The pinning observed in Fig. 4 taken from Freyhardt [16] compares the Ullmaier *et al.* [14] electron-irradiation results with this neutron-induced pinning. The resistivity increase is equivalent in both cases, but the disparity of increasing pinning force in favor of the neutron-irradiated sample is evident. The neutron-damaged samples, in which many of the defects produced lie in large cascades, exhibit much larger pinning than for the more homogeneous distribution in the electron-irradiated case. A quantitative evaluation of these two cases can be found in the review by Ullmaier [17].

The pinning curve of the neutron-irradiated specimen in Fig. 4 exhibits

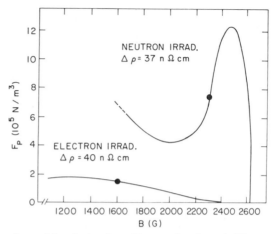

Fig. 4. Comparison of the pinning force F_p per unit volume in Nb samples irradiated at 4.5 K with electrons and neutrons, respectively. The mean concentration of Frenkel defects is about equal for both cases.

a peak close to H_{c2} in the flux pinning (here the volume pinning force F_p) as a function of applied magnetic field. A similar peak is often observed in the plot of critical current versus field following fast-particle irradiation [18]. Niobium and its alloys, particularly when they are doped with interstitial impurities, show the most pronounced peaks. Several explanations of this "peak effect" have been put forward [19]. Van der Klein *et al.* [18] showed that their data on neutron-irradiated niobium agree with the model proposed by Campbell and Evetts [19] in which the flux lines are strongly pinned by line defects that are parallel to the flux lines. The line defects in this case were the edges of dislocation loops. Again, as has already been seen, the *kind* of defect is strongly involved in determining the observed changes in critical properties.

Other neutron-irradiation studies on Nb and Va [20], in which pinning measurements could be coupled with transmission electron microscopy, advanced the qualitative understanding of the role played in pinning by dislocation lines and loops. Brown [21] and Brown *et al.* [22] investigated the pinning forces in Nb following fast-neutron irradiation at 4.6 K. They interpreted their data with the assumption that the elementary interaction force between a cascade and a flux line is predominantly magnetic in character. The order-of-magnitude consistency they achieved in their analysis is quite good considering the many experimental (e.g., distribution of cascade volumes, flux-line-lattice elastic constants) and theoretical (microscopic pinning forces f_p, summation problem) unknowns.

The pinning of fluxoids in Nb and Nb alloys by voids produced by nickel-ion irradiation has recently been investigated by Freyhardt *et al.* [23, 24]. Electron microscopy of similarly irradiated specimens measured the number and size distribution so that quantitative evaluations could be made of the effects of the size of defects on volume pinning forces as determined from changes in critical current. Large increases in J_c indicated that the voids acted as strong pinners. Where the samples were doped with oxygen, the voids arranged themselves in a void lattice. When the flux-line lattice is a multiple of the void lattice spacing (''matching''), there is a considerable increase of J_c, indicating increased pinning for this condition. These interesting investigations are being continued by those authors.

III. Body-Centered-Cubic Alloys

The data reported to date indicate that the T_c's of the bcc alloys of Nb are affected very little by irradiation, as might be expected from the results on pure Nb. Swartz *et al.* [25] reported no change in T_c for a Nb 30 at. % Ta alloy irradiated to a fluence of $5 \times 10^{16} \, n/cm^2$ at reactor temperature, and Okada *et al.* [26] reported small or no decreases in T_c for Nb–Ti alloys irradiated to $1.3 \times 10^{18} \, n/cm^2$ ($E > 0.1$ MeV), also irradiated at ambient temperatures. Schmelz *et al.* [27] irradiated Nb–Ti with 25-MeV oxygen ions at $T < 30$ K and found a T_c depression of only 0.14 K for an oxygen fluence of $3 \times 10^{15} \, O/cm^2$. Changes in H_{c2} are similarly negligible in these alloys. Coffey *et al.* [5] observed changes in T_c of ~ 0.2 K in Nb–Zr and Nb–Ti alloys following doses of 1×10^{17} 15-MeV deuterons/cm^2 at low temperature. Decreases of $\sim 10\%$ in H_{c2} were observed in these alloys, and recovery was found to take place following anneals to 300 K. These authors attributed these changes in H_{c2} to the decreasing T_c and $H_c(0)$, but also point out that changes in κ should be included. Wohlleben [28] found that irradiation of Nb 66 at. %-Ti by 1×10^{17} 3.1-MeV p/cm^2 produced a decrease in T_c of 0.17 K and a decrease in H_{c2}. Annealing the samples to room temperature produced recovery of $\sim 50\%$ in T_c and a similar recovery in H_{c2}. This annealing behavior following low-temperature irradiation will be discussed further in the consideration of critical-current changes in Nb–Ti.

The problem of interpreting the effects of irradiation on critical currents, as stated before, is one of differentiating between the mechanical pinning effects caused by microstructural changes of the pinning centers and the electronic changes that can alter the volume pinning forces. This uncertainty is enhanced in alloy superconductors such as Nb–Ti because of the presence of many, usually intentionally included, defects in the material. These defects, such as precipitates, inclusions, cell walls, dislocation tangles, and second-phase regions, have been engineered into the conductor to maximize one or more of the critical parameters (usually the critical-current density J_c). These effects are discussed in detail in the chapter by McInturff, this volume. The interactions of radiation-produced defects with these preexisting defects and the assignment of critical property changes observed to unique mechanisms associated with particular pinners is a problem that is continuing to receive attention.

The radiation-induced changes in J_c for Nb–Ti have been recently tabulated by Brown [20]. He points out that most of the neutron-irradiation experiments are consistent with each other. Also, there is a general rule that the greater the J_c of the starting material, indicating better metallurgical optimization of the flux pinning, the greater are the decreases in the critical current for a given fluence of neutrons. Söll *et al.* [29] showed that for a fluence of $3.6 \times 10^{18} \, n/cm^2 \, (E > 0.1$ MeV) at 4.6 K for values of J_c (initial critical current measured at 4 T) below a value of $\sim 1.7 \times 10^4$ A/cm^2, increases in J_c were observed. For J_{c0} values above this value, decreases were observed, while at this value no change took place. This work very nicely demonstrated the strong correlation of the metallurgical state with the radiation-induced changes in critical properties to be expected in these alloy superconductors.

Söll [30], following Narlikar and Dew-Hughes [31], Neal *et al.* [32], and Hampshire and Taylor [33], showed that for Nb–Ti the major metallurgical factor in maximizing J_{c0} was the dislocation-cell-wall density. As the cell-wall density increased, J_{c0} increased and the reduced field h at which maximum flux pinning $F_p(max)$ occurred increased. These two parameters, h and $F_p(max)$, were then shown to be the yardsticks to use in predicting how radiation will affect J_c. For cell-wall densities sufficiently high that optimal pinning obtains, further introduction of pinning centers such as cascades (as in fast-neutron irradiation) will not further enhance the pinning. Figures 5 and 6 from Söll [29] demonstrate the importance of the cell-wall density. Samples 1 and 2, involving either cold work or cold work followed by a 380°C anneal, produce the highest density of cell walls and consequently the highest values of J_{c0}. Sample 5 had a higher content of Ti than the others and had precipitates of the α phase present. These precipitates could have acted as additional flux-pinning centers, thereby raising J_{c0}. It also seems reasonable that the precipitates would act as dislocation pinning centers during the cyclic cold working of this sample, thereby providing additional cell-wall nucleation sites. The increase of cell-wall density would then account for the increased J_{c0}.

Annealing to 520°C and 600°C produced lower J_{c0} values. The conclusion here is that above 380°C cell growth takes place, which decreases the

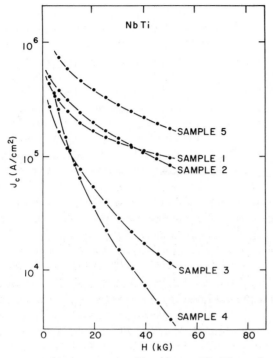

Fig. 5. Critical current densities of some Nb–Ti samples.

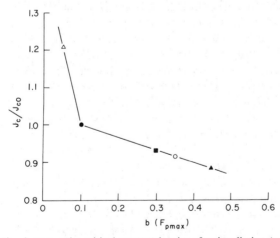

Fig. 6. Relation between the critical current density after irradiation (normalized to the values before irradiation) and the induction value at the maximum pinning force: ▲, sample1, cold worked; ■, sample 2, cold worked then 380°C; ●, sample 3, cold worked then 520°C; △, sample 4, cold worked then 600°C; sample 5, cold worked then α precipitations.

cell-wall density, thereby decreasing J_{c0}. This dependence of J_{c0} on the cell-wall density has been demonstrated by Neal et al. [32], who used transmission electron microscopy to show that J_{c0} scaled as $1/d$ for d between 400 and 1000 Å, where d is the mean diameter of the cell. It can be seen in Fig. 6 that fluences of up to $4 \times 10^{18} n/\text{cm}^2$ ($E > 0.1$ MeV) produced no change in J_c for sample 3, annealed at 520°C, but that J_c for the same fluence increased by $\sim 20\%$ for sample 4 annealed at 600°C. For this sample, whose cell-wall density was lower than for the rest, the implication is clear that flux-pinning centers in the form of neutron-induced cascades are being created whose effect is to cause an increase in J_c. For samples 1 and 2 with high initial J_{c0}'s, decreases in J_c are observed due to the irradiation. Sample 3 exhibits behavior intermediate to the others where no change in J_c is observed for a neutron fluence up to 4×10^{18} n/cm^2. In this sample there has obviously been a balance achieved between the two mechanisms that change J_c, pinning centers (cascades), which increase J_c, and radiation defects, which *decrease* J_c. The latter are now considered.

Other than by directly changing the size or number density of flux-pinning centers, the volume pinning force F_p can also be altered by changing the electrical properties of the superconductor. In the case of irradiated Nb–Ti, the radiation-induced defects produced in the interior of the cells raise the resistivity of the interior cells, thereby causing a decrease of the pinning strength of the cell walls. This is exhibited by a decrease in J_c following irradiation. This type of explanation was first used by Dew-Hughes and Witcomb [34] to explain the effects of deformation on the pinning force of Mo 34 at. % Re alloys, and later applied to low-temperature neutron-irradiated Nb–Ti by Söll et al. [35].

The physical basis for this argument rests with the reduction of the core energy of flux lines in regions of high κ. The κ of the cell wall is higher than that of the cell interior, thereby producing pinning of the flux lines in the cell walls [31]. Dew-Hughes and Witcomb calculate the pinning force due to the variation of κ in crossing a cell wall and get an expression

$$F_p \propto \Delta\kappa/d^3 \qquad (9)$$

where $\Delta\kappa$ is the difference in κ between the cell wall and the cell, and the number of cells per unit volume is $1/d^3$. Figure 7 from Dew-Hughes and Witcomb demonstrates this variation of κ from cell wall to cell, and shows the dependence of $\Delta\kappa$ on the degree of deformation. It is pointed out that many factors can alter this effect; i.e., cell-wall density $1/d^3$, thickness of the walls, the width of the boundary between cell and cell wall, and the density and nonuniform distribution of dislocations within the cell wall, to mention the more important items. (See also the introductory chapter by

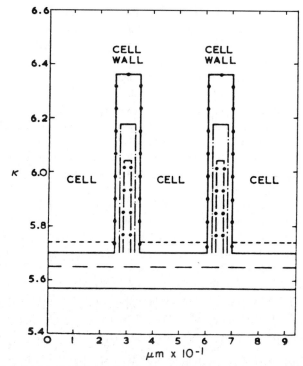

Fig. 7. Variation of κ with distance through specimens subjected to increasing deformation by longitudinal rolling. Note dislocation rearrangement effects; cell walls of higher κ, and a κ value for the cell lower than the bulk value of 8% deformed specimen.

Dew-Hughes, and, specifically for pinning in Nb–Ti, that by McInturff, this volume.) Obviously, the higher the dislocation density in the wall, the higher the κ and the greater the pinning.

Consider the effect of irradiation on this system. The Frenkel pairs created will increased the resistivity. In the cell wall, however, where there is a high concentration of dislocations and associated strain fields, the production of Frenkel pairs will be curtailed by enhanced spontaneous recovery at the dislocations, reducing the numbers created relative to the dislocation-free cell interior. The relative increase in resistivity owing to the irradiation will be much greater in the cell interior than in the cell walls. For highly tangled cell walls, one could, in first approximation, consider the irradiation as completely ineffective in altering the resistivity in the cell walls and only consider the radiation-induced resistivity in the the cell interior.

This effect on κ with a changing ρ_n is shown schematically in Fig. 8. The $\Delta\kappa$ (the difference between cell-wall κ and cell-core κ) before irradiation is

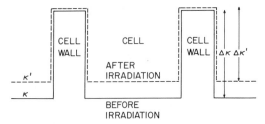

Fig. 8.   Effective defect density for flux pinning across a cell.

greater than is $\Delta\kappa'$ after irradiation, and since $F_p \propto \Delta\kappa$, the result of the irradiation is to decrease the pinning forces attributable to the cell walls. In the previously discussed work of Söll *et al.* [29] (see Figs. 5 and 6), samples 1, 2, and 5 are then considered to be sufficiently cell-wall pinned that the increasing resistivity of the interiors could effect a decrease of the cell-wall F_p's. In sample 4, where recrystallization and grain growth at the 600°C anneal has reduced the density of the cell walls, the cascades introduced by the neutron irradiation act as stronger pinners than (or additions to) the existing ones, and a net increase in pinning is seen. (Note that this effect should saturate at high fluences and the pinning force then start to decrease if ρ_n is still increasing.) In sample 3, where no change in pinning force was observed, apparently the increase in F_p being produced by cascade pinning is just offset by the decrease in F_p being produced by the increasing resistivity. At higher values of fluence, it is probable that this balance would change, most likely in favor of decreasing J_c. Based upon the work of Neal *et al.* [32], an estimate of ~ 850 Å can be made for d for sample 3. This is taken to be the cell size below which little further pinning can be achieved by radiation, but rather resistivity-induced F_p reduction is the dominant effect on J_c.

For reductions of F_p caused by resistivity due to irradiation-produced defects one would expect that the temperature of irradiation would be an important parameter affecting the magnitude of the decreases observed. This is because of thermal recovery of defects that are stable at helium temperature, but become mobile, migrate, and annihilate or become trapped at sinks as the temperature is raised. As the defects recover, so does the resistivity due to them. A greater change in J_c would then be predicted for low-temperature irradiation as opposed to, say, room-temperature irradiation for equal fluences with identical particles. Also, annealing of the low-temperature-irradiated Nb–Ti samples would produce recovery of J_c.

The work of Söll *et al.* [35] bears out these contentions. They irradiated Nb 66 at. % Ti at 5 K with reactor neutrons and observed changes in J_c in fields up to 3 T after irradiation and then after annealing. The Nb–Ti was well optimized metallurgically, as evidenced by a J_c of 1.3×10^6 A/cm² at zero field and 4.2 K. Figure 9 shows the results of the irradiations and the anneals. Irradiation to a fluence of 7.5×10^{18} n/cm² ($E > 0.1$ MeV) gives a reduction of J_c of ~40% (curve c). Annealing to 100 K produces the recovery seen in curve d, and finally annealing to 270 K (curve e) produces ~70% recovery of the change in J_c. Other samples reported in this work irradiated to 4.5×10^{18} n/cm² and annealed showed recovery at room temperature from 65 to 100%. (Note that the reduction of J_c and the recovery show roughly the same relative behavior independent of field, up to ~3 T.) The model of reduced F_p owing to an increasing κ in the cell caused by radiation-induced resistivity fits nicely the decreases in J_c observed. Also, the recovery of J_c is predicted by the expected recovery in $\Delta\rho_n$ on warming from helium to room temperatures. The samples used in the work above by Söll et al. [35] were multifilament wires, but with a gauge length of sample 1.5 mm long from which the copper matrix has been removed and all but one Nb–Ti filament removed.

Short samples of various types of Nb–Ti multifilament conductors have been irradiated at helium temperature with ~30-GeV protons from the Brookhaven Alternating Gradient Synchrotron and annealed similarly to those above. The primary purpose of these experiments was to determine the effects of beam spill on the conductors for superconducting

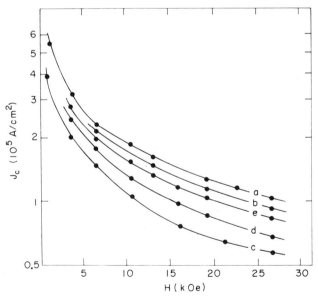

Fig. 9. Critical current of an Nb–Ti wire versus a transverse magnetic field at 5.3 K before and after neutron irradiation. (The zero field value of curve a is 1.2×10^6 A/cm².)

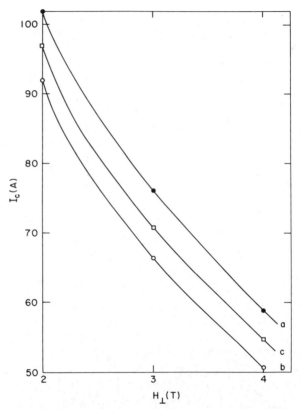

Fig. 10. Effect of irradiation on critical currents in Nb–Ti conductor (379 core), up to 40 kG: ●, unirradiated; ○, 1.5×10^{18} p/cm^2, 30 GeV; □, irradiated and annealed to 290 K.

bending magnets envisioned for intersecting-storage-ring applications. Several different types of multifilament wire were examined that were state of the art for current optimization and that were considered prime candidate conductors for this application. The results [36] for one of these conductors, Nb 45 wt. % Ti 379-core 0.0116-in.-diam cupronickel-jacketed wire with a copper-matrix-to-superconductor ratio of 1.25:1, are shown in Fig. 10. For a total fluence of 1.5×10^{18} p/cm^2 a reduction (curve b) of 14% in the critical current was observed measured at 4.2 K and 4 T, and 13% at 3 T. (Comparison at 2 T is invalid since the I_c determinations for currents in excess of 100 A were in error for these experiments.) The independence of field for I_c reductions seen in the neutron irradiations of Söll et al. in applied magnetic fields up to 4 T does, however, seem to be the case for these high-energy proton irradiations also. Curve c shows the results of annealing the irradiated samples to room temperature, where 50% recovery is observed.

It is interesting to compare this experiment involving low-temperature irradiation to a fluence of 1.5×10^{18} p/cm^2 with one on an identical sample to 1.6×10^{18} p/cm^2, but with an anneal to room temperature at an intermediate fluence (7×10^{17} p/cm^2). The results [37] of I_c versus field for the latter irradiation are shown in Fig. 11. Curve a is for the unirradiated sample, curve b for the sample after the first irradiation segment to 7×10^{17} p/cm^2, curve c the result of a room-temperature anneal following the first segment, and d the result of the additional 9×10^{17} p/cm^2 irradiation. The result of the first anneal (curve c) was a recovery of 60% of the decrease in I_c induced by the first-segment irradiation. This is a 10% greater recovery than that seen for the 1.5×10^{18} p/cm^2 experiment shown in Fig. 10. Following the additional 9×10^{17} p/cm^2 segment (curve

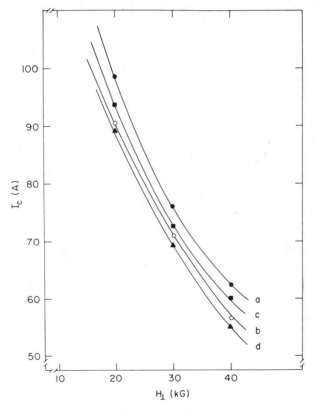

Fig. 11. Critical current versus applied magnetic field for an optimally treated Nb–Ti conductor (cupronickel jacket) prior to and after irradiation: ●, unirradiated; ○, 7×10^{17} p/cm^2; ■, ~273 K anneal; ▲, additional 9×10^{17} p/cm^2.

d of Fig. 11) the sample was again annealed to room temperature and subsequent measurement made (not shown in the figure). This anneal caused
the recovery of 70% of the I_c decrease due to the second-segment irradiation, again a higher recovery percentage than for the single-high-dose
case.

The comparison between these two irradiation and annealing procedures [36] is presented in Fig. 12. Here the relative change in I_c measured
at 40 kG is plotted for both samples as a function of fluence. The vertical
dashed lines represent the changes in I_c following anneals to 300 K. The
lines drawn through the data points are only guides for the eye. The sa-
lient point is that even though the high-total-fluence segmented run had
a larger total I_c decrease than the unsegmented one, the effect of the two
anneals in the segmented case is to cause a greater relative recovery in I_c
at the completion of the final anneal.

This dependence of recovery on fluence is understandable in the interpretation that the critical-current decreases are due to radiation-induced
resistivity. The recovery of the resistivity change is due mainly to the
mutual annihilation of interstitials and vacancies and to a lesser extent to
these migrating defects finding other sinks of annihilation such as dislocations or grain boundaries. As the concentration of the defects increases
owing to the irradiation, the probability increases that mutual interaction
between like species of defects during migration of that species can produce trapping of pairs (or higher-order clusters as the trapping process
proceeds), thereby retaining these defects and their associated resistivity

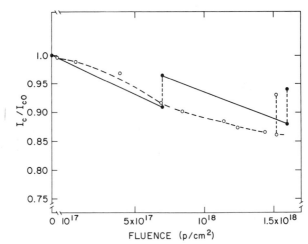

Fig. 12. Relative changes in critical currents of Nb–Ti samples (379 core) as a function
of fluence 30-GeV proton irradiation, $H_\parallel = 4$ T.

contributions to higher temperatures (room temperature). (Annealing experiments on irradiated Nb–Ti to sufficiently high temperatures above room temperature to remove these clusters and to produce complete recovery in I_c have not been carried out.) The lower the concentration, then, the less the mutual interaction of defects to reduce the recovery. This could have a large bearing on the operation of magnets in a radiation field. One desires to keep the cumulative damage to a minimum in, say, a fusion reactor magnet to keep the degradation of I_c as low as possible over its radiation lifetime to minimize resistivity buildup caused by the irradiation. The same arguments can also be applied to the stabilizer material of the superconducting magnet to keep its resistivity minimized.

The most-definitive results on the resistivity recovery of Nb 48 wt. % Ti following low-temperature irradiation are those of Brown *et al.* [38]. For samples with an initial resistivity ρ_0 of 36 $\mu\Omega$ cm, fast-neutron irradiation of 1.8×10^{18} n/cm^2 ($E > 0.1$ MeV) produced a change of 14.1 $\mu\Omega$ in the measured resistivity at ~ 14.5 K. Isochronal annealing of these irradiated specimens produced 40% recovery of the resistivity change at 100 K and 70% recovery at 290 K. This recovery is seen to agree with that of I_c very closely.

At high fluences there appears to be a saturation in the decrease of I_c. Ambient reactor irradiations [36, 39, 40] in Nb 45 wt. % Ti multifilament wires to fluences in excess of 10^{20} n/cm^2 ($E > 1$ MeV) showed a flattening at $\sim 20\%$ reduction of I_c (measured at 4 T). This is shown in Fig. 13. Plotted for comparison are the 4.2 K reductions for similar Nb–Ti short samples irradiated with 30-GeV protons and plotted as a function of "equivalent" neutrons based upon "damage energy" calculations [40].

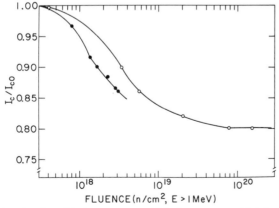

Fig. 13.   Saturation in critical currents at high fluences: ○, high-flux-beam reactor, 373 K, 402 core; ●, 30 GeV protons, 4 K, 379 core, equivalent neutron fluence, $H_\perp = 4$ T.

The low-temperature data do not extend to high enough fluence to determine if and at what level a saturation effect is present for 4.2 K irradiations. The fast decrease of the proton-irradiated samples does, however, reflect again the fact of greater resistivity induced at 4.2 K vis-à-vis room-temperature irradiations, and thus a larger reduction of I_c. The annealing of these proton-irradiated samples has been discussed previously.

IV. Non-A15 Compounds

A. Laves Phases

Several superconducting compounds crystallize with the C15 Laves phase structure. The highest transition temperatures are found in compounds with the cubic $MgCu_2$ (C15) type of structure. Of this class of compounds, HfV_2 and pseudobinaries based on HfV_2, such as $(Hf,Zr)V_2$ have attracted some attention in recent years due to their rather high critical fields, critical currents, and moderate transition temperatures, ~ 13 T, 10^5 A/cm^2, and 10 K, respectively [41]. Tapes have also been prepared, making their use in magnets feasible. These materials are discussed in more detail in the chapter by Dew-Hughes and Luhman, this volume.

The effect of neutron irradiation of HfV_2 has recently been reported [42, 43]. The depression of T_c is greater than that observed for bcc alloys, but less than for other intermetallic compounds. Starting materials with T_c's of ~ 9.4 K had their T_c's decreased to only 8.6 and 8.4 K for fluences of 6×10^{18} and 1×10^{19} n/cm^2, respectively. This change in T_c, amounting to about 10% for a fluence of 1×10^{19} n/cm^2, may be compared to the $\sim 50\%$ decrease observed in Nb_3Al. At present, no information is available concerning the effects of irradiation on the lattice parameter, degree of order, H_{c2}, or J_c in these very interesting materials.

B. Chevrel Phases

The Chevrel phases [44] can be written with the ideal formula $M_xMo_6S_8$ with $0 \le x \le 4$ depending on the size of the metal atom M. The ternary lead molybdenum sulfide with the stoichiometric formula $PbMo_6S_8$ has been reported to have extremely high critical fields, ~ 50 T [45], and reasonable T_c's of ~ 10–13 K [46]. The ideal structure consists of a cubic network of S atoms with the vertices of the Mo_6 octahedra lying on the face centers of the S cubes. Four cubes are loosely linked through a Pb atom, yielding open channels running through the structure parallel to the cube edges [47]. The possibility of fabricating conductors from these materials is discussed in the chapter by Dew-Hughes and Luhman.

Sweedler *et al.* [42] and Brown *et al.* [43] have studied the effects of fast-neutron irradiation on $PbMo_6S_8$, $PbMo_6S_7$, and $SnMo_5S_8$. Irradiation at reactor temperatures of $PbMo_6S_8$ resulted in very large decreases in T_c, from 12.8 K to below 4.2 K for a fluence of $1 \times 10^{19} n/cm^2$. Low-temperature irradiation followed by a room-temperature anneal also yielded large decreases in T_c for $PbMo_6S_7$ and $SnMo_5S_8$. Depressions of 61 and 51% for a fluence of 1.5×10^{19} n/cm^2 ($E > 0.1$ MeV; approximately equivalent to an $E > 1$ MeV fluence of $\sim 6 \times 10^{18}$ n/cm^2), respectively, were observed. No structural studies have been reported for the irradiated Chevrel phases, and the nature of the defects responsible for the large depressions is not currently known. However, considering the sensitivy of T_c of these materials to structural effects [48], the large neutron-induced decreases in T_c are not surprising. As with the Laves phases, no work has been reported on the effects of irradiation on H_{c2} or J_c.

C. NbSe₂

$NbSe_2$, although not of much promise as a magnet material, is interesting from the point of view of superconductivity in layered compounds. These layered compounds consist of two close-packed planes of metalloid, Se in this case, with Nb atoms in between in trigonal prismatic space lattices. The bonding between Nb and Se atoms within the layers is of the strong covalent type while the bonding between layers is of the weak Van de Waals type. The effect of neutron irradiation on the T_c of double hexagonal $NbSe_2(2H-NbSe_2)$ has been determined [42] to a fluence of 3×10^{18} n/cm^2. As with the Chevrel phases, this material is also extremely sensitive to neutron irradiation, a reduction in T_c of almost 50% being observed, compared to the A15s, where T_c at the same fluence is decreased only 20%.

The rapid decrease in T_c for $2H-NbSe_2$ upon irradiation with high-energy neutrons may reflect the sensitivity of T_c to the occupation of the Nb sublattice, as shown by Antonova *et al.* [49] These workers prepared $2H-NbSe_2$ with different compositions, and T_c was determined as a function of changes in the occupation of the Nb and Se sublattice sites as the concentration was varied. A correlation between T_c and the degree of ordering of the Nb sublattice was observed, suggesting that occupation of the Nb sublattice by Nb is important in determining the superconducting properties. The depressions in T_c observed by Antonova *et al.* from ~ 6 to

2 K are similar to these observed for the neutron-irradiated samples, and most likely arise from the same cause, i.e., disruption of the Nb sublattice. In the case of the experiments of Antonova *et al.*, disorder in the Nb sublattice was produced by a change in composition, whereas in the neutron-irradiation study disorder was brought about by displacement collisions resulting from the high-energy irradiation without compositional changes. Again, no data are available for the irradiation effects on H_{c2} or J_c. The fractional decrease in T_c for neutron-irradiated non-A15 compounds is shown in Fig. 14 (the A15 compound Nb_3Ge is included for comparison). Viewed in this way the relative change in T_c for ambient-temperature neutron irradiation is greatest for $NbSe_2$, followed by the Chevrel phases, Nb_3Ge, HfV_2 (Laves phase), and the Nb-based bcc alloys.

V. A15 Compounds

A. *Introduction*

In the alloy superconductors discussed so far, the effects of irradiation on the critical current are either through the generation (or possibly

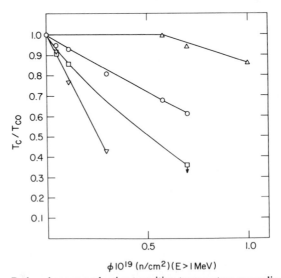

$\phi 10^{19} \, (n/cm^2)(E > 1 \, MeV)$

Fig. 14. Reduced superconducting transition temperature normalized to the unirradiated value T_{c0} as a function of high-energy ($E > 1$ MeV) neutron fluence Φ for various types of superconducting compounds: $\triangle$, HfV_2, Laves phase; $\bigcirc$, Nb_3Ge, A15; $\square$, $Mo_3 Pb_{0.5} Sn_4$, Chevrel phase; $\triangledown$, $NbSe_2$.

destruction) of pinning centers (microstructural changes) or by the alteration of the effectiveness of preexisting pinning centers through electronic properties such as a changing κ. For high fluences in these alloys, changes in T_c of only a few tenths of a kelvin were observed, so that ΔT_c effects on other critical properties were small.

For the compound superconductors, particularly the Nb- and V-base A15 materials, the situation is quite different. Here, T_c is quite sensitive to irradiation and at fluences above $\sim 10^{18}$ n/cm^2 large decreases relative to the unirradiated value, are observed. Thus ΔT_c effects become an important parameter in determining the other superconducting properties.

For the A15 compounds the large decreases in T_c, up to ~ 15 K for Nb_3Ge, are generally accompanied by a decrease in the degree of long-range order and an expanded lattice parameter relative to the unirradiated values. Other defects, such as static displacements of the A-site atoms and highly disordered or amorphous regions, may also result from neutron or heavy-ion irradiation. In addition to the defects in A15 compounds as prepared, the complex nature of the chemical bonding must also be considered. Thus it is extremely difficult to separate out one particular defect as being responsible for the observed effects. Most likely, various defect configurations interact in a complex way to bring about changes in the electronic density of states, which in turn directly affect the superconducting critical parameters.

In the following sections, we discuss how T_c responds to irradiation by neutrons and charged particles, and attempt to elucidate the nature of the resulting defect state by annealing studies and by use of diffraction measurements with x rays and neutrons. Finally we discuss the several models that have been proposed to explain the connection between the defect state and the electronic properties.

B. Transition Temperatures

1. Neutron Irradiation

The A15 compounds are the single most important group of superconducting materials currently available. Their high T_c's (up to ~ 23 K for Nb_3Ge [50]), high H_{c2}'s, and J_c's make them suitable for many applications. Fabrication techniques, described in the chapter by Luhman, this volume, have been developed to prepare these very brittle compounds into wire and tape conductors for use in superconducting magnets and power transmission lines. A wide-scale application of the A15 compounds will be for use as bending magnets in large particle accelerators and for plasma confinement in future fusion power reactors. In such applications, particularly fusion, the magnets will inevitably be exposed to various levels of radiation, depending on the particular application and design. Various estimates [51] indicate total integrated neutron fluences ranging from 10^{18} to 10^{20} n/cm^2 over the lifetime of a fusion reactor. Such high doses will clearly have an effect on the superconducting properties and a great deal of recent research has been devoted to studying this problem.

Swartz and co-workers [25] were the first to irradiate a series of Nb-based A15 compounds with high-energy neutrons at ambient reactor temperatures to a fluence of about 1.5×10^{18} n/cm^2. Small decreases in T_c (~ 0.2 K) were observed for several A15 compounds. Bean and co-workers [52] doped V_3Si and Nb_3Al with fissionable impurities and observed decreases in T_c of 0.22 K and 0.45 K, respectively. Coffey et al. [5] observed a drop of ~ 0.1 K in the T_c of Nb_3Sn irradiated with 15-MeV deuterons at 30 K, and Cooper [53] observed decreases from 0.16 to 0.2 K for neutron irradiations to 2.7×10^{18} n/cm^2 in Nb_3Sn.

More recently, Bett [54] irradiated Nb_3Sn tapes to higher fluences than previous workers ($\sim 5 \times 10^{19}$ n/cm^2), and observed very large depressions in T_c, ~ 13 K. He also noted that T_c could be restored to the unirradiated value by annealing at 900°C. Following Bett's work, Sweedler et al. [55] irradiated various A15 compounds to fluence levels of 5×10^{19} n/cm^2 ($E > 1$ MeV) and found that the large depressions of T_c observed by Bett for Nb_3Sn were also seen in Nb_3Al and Nb_3Ga. A systematic study was undertaken [42, 56–59] and some of the results are summarized in Fig. 15, which shows the decrease in T_c against neutron fluence for Nb_3Ge, Nb_3Al, Nb_3Sn, Nb_3Pt, V_3Si, and Mo_3Os.

The irradiations were carried out at the reactor ambient temperature of ~ 150°C, and the samples were generally in powdered form. For a complete description of the sample preparation, irradiation procedures, and T_c determination, the references listed above may be consulted.

With the exception of Mo_3Os, all of the compounds show very large depressions of T_c, up to ~ 16 K in the case of Nb_3Ge. The rate of depression for the samples shown is largest at low fluences ($\sim 10^{18}$ n/cm^2) and steadily decreases until a leveling off is observed at high fluences ($\sim 3 \times 10^{19}$ n/cm^2). This leveling off, or saturation, appears to be characteristic of all of the systems studied to date, and was first observed by Poate et al. [60] for alpha-particle-irradiated films of Nb_3Ge. The saturation value of T_c is ~ 2–4 K, depending on the particular material.

The effect of irradiation on Mo_3Os is quite different from that on the other A15 superconductors. There is only a small decrease in T_c with increasing fluence, even at quite high irradiation levels ($\sim 1.5 \times 10^{20}$ n/cm^2), where T_c has decreased only $\sim 16\%$ compared to about 60–80% for the other A15 compounds shown. It is believed that this striking dif-

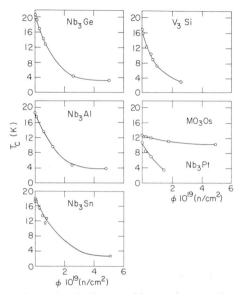

Fig. 15. Midpoints of superconducting transition temperature T_c as a function of high-energy ($E > 1$ MeV) neutron fluence Φ for A15 compounds. Irradiations carried out at reactor temperature ($\sim 150°$C). Solid lines are for guiding the eye.

ference in behavior is due to the fact that Mo_3Os resembles the Nb- and V-based A15 compounds in which the B element is a transition metal, and would accordingly be expected to be less sensitive to the degree of long-range order [61, 62] than for the case where the B element is a non-transition metal. Even though Pt is also a transition metal, there is some evidence [62, 63] that it behaves more like a non–transition metal, such as Al or Ga, when combined with Nb in the Nb_3Pt phase. This may be due to the fact that Pt is at the end of the transition series and behaves "almost" like a non–transition metal when in the A15 structure, whereas Os, with only six d electrons, behaves unambiguously as a transition metal in the A15 phase.

For purposes of comparison with the literature, the data in Fig. 15 are plotted on a semilogarithmic scale in Fig. 16, with T_c normalized to its unirradiated value T_{c0}. While it is possible to display the data in this fashion, some important trends may be masked. The decreasing slope in T_c versus fluence curve with increasing fluence is not evident in the semilogarithmic plot and the suggestion of a plateau is misleading. Although the actual magnitudes of the T_c depression are small at fluences $< 10^{18}$ n/cm², the *rate* of depression is large.

Irradiation with high-energy particles is, of course, only one of several ways in which T_c may be altered. Varying the composition within the A15

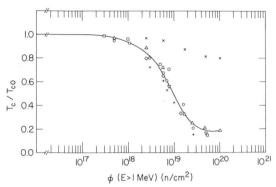

Fig. 16. Reduced superconducting transition temperature normalized to the unirradiated value T_{c0} as a function of high-energy ($E > 1$ MeV) neutron fluence Φ for A15 compounds: O, Nb_3Ge; △, Nb_3Al; ◇, Nb_3Pt; □, Nb_3Sn; ▽, Nb_3Ga; ×, Mo_3Os; +, V_3Si. Data are displayed on a semilogarithmic plot for easy comparison with other published data. Solid line is for visual aid.

phase will also, as a rule, strongly affect T_c. In order to gain a better understanding of the relationship between radiation-induced and compositionally induced changes in T_c, several A15 systems have been studied as a function of both irradiation and composition.

Figure 17 shows the results of such a study for Nb–Al alloys in the A15

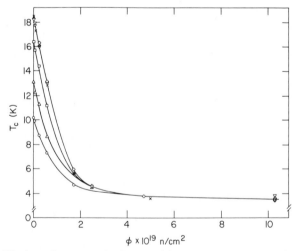

Fig. 17. Midpoints of superconducting transition temperature T_c as a function of high-energy ($E > 1$ MeV) neutron fluence Φ for alloys in the Nb–Al system: O, 18.7 at. % Al; △, 20.2 at. %. Al; □, 21.9 at. % Al; ◇, 23.2 at. % Al; ▽, 24.5 at. % Al; ×, 26 at. % Al; ⊕, 27 at. % Al. (After Sweedler *et al.* [57] and Moehlecke [59].) Solid lines are for guiding the eye.

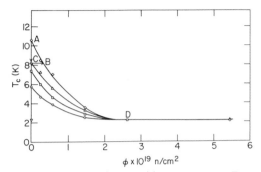

Fig. 18. Midpoints of superconducting transition temperature T_c as a function of high-energy ($E > 1$ MeV) neutron fluence Φ for alloys on the Nb-rich side of stoichiometry in the A15 phase of Nb–Pt: $\Diamond$, 25.1 at. % Pt; $\triangle$, 24.1 at. % Pt; $\square$, 23.1 at. % Pt; $\bigcirc$, 22.1 at. % Pt. (After Sweedler *et al.* [57] and Moehlecke [59].) Letters refer to discussion in text. Solid lines are for guiding the eye.

phase. Several interesting features are observed. As the unirradiated value of T_c (T_{c0}) increases, the initial rate of depression (slope) for the T_c versus fluence curve also increases. T_{c0} increases as we approach the stoichiometric composition, suggesting that T_c is the most sensitive to the structural defects induced by the irradiation at the stoichiometric composition.

This point is more clearly seen in the Nb–Pt system illustrated in Figs. 18 and 19. For this case, it is possible to prepare samples on both sides of stoichiometry, with compositions ranging from 22.1 to 28 at. % Pt [64], which permits the study of the depression of T_c as a function of composi-

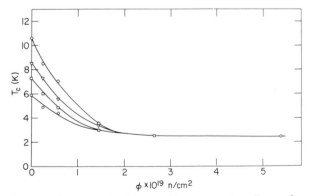

Fig. 19. Midpoints of superconducting transition temperature T_c as a function of high-energy ($E > 1$ MeV) neutron fluence Φ for alloys on the Pt-rich side of stoichiometry in the A15 phase of Nb–Pt: $\Diamond$, 25.1 at. % Pt; $\triangledown$, 26 at. % Pt; $\square$, 27 at. % Pt; $\bigcirc$, 28 at. % Pt. (After Sweedler *et al.* [57] and Moehlecke [59].) Solid lines are for guiding the eye.

tion and irradiation on either side of stoichiometry. We note that both the highest T_c and the greatest rate of T_c depression occur at the stoichiometric composition, as is also seen from the T_c versus composition curves shown in Fig. 17. The trend of increasing initial rates of T_c depression with increasing T_c is observed not only within a given A15 system, but also in the case of different Nb-based A15 compounds. This is seen in Fig. 20, where the initial slopes are shown as a function of T_c for several such compounds. It is seen that as T_c increases, so does the initial depression rate. The significance of these observations will be discussed in Section V,B,6.

It should be noted that V_3Ga, another A15 compound that has its maximum T_c at the stoichiometric composition, does not follow the foregoing trend upon irradiation. Francavilla $et\ al.$ [65] have measured T_c for irradiations up to $\sim 10^{19}\ n/cm^2$ ($E > 1$ MeV) for three samples, encompassing both sides of stoichiometry, and found an increasing slope as a function of increasing Ga content, in contrast to the Nb–Pt results. This difference is not understood at present.

We also note that in Fig. 17, the data for Al compositions around 24 at. % fall on essentially the same curve. This is because the Al-rich phase boundary of the A15 phase occurs at approximately this point. Furthermore the saturation value of ~ 3.5 K is seen not to depend on the initial value of T_c.

Bauer $et\ al.$ [66] have also observed large T_c depressions for the A15 compounds Nb_3Sn and Nb_3Al when doped with fissionable U^{235} and B^{10} and irradiated with thermal neutrons. Irradiations with fast neutrons ($E > 1$ MeV) were also carried out. T_c decreases of 6.4 and 5.18 K

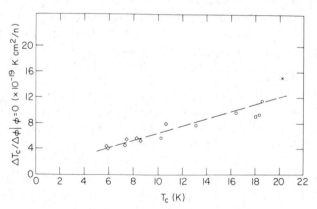

Fig. 20. Initial depression rate of the superconducting transition temperature $\Delta T_c/\Delta\Phi|_{\Phi=0}$ as a function of T_c for neutron-irradiated Nb-based A15 compounds: ×, Nb_3Ge; ○, Nb–Al; ◇, Nb–Pt; □, Nb_3Sn. (After Moehlecke [59].)

were observed for Nb_3Sn and Nb_3Al, respectively. $Nb_3(Al,Ge)$ was also irradiated to a fast-neutron fluence of 5×10^{18} n/cm^2, with a resultant decrease in T_c of 5.1 K, and doped $Nb_3(Al,Ge)$ showed a T_c drop of 7.2 K. The T_c of V_3Si was decreased by 2.7 K for a fast-neutron fluence of 5×10^{18} n/cm^2. Voronova et al. [67] also doped Nb_3Sn with fissionable U^{235} and observed decreases similar to those noted by Bauer et al.

Söll et al. [68] have carried out neutron irradiation at low temperatures (10 K) for Nb_3Sn and have also observed the large T_c decreases observed for ambient-temperature irradiations noted above. Decreases of ~ 3.5 K were seen for a fluence of 10^{19} n/cm^2 ($E > 0.1$ MeV). Significantly, very little recovery of T_c was seen upon annealing to room temperature, indicating that the defects responsible for the depression of T_c are stable to at least this point. They are in fact stable to temperatures considerably higher than room temperature, as discussed in more detail in Section V,B,3.

At fluences below 10^{18} n/cm^2, Söll et al. observed little or no change in T_c for Nb_3Sn. Viswanathan et al. [69] also observed no change in T_c for fluences below 10^{18} n/cm^2 on certain samples of V_3Si but noted that this effect was strongly sample dependent. Brown et al. [43], however, did not observe this "plateau" region in the T_c versus fluence curve for their low-temperature neutron irradiation of Nb_3Al, and Wiesmann et al. [70] have suggested, based on their alpha-particle irradiations of Nb_3Sn and Nb_3Ge, that this behavior may be due to the initial defect state of unirradiated material.

Couach et al. [71] reported low-temperature neutron irradiation results for V_3Ga irradiated to fluences of 3×10^{18} n/cm^2. A drop in T_c of about 2 K was observed, which is similar to that reported for the irradiations of Francavilla et al. [65].

2. CHARGED-PARTICLE IRRADIATION

Besides exposure to high-energy neutron irradiation, the A15's have also been irradiated with energetic charged particles, including alpha particles, heavy ions such as oxygen and argon, electrons, deuterons, and protons. Broadly speaking, the effect of charged-particle irradiations on T_c is similar to that observed for neutron irradiation. Large depressions of T_c have been observed with alpha particles [60, 72] and oxygen ions [73], with the same saturation value of T_c of ~ 2–4 K as observed for neutron irradiation. Smaller reductions in T_c have been observed for electron [74], deuteron [5], and proton [36, 67] irradiations.

The most extensively used charged-particle irradiations have been carried out with 2-MeV He^4 particles on a variety of A15 compounds. The resulting effect on T_c is shown in Fig. 21, after Poate *et al.* [72]. The same general trend is seen as previously noted for the neutron irradiations (see Fig. 16), i.e., substantial decreases in T_c at low fluences with a stauration value of ~ 1–3.5 K, depending on the sample. Although the T_c behavior for alpha-irradiated A15 films is similar to that of neutron-irradiated A15 bulk material, there have been differences observed as regards structural and lattice parameter changes. This point will be more fully discussed at the end of this section, where the various models are described.

We also note that alpha-particle irradiations carried out at low tempera-

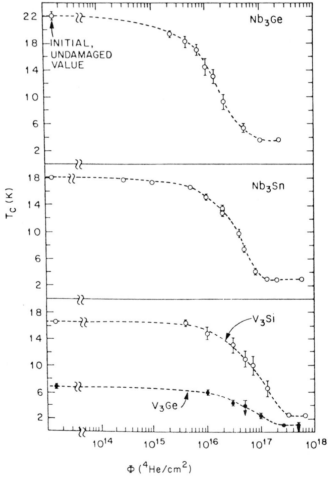

Fig. 21. Superconducting transition temperature T_c as a function of ^{4}He fluence ($E = 2$ MeV) for some A15 compounds. (After Poate *et al.* [72].)

tures also show large reductions in T_c [74]. However, the behavior observed by Ghosh *et al.* at low fluences is somewhat different from that seen by Poate *et al.* In the former case, a "plateau" region, where little or no change in T_c occurs, exists at low fluences for Nb_3Sn, similar to that observed for neutron-irradiated single-crystal V_3Si by Viswanathan *et al.* [69]. This plateau region was not observed for Nb_3Ge, which, however, had a higher residual resistivity than Nb_3Sn, suggesting that the initial defect state of the unirradiated sample may determine the initial rate of T_c depression. However, more work will be needed on well-characterized samples before a definitive statement can be made.

3. ANNEALING STUDIES

An important feature of the radiation-induced changes is their reversibility upon annealing. All of the superconducting and normal-state properties, such as T_c, the lattice parameter, degree of order, and crystallinity, assume essentially their preirradiated values following annealing of the irradiated material. This reversibility, or the elimination of the radiation-induced defects, has made irradiation an extremely useful probe in understanding the role of defects and instabilities in the A15 compounds.

Below we discuss various annealing studies that have been reported on the A15 compounds. We should bear in mind that interpretation of annealing studies is difficult due to the complex nature of the A15 compounds, the relative absence of data concerning diffusion in these materials, and the complex nature of the radiation-induced defects. A model for the mechanism of recovery upon annealing has recently been suggested [75].

Several parameters need to be considered in discussing annealing experiments. First, since the number of induced defects depends upon the fluence, the latter is clearly an important variable. Second, the initial defect state as manifested by the resistivity, T_{c0}, lattice parameter, and order parameter must also be considered. Third, the temperature and time of the anneal are external variables that may be varied, and that in combination yield useful information concerning activation energies and shed some light on the nature of the defects.

Several systems have been investigated, including Nb_3Ga [58], Nb_3Sn [58, 76], Nb_3Al [57], Nb_3Ge [57, 77], Nb_3Pt [59], V_3Si [57], V_3Ga [65, 78], and Mo_3Os [59]. Annealing effects as a function of initial composition, fluence, time, and temperature have been reported. Figures 22 and 23 show the results for isothermal anneals at 750°C for Nb_3Ga and Nb_3Sn, respectively. In both cases, T_c recovers to its preirradiated value, but considerably longer times are required for Nb_3Sn than for Nb_3Ga. This has

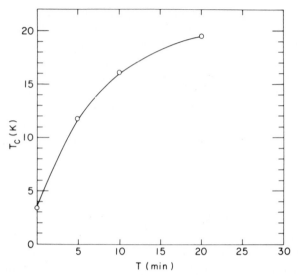

Fig. 22. Isothermal annealing of Nb₃Ga ($T = 750°C$) after neutron irradiation to a fluence of 5×10^{19} n/cm^2 ($E > 1$ MeV). T_{c0} was 20.3 K, irradiation temperature $\sim 150°C$, annealing temperature 750°C. (After Sweedler *et al.* [58].)

been attributed to the difference in radius ratio of Sn and Ga with respect to Nb, resulting in a greater driving force for reordering in the case of Nb₃Ga [79]. This greater driving force is reflected by a shorter time required for reordering and therefore a faster T_c recovery time. The results of isothermal annealing studies on Nb₃Ge, prepared by chemical vapor deposition and irradiated to a fluence of 2.6×10^{19} n/cm^2, are shown in Fig. 24, where T_c is plotted against $t^{1/2}$. We see that over a considerable

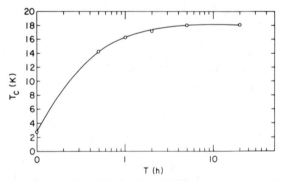

Fig. 23. Isothermal annealing of Nb₃Sn ($T = 750°C$) after neutron irradiation to a fluence of 5×10^{19} n/cm^2 ($E > 1$ MeV). T_{c0} was 18.1 K, irradiation temperature $\sim 140°C$, annealing temperature 750°C. (After Sweedler *et al.* [58].)

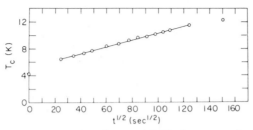

Fig. 24. Isothermal annealing of Nb$_3$Ge ($T = 550$°C) after neutron irradiation to a fluence of 2.6×10^{19} n/cm^2 ($E > 1$ MeV). T_c is plotted as a function of $t^{1/2}$. (After Sweedler *et al.* [77].)

time interval at 550°C, a $t^{1/2}$ dependence is followed, which is suggestive of a diffusion process during reordering rather than the diffusing species being short-circuited by dislocations or grain boundaries [79]. T_c for chemically vapor-deposited and irradiated Nb$_3$Ge is completely recovered following a 30-d anneal at 625°C to its preirradiated value of 21.7 K [59].

Isochronal anneals have also been carried out on neutron-irradiated Nb$_3$Ge for different fluences, as shown in Fig. 25 [57]. Very little recovery of T_c is seen up to temperatures of ~450°C, an indication of the stability of the related defect. It is also interesting to note that this is independent of the fluence and hence the number of induced defects. Above this temperature, rapid recovery begins and is close to completion by 700–800°C. T_c begins to drop due to precipitation of second-phase material and an accompanying irreversible shift in composition to the Nb-rich region of the A15 phase [77].

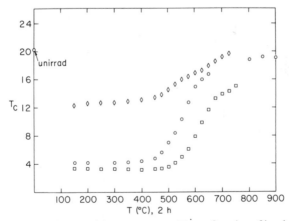

Fig. 25. Superconducting transition temperature T_c as a function of isochronal annealing temperature for neutron-irradiated Nb$_3$Ge: ◇, 7.0×10^{18} n/cm^2; ○, 2.6×10^{19} n/cm^2; □, 5.0×10^{19} n/cm^2.

The effect of annealing on the T_c of neutron-irradiated Nb–Pt alloys is quite different from the behavior described above for Nb_3Ge, Nb_3Sn, and Nb_3Ge. This is interesting in that the dependence of T_c on irradiation and composition is very similar to that of other Nb-based A15 compounds. Figure 26 shows the effect of a 2-h isochronal anneal up to 700°C for two alloys in the Nb–Pt system containing 25.1 and 23.1 at. % Pt, respectively [59]. We note that even though T_c is severely degraded to the saturation value of ~3 K, there is very little recovery of T_c, even up to 700°C. This is in marked contrast to the previously discussed Nb-based A15 compounds, where rapid recovery began to occur at ~450°C and complete

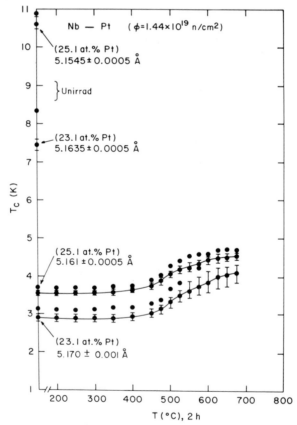

Fig. 26. Superconducting transition temperature T_c as a function of isochronal annealing temperature for two Nb–Pt alloys neutron irradiated to a fluence of 1.44×10^{19} n/cm^2 ($E > 1$ MeV). Data points represent successive 2-h anneals at each temperature, the upper circles corresponding to transition onsets and the bars indicating transition widths (10–90%). Lattice parameters are also shown. (After Moehlecke [59].)

recovery was attained by about 700–800°C, depending on the system. Thus, the radiation-induced disorder is considerably more stable in the Nb–Pt system than in the Nb-based materials where the B atom is a non–transition metal.

This is consistent with the concept of Nb_3Pt being of an intermediate nature between an A15 compound in which the B atom is, respectively, a transition or non–transition metal. That is, although Pt is a transition metal, when combined in the A15 phase it behaves more like a non–transition metal with respect to many of its properties (T_c versus composition, T_c versus fluence, homogeneity range of the A15 phase). However, with respect to its annealing behavior, it behaves more like an A15 compound such as Mo_3Os. This is seen in Fig. 27, where the recovery behavior under isochronal annealing conditions for neutron-irradiated Mo_3Os is shown [59]. As for the Nb–Pt system, we see very little recovery of T_c up to 700°C, again attesting to the remarkable stability of the radiation-induced disorder in these materials when compared to the Nb-based A15 materials containing B atoms of the s–p element.

Another interesting case is V_3Si, whose annealing behavior is different from that of the two cases discussed above. Figure 28 shows the

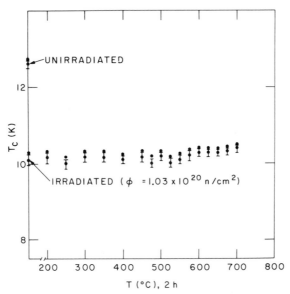

Fig. 27. Superconducting transition temperature T_c as a function of isochronal annealing temperature for Mo_3Os neutron irradiated to a fluence of 1.03×10^{20} n/cm^2 ($E > 1$ MeV). Data points represent successive 2-h anneals at each temperature, the upper circles corresponding to transition onsets and the bars indicating transition widths (10–90%). (After Moehlecke [59].)

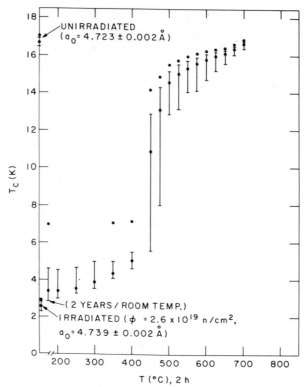

Fig. 28. Superconducting transition temperature T_c as a function of isochronal annealing temperature for V_3Si neutron irradiated to a fluence of 2.6×10^{19} n/cm^2 $(E > 1\ MeV)$. Data points represent successive 2-h anneals at each temperature, the upper circles corresponding to transition onsets and the bars indicating transition widths (10–90%). (After Moehlecke [59].)

isochronal annealing curves for V_3Si irradiated to a fluence of 2.5×10^{19} n/cm^2 [59]. In this material, recovery of T_c begins at relatively low temperatures, $\sim 250°C$, and is essentially complete by $\sim 550°C$. The region of the most rapid recovery, 450–500°C, is considerably lower than for Nb_3Al or Nb_3Ge, where the corresponding region is about 500–700°C. Thus, for V_3Si, defects associated with the drop in T_c appear to be quite mobile compared to the Nb-based materials (viz., a lower self-diffusion energy for the V compound).

Other interesting features are observed in the isochronal annealing curves of the Nb–Al A15 alloys shown in Fig. 29. It is seen that for off-stoichiometric alloys, T_c after annealing has been enhanced with respect to its preirradiated value. This increase in T_c is about 1 and 0.3 K for the 20.2 and 23.2 at. % Al sample, respectively [57]. In order to be sure

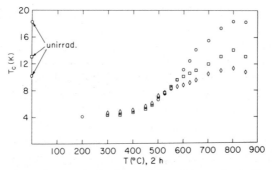

Fig. 29. Midpoints of superconducting transition temperatures T_c as a function of isochronal annealing temperature for Nb–Al alloys neutron irradiated to a fluence of 2.6×10^{19} n/cm^2 ($E > 1$ MeV): $\bigcirc$, 23.2 at. % Al; $\square$, 20.2 at. % Al; $\diamond$, 18.7 at. % Al. Data points represent successive 2-h anneals. (After Sweedler *et al.* [57] and Moehlecke [59].)

that the observed enhancement was indeed due to the irradiation, an unirradiated sample was given an identical treatment, with no observed increase in T_c. The enhanced T_c is believed to be due to a process similar to radiation-enhanced diffusion. One way in which this could occur is for the second (Nb-rich) phase that is present prior to the irradiation to be disordered by the radiation just as is the A15 phase. As the temperature is raised, only the A15 phase is reordered until the phase boundary with the Nb-rich phase is crossed (in Nb–Al ~800°C for 20% Al). The precipitation of this phase (Nb_5Al_3) pushes the A15 composition further from stoichiometry. This results in a turning back down of the T_c toward the unirradiated value for annealing temperatures above 800°C.

A similar situation has been observed in neutron-irradiated V_3Ga samples by Francavilla *et al.* [65, 78] and is shown in Fig. 30. These workers observed that a temperature of only 450°C was required to fully

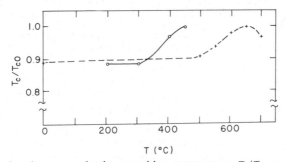

Fig. 30. Reduced superconducting transition temperature T_c/T_{c0} as a function of isochronal annealing temperature for V_3Ga. Open circles, V 25.6 at. % Ga neutron irradiated to a fluence of 0.6×10^{19} n/cm^2 ($E > 1$ MeV), annealing time 4 d; crosses, as-cast sample, 24.3 at. % Ga, annealing time 7 d. (After Francavilla *et al.* [78].)

order a V_3Ga sample that had been irradiated to a fluence of 6×10^{18} n/cm^2, whereas 650°C would normally have been needed to achieve complete ordering for an unirradiated sample of similar composition. Presumably, some form of radiation-enhanced diffusion accounts for the lower ordering temperature.

4. LATTICE PARAMETER

Until now we have considered only the changes in the critical currents, critical field, and transition temperature due to the radiation-induced defects, without inquiring too deeply into the nature of the defects or their effect on the crystal lattice. Clearly, in order to better understand the effect on the superconducting properties brought about by the irradiation, the changes in lattice properties will be important. In this section we discuss the changes in lattice parameter a_0 and in Section V,B,5, the effects of order are considered.

The most extensive lattice-parameter studies have been carried out as a function of neutron irradiation [54, 57, 59, 66]. Similar studies of the effects of alpha-particle irradiation on thin films have also been reported [72]. It has generally been found that the A15 cell dimension a_0 expands upon irradiation regardless of the type of incident radiation.

Typical lattice expansions for neutron-irradiated A15 compounds are shown in Figs. 31 and 32. We note that the rate of lattice-parameter increase as a function of fluence is greater at low fluences and begins to level off at higher fluences. Also, the magnitude of the increase can be quite large. Bett [54] has reported increases in a_0 for Nb_3Sn up to 1% for irradiation levels of 9×10^{19} n/cm^2, and Skvortsov et al. [76] observed a 1.25% expansion for Nb_3Sn irradiated to an equivalent dose of $\sim 10^{22}$ n/cm^2, while Bauer et al. [66] noted a 0.5% increase for a fluence of 5×10^{18} n/cm^2.

For other A15 compounds, the magnitude of the expansion varies. Nb_3Ge (Fig. 32) increases 1.0%, while Nb_3Al showed an increase of 0.3% for an irradiation of 5×10^{19} n/cm^2. Mo_3Os and Nb_3Pt show the smallest relative expansions, 0.16% and 0.18% for fluences of 5×10^{19} and 3×10^{19} n/cm^2, respectively.

Thus, the defects introduced by irradiation affect the lattice in such a manner as to bring about a rather large increase in the size of the A15 unit cells relative to the unirradiated value. However, no accompanying change in the basic integrity of the structure is apparent in the majority of compounds at irradiation levels where the most drastic changes in T_c occur. The A15 structure is maintained, with little or no evidence for line broadening, the appearance of new phases, or a change to an amorphous

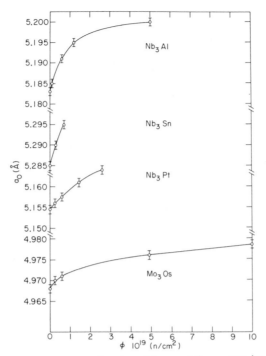

Fig. 31. Lattice parameter a_0 as a function of fluence Φ for neutron-irradiated A15 compounds.

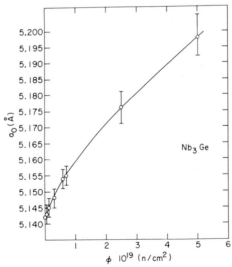

Fig. 32. Lattice parameter a_0 as a function of fluence Φ for neutron-irradiated Nb_3Ge. (After Moehlecke [59].)

structure for Nb_3Al, Nb_3Pt, Nb_3Si, V_3Si, and Mo_3Os to fluences of up to 2×10^{19} n/cm^2.

A typical example of the stability of the A15 phase under irradiation is shown in Fig. 33, which shows Debye–Scherrer x-ray patterns for neutron-irradiated Nb_3Al. Note that all the lines of the A15 phase are still visible and quite sharp even in the highly irradiated material. Figure 34 is a more quantitative indication of the structural stability, showing the absence of line broadening for irradiated Nb_3Pt relative to the unirradiated sample. A similar degree of stability has also been observed in off-stoichiometric Nb–Pt alloys and Nb–Al alloys.

Nb_3Ge, however, behaves somewhat differently from the other A15 compounds with regard to its crystal structure after irradiation. Figure 35 shows a portion of the x-ray diffraction pattern for unirradiated and irradiated Nb_3Ge before irradiation and after irradiation to a fluence of 5×10^{19} n/cm^2. The sample was prepared by chemical vapor deposition [80] and, as seen in the diffraction pattern, is predominantly A15, but contains a number of impurity phases. Although the pattern is complex every line can be accounted for. Upon irradiation to a fluence of 5×10^{19} n/cm^2, the diffraction peaks corresponding to the A15 phase are seen to be greatly diminished in intensity, whereas the impurity phases, NbO, hexagonal Nb_5Ge_3, and tetragonal Nb_5Ge_3, are still present. A large diffuse background, not present in the diffraction pattern of the unirradiated

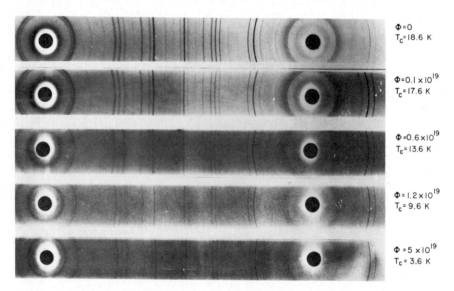

$\Phi = 0$
$T_c = 18.6$ K

$\Phi = 0.1 \times 10^{19}$
$T_c = 17.6$ K

$\Phi = 0.6 \times 10^{19}$
$T_c = 13.6$ K

$\Phi = 1.2 \times 10^{19}$
$T_c = 9.6$ K

$\Phi = 5 \times 10^{19}$
$T_c = 3.6$ K

Fig. 33.   Debye–Scherrer x-ray patterns for Nb_3Al neutron irradiated to various fluences. (After Sweedler and Cox [81].)

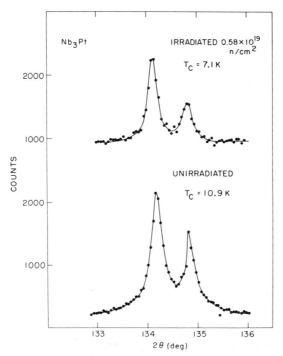

Fig. 34. X-ray (Cu–Kα) diffractometer scan of the (611) reflection from Nb_3Pt. Bottom, unirradiated; top, irradiated to a fluence of 0.58×10^{19} n/cm^2 ($E > 1$ MeV).

sample, is clearly visible in that of the irradiated one, as indicated by the broken line. The regularity of the A15 lattice is thus considerably reduced and the pattern is more indicative of an amorphous or microcrystalline phase than a crystalline structure. This is in marked contrast to the other A15 compounds, as noted above. However, the preirradiated lattice parameter and A15 structure can readily be restored by appropriate annealing.

The lesser stability of the Nb_3Ge crystal structure to irradiation is undoubtedly related to the metastable nature of the A15 phase. Stoichiometric single-phase material is very difficult or impossible to prepare and metastable techniques, such as rf sputtering or coevaporation, are required. Near-stoichiometric phases are unstable above $\sim 1000°C$, irreversibly decomposing to a Nb-rich A15 phase and tetragonal Nb_5Ge_3 [77]. We also note that Nb_3Ge has the largest initial depression rate of T_c upon irradiation (Fig. 15), indicating the high degree of sensitivity of T_c to small changes in lattice structure. In view of the foregoing discussion, it is not surprising that the structural stability of Nb_3Ge is more sensitive to the in-

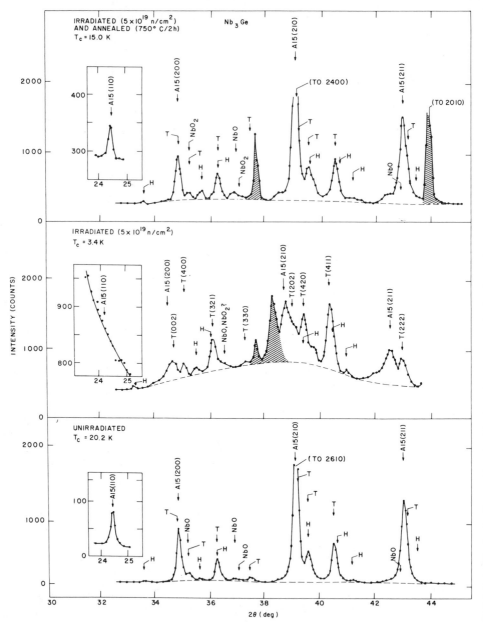

Fig. 35. Part of the x-ray (Cu–Kα) diffractometer scan for Nb_3Ge. Bottom, unirradiated; middle, irradiated to a fluence of $5.0 \times 10^{19}\ n/cm^2\ (E > 1\ MeV)$; top, irradiated to a fluence of $5.0 \times 10^{19}\ n/cm^2$ and annealed at 750°C for 2 h. Lines are identified as A15, hexagonal (H) and tetragonal (T) Nb_5Ge_3, NbO_2, and NbO. The Al lines (shaded peaks) come from the sample holder. Note the greatly diminished intensity of the A15 peaks in the irradiated sample and the appearance of a broad diffuse peak indicated by the broken line.

troduction of radiation-induced defects than in the case of the other A15 compounds.

Although the behavior of T_c is similar with neutron or charged-particle irradiation, the effect on the lattice appears to be somewhat different. Poate *et al.* [60, 72] have noted a diminishing of the intensity of the high-angle lines on x-ray patterns for alpha-irradiated A15 films, a feature not observed in the case of neutron irradiation. At present, the origin of this difference is not understood.

The lattice parameter returns to its preirradiated value upon annealing, but the variation as a function of T_c does not necessarily follow the same path. This is illustrated by some data of Francavilla *et al.* [65] for V₃Ga shown in Fig. 36. Annealing at low temperatures (180–360°C) leads to substantial recovery of a_0 but little recovery in T_c. Subsequent annealing at 400°C and 475°C produces a large recovery of T_c but virtually no change in a_0. It is suggested that the low-temperature stage involves the move-

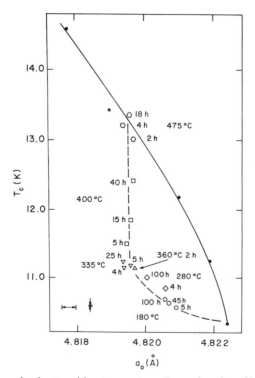

Fig. 36. Superconducting transition temperature, T_c as a function of lattice parameter a_0 for V₃Ga (24.9 at. % Ga) after neutron irradiation to successively higher fluences (solid line) followed by annealing at successively higher temperatures (broken lines). Note the two distinct recovery stages. (After Francavilla *et al.* [65].)

ment of interstitial loops, while in the second stage, vacancies become mobile and the process of reordering occurs. It may be inferred from this work that correlations between a_0 and T_c must be regarded with caution.

5. ORDER PARAMETER

In order to gain further insight into the nature of the radiation-induced defects in the A15 compounds, several diffraction studies have been carried out to determine the degree of long-range order (LRO).

The degree of long-range order is conveniently expressed by the Bragg–Williams order parameter S, which may be written in a generalized form as

$$S_A = (P_a - \Gamma_a)/(1 - \Gamma_A) = 1 - 4x/3(1 - y) \qquad (10)$$

and

$$S_B = (P_B - \Gamma_B)/(1 - \Gamma_B) = 1 - 4(x + y)/(3 + y), \qquad (11)$$

where $P_A = \frac{1}{3}(3 - x)$ is the fraction of A sites occupied by A atoms, $P_B = 1 - x - y$ the fraction of B sites occupied by B atoms, $\Gamma_A = \frac{1}{4}(3 + y)$ and $\Gamma_B = \frac{1}{4}(1 - y)$ is the fraction of A and B atoms in the alloy, respectively. The parameters x and y are related to the site occupation and composition, respectively, in the generalized A15 phase, which may be written as $(A_{3-x}B_x)[B_{1-x-y}A_{x+y}] = A_{3+y}B_{1-y}$, where the round and square brackets represent the A and B sites, respectively. At the stoichiometric ratio of $3A:1B$, $y = 0$ and the order parameters reduce to the more familiar definition with $S_A = S_B$. For off-stoichiometric compositions, ($y \neq 0$) $S_A \neq S_B$ and two order parameters are needed to completely describe the state of order for the A15 phase. Another way to describe the degree of site occupation is to consider the total number of "wrongly occupied sites," which incorporates both A and B sites and which can be expressed as the fraction of atoms not on their normal lattice sites. For the generalized A15 phase this is equal to $\frac{1}{4}(2x + y)$.

Order-parameter measurements have been reported for neutron-irradiated Nb_3Al [81], Nb_3Ge [57], Nb_3Sn [82], Nb_3Pt [59], and V_3Si [83]. In each case, a reduction in the LRO was observed as a function of increasing fluence, and hence decreasing T_c. The diffraction results are well accounted for by a simple site-exchange model, which is characterized by the order parameter S. Site-exchange disorder, where A and B atoms exchange lattice sites, has also been referred to as "antistructure" disorder.

In order to extract an order parameter from diffraction data, the intensities of the Bragg peaks of powdered or single-crystal specimens must be accurately measured and then compared with calculated values. In the literature this is often reported on the basis of the first few low-angle peaks or even the ratio of two adjacent peaks. However, a much more systematic and reliable procedure is to carry out a least-squares analysis of all the observable data with appropriate variable parameters, which include the order and parameter and a Debye–Waller temperature factor. This procedure is not restricted to single-phase samples. For further details the reader may consult Sweedler and Cox [81] and Cox *et al.* [84].

In the A15 structure, with the space group $Pm3n(O_h^3)$, certain Bragg peaks are more sensitive to changes in the order parameter than are others. These peaks have intensities that involve the square of the difference of the scattering factors or amplitudes of the A and B atoms. Since for x rays the scattering factors are roughly proportional to the atomic number Z, the greater the difference in Z between two elements, the easier it will be to observe those peaks that are sensitive to the degree of order, the so-called difference or superstructure peaks. For the A15 structure these difference lines are the (110), (220), (310), (411), (422), (510), (530), and (620) reflections, for example. Since the intensity of the diffracted peak also falls off as a function of $(\sin\theta)/\lambda$, where 2θ is the diffraction angle and λ the incident wavelength, the low-angle difference peaks are generally stronger than the high-angle peaks. The (110) reflection is usually the strongest and is particularly useful when the difference in Z is small. For Nb_3Sn and Nb_3Ge, where this is the case, this may be the only observable difference peak. However, for Nb_3Al and V_3Si a number of difference peaks can be observed, and it is easy to get a qualitative estimate of the degree of order by visual inspection.

The first comprehensive study of this type was a neutron-diffraction investigation of neutron-irradiated Nb_3Al [81]. Neutron-diffraction data have the advantage that there is no angular falloff of the scattering amplitudes, and preferred orientation and particle size effects that can be very troublesome in x-ray studies are usually negligible. Furthermore, the neutron counters are not sensitive to the high γ-ray background from the irradiated samples. However, much larger samples are required, and this can pose handling problems. A portion of the diffraction pattern from Nb_3Al is shown in Fig. 37 for unirradiated and irradiated material. As noted above, the (110) reflection is particularly sensitive to the degree of order and it is clearly seen that upon successive irradiations, the intensity of this peak is decreased relative to the unirradiated value. The intensities of the (200) and (211) reflections are much less sensitive to disorder, and their ratio is independent of the order parameter. A least-squares analysis of the complete diffraction patterns yielded the long-range order parameters shown in Fig. 38.

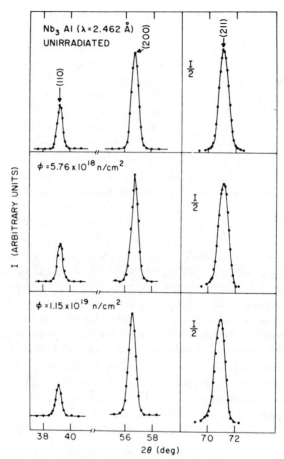

Fig. 37. Part of the neutron diffraction scans for Nb_3Al as a function of neutron fluence. Note the decrease in the intensity of the (110) peak, corresponding to a decrease in order parameter S. (After Sweedler and Cox [81].)

One feature of these least-squares analyses is that the order parameter is strongly correlated with the assumed composition [85]. In the analysis reported in Sweedler and Cox [81], a slightly Al-rich composition $Nb_{74}Al_{26}$ was assumed. However, recent studies of the Nb–Al system [59, 86] show that the maximum equilibrium Al composition lies between 24 and 25 at. % Al. The order parameters shown in Fig. 38 have accordingly been recalculated on the basis of the stoichiometric composition Nb_3Al. This has the effect of increasing each of the reported values of S by 0.03.

It should be pointed out that x-ray diffraction measurements might equally well be interpreted on the basis of vacancy formation rather than

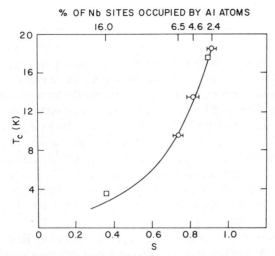

Fig. 38. Superconducting transition temperature T_c as a function of order parameter S for neutron-irradiated Nb_3Al. Also shown is the fraction of Nb sites occupied by Al atoms. The open circles correspond to directly measured values of S, and the squares represent values of S derived from Aronin's expression as described in text. The solid line depicts the relationship $T_c = T_{c0} \exp[-3.2(1 - S/S_0)]$, where $T_{c0} = 18.6$ K and $S_0 = 0.915$. (After Sweedler and Cox [81], but note slight change in assumed composition as discussed in text.)

of antistructure defects. Cox *et al.* [84] have discussed this possibility and concluded that the concentration of vacancies necessary to explain the observed diffraction intensities would be several percent, an unreasonably large figure. Also, direct density measurements on neutron-irradiated single-crystal V_3Si yielded a maximum vacancy content of less than 0.5% [83].

It can be seen from the figure that there is a small amount of disorder present in the unirradiated sample. This implies that the low-temperature anneal was not completely effective in ordering the arc-melted sample [87].

From the three measured values of S, it is possible to use the Aronin relationship [88] between S and fluence Φ, $S = S_0 e^{-k\Phi}$, where S_0 is the initial order parameter and k is a proportionality constant. This gives a value for k of 1.86×10^{-20} cm^2/n, which has been used to calculate S for the two points represented by squares in Fig. 38.

The initial depression of T_c with site-exchange disorder is roughly linear, but T_c is approaching its saturation value when S has fallen to slightly below 0.4 ($\sim 5 \times 10^{19}$ n/cm^2). The solid line in Fig. 38 corresponds to an exponential dependence of the type $T_c = T_{c0} \exp[-\alpha(1 - S/S_0)]$, where T_{c0} is the preirradiation value of T_c (18.6 K) and α is a constant proportional to the initial rate of T_c depression, and is equal to 3.2. Order measurements to considerably higher fluence of $\sim 10^{22}$ n/cm^2 for Nb_3Sn yielded an $S = 0$ with a T_c in the 3–4 K range [76], indicating that for low values of the order parameter, T_c is relatively insensitive to changes in S. For S close to 1, however, the sensitivity of T_c to small changes of order is quite high, being about -2.0 K/% of Nb sites occupied by Al atoms, or alternatively -1.3 K/% of wrongly occupied sites. This is seen to be a rather large value when it is recalled that nonmagnetic impurity atoms tend to depress T_c only a few tenths of a kelvin/atomic percent. If the small amount of residual disorder in Nb_3Al could somehow be eliminated, one might expect a significantly higher T_c. It is tempting to extrapolate the T_c versus S curve to perfect order, i.e., $S = 1$, to obtain an estimate of T_c for fully ordered material. A linear extrapolation yields a T_c of ~ 22 K. However, it must be emphasized that there is no *a priori* justification for such an extrapolation. In fact, there is reason to believe that one would encounter a plateau region close to $S = 1$, as inferred from data for Nb_3Sn [68, 70] and V_3Si [69] described in Section V,B,1.

The disorder present in the unirradiated material is believed to promote greater stability of the A15 phase relative to the σ phase and is a frequently encountered feature of the high-T_c A15 compounds. It is possible that the enhanced stability relative to the perfectly ordered A15 phase originates from the increased lattice parameter resulting from the larger Nb atoms occupying the smaller Al sites (antisite disorder). This increased lattice parameter leads to a greater Nb–Nb separation along the A chains, decreasing the repulsive interaction term and thereby increasing the stability. It is interesting to note that Nb_3Sn has a considerably larger lattice parameter (5.29 Å) than those of the other high-T_c Nb-based A15 compounds, which range between ~ 5.14 and 5.18 Å. This results from the large Sn radius (1.45 Å) compared to Al (1.39 Å) and Ge (1.36 Å) [89] and leads to Nb_3Sn being quite stable. In fact, Nb_3Sn is the only stable compound above 1200°C in the Nb–Sn phase diagram [90] and also appears to be fully ordered at the stoichiometric composition, unlike Nb_3Al or Nb_3Ge.

The same trend observed for Nb_3Al is also seen for neutron-irradiated Nb_3Ge. Figure 39 shows a portion of the x-ray diffraction pattern for Nb_3Ge unirradiated and irradiated to a fluence of 2.5×10^{19} n/cm^2. Again, we see a reduction in the intensity of the (110) peak upon irradiation, indicating a decrease in the degree of order. In the case of Nb_3Ge, additional interesting features are observed. The unirradiated sample was prepared by a chemical vapor deposition process [80] and the presence of impurity phases is evident in the diffraction pattern. The H and T refer to the hexagonal and tetragonal forms of the primary impurity phase,

Nb_5Ge_3. Upon irradiation at reactor temperature, the relative amount of impurity phase has increased somewhat, due most likely to radiation-enhanced precipitation process. An increased amount of Nb_5Ge_3 in the irradiated sample could cause a slight shift in the composition of the A15 phase to the Nb-rich side of the phase field. However, this shift would be less than 1 at. % and could not account for the large drop in T_c.

Another interesting feature in Fig. 39 is that the A15 lines of the irradiated material are only slightly broadened with respect to those of the unirradiated sample. This indicates that, to this fluence, the basic integrity of the A15 structure is retained. However, as previously noted, at 5×10^{19} n/cm^2 (Fig. 35) the lattice is much less crystalline, a feature not observed for the other A15 compounds for this range of fluences.

The dependence of T_c or S for neutron-irradiated Nb_3Ge is shown in Fig. 40. The same general trend is observed as in the case of Nb_3Al (Fig. 39), i.e., a roughly linear initial drop and a leveling off at low S values. For Nb_3Ge the initial decrease in T_c is larger than for Nb_3Al, about -2.7 K/% of Nb sites occupied by Ge atoms (-1.8 K/% of wrongly occupied sites). Once again, elimination of the residual disorder (should this be possible) should lead to a significantly higher T_c. With the same caveat as for Nb_3Al, linear extrapolation leads to a fully ordered T_c of ~28–30 K.

The data of Fig. 40 can also be approximated by an exponential of the form $T_c = T_{c0} \exp[-\alpha(1 - S/S_0)]$ with α equal to 3.6. This may be compared with the corresponding values of 3.2 and 2.8 for Nb_3Al and Nb_3Sn [73], which reflect the greater sensitivity of T_c to disorder the higher the T_c. However, it is to be recalled that this does not hold true for V–Ga alloys [65].

Another interesting A15 compound is Mo_3Os. This material exhibits behavior different from that of the Nb- and V-based A15 compounds containing non–transition B atoms, the effect of irradiation upon T_c being considerably less (Fig. 15). There is a narrower homogeneity range for the A15 phase [91] and the dependence of T_c on thermally induced disorder is less [61]. This system provides an interesting example of how neutron irradiation may be used as a tool to complement thermal studies of ordering effects, and thus broaden our insight into the mechanisms of superconductivity in these materials.

Flükiger and co-workers have studied the effects of ordering on T_c of Mo_3Os [61]. Their approach was to freeze-in thermal disorder by rapidly quenching from high temperatures. The amount of frozen-in disorder is limited by the thermal disorder initially present at the quenching temperature and the rate of quenching. By using this technique, order parameters in the range $0.78 < S < 0.88$ were obtained. This range has been considerably extended using irradiation-induced disorder, which circumvents

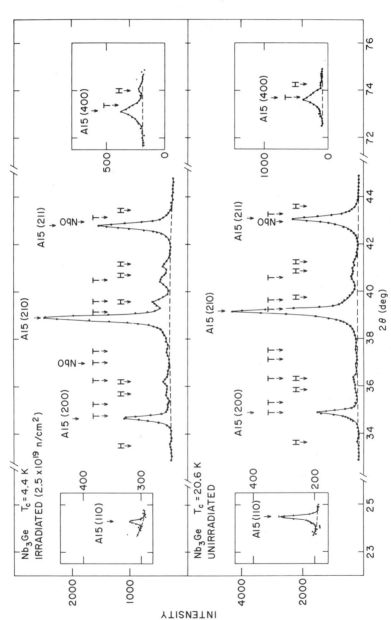

Fig. 39. Part of the x-ray (Cu–Kα) diffraction scans for Nb₃Ge. Bottom, unirradiated; top, irradiated to a fluence of 2.5 × 10¹⁹ n/cm² ($E > 1$ MeV). Lines are identified as A15, hexagonal (H) and tetragonal (T) Nb₅Ge₃, and NbO. Note the decrease in the intensity of the (110) reflection shown in the two insets, as well as the slight enhancement of the impurity phase. The line width of the (400) reflection remains roughly the same. See also Fig. 35. (After Sweedler *et al.* [57] and Moehlecke [59].)

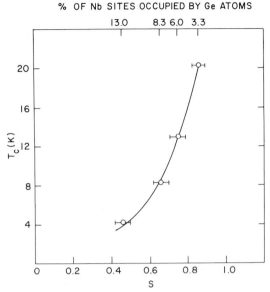

Fig. 40. Superconducting transition temperature T_c as a function of order parameter S_A for neutron-irradiated Nb_3Ge (actual composition 76 ± 1 at % Nb). Also shown is the fraction of Nb sites occupied by Ge atoms. The solid line depicts the relationship $T_c = T_{c0}$ exp[−3.6(1 − S/S_0)] where $T_{c0} = 20.3$ K and $S_0 = 0.86$. (After Sweedler *et al.* [57] and Moehlecke [59].)

the thermal considerations inherent in the quenching process, to values as low as $S \approx 0.2$ [59, 92].

Figure 41 shows the results for T_c versus S for thermally quenched and neutron-induced disorder in Mo_3Os. The squares were obtained from order measurements on thermally quenched samples, while the other points were inferred from neutron-irradiated samples using the previously mentioned relationship between order parameter and fluence developed by Aronin [88], $S = S_0 e^{(-k\Phi)}$, with $k = 2.7 \times 10^{20}$ cm²/n.

For Mo_3Os, Fig. 41 indicates a smooth extrapolation between the thermally and neutron-induced disorder and shows, as expected, that T_c for Mo_3Os is considerably less sensitive to disorder than for the Nb- and V-based A15 compounds were the B element is a non–transition metal, as in Nb_3Al. Data for the latter are shown for comparison. The initial T_c decrease is only −0.3 K/% of Mo sites occupied by Os atoms (or alternatively −0.2 K/% of wrongly occupied sites), about an order of magnitude less than that for Nb_3Ge.

As noted above, Skvortzov *et al.* [76] made order measurements on Nb_3Sn irradiated to an equivalent fluence of ∼10^{22} n/cm², using fission-fragment damage. At these very high irradiation levels, S has been re-

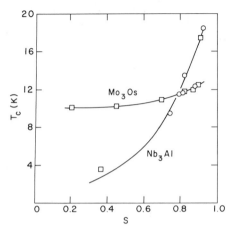

Fig. 41. Superconducting transition temperature T_c as a function of order parameter S for Mo_3Os. The open circles correspond to the directly measured values of S of Flükiger *et al.* [61] and the squares represent values derived from Aronin's expression as described in text. Data for Nb_3Al are shown for comparison. (After Moehlecke [59] and Sweedler *et al.* [92].)

duced to zero (a value of $S = 0$ for a stoichiometric A15 compound implies that $\frac{3}{4}$ of the B sites are occupied by A atoms, $\frac{1}{4}$ of the A sites are occupied by B atoms, and $\frac{3}{8}$ of the total number of sites are "wrongly" occupied). They found that at higher degrees of disorder the order parameter and lattice parameter were approximately related to an expression of the type

$$(\Delta a / a_0) = 1.25(1 - S^2) \qquad (12)$$

where Δa is the lattice-parameter increase upon irradiation, and suggested that the elastic energy, proportional to a_0^2, is equal to the disordering energy, proportional to $1 - S^2$, and that a simple hard-sphere model is insufficient to explain a relationship of this nature.

6. DEFECT STATE IN A15 COMPOUNDS

The nature of the defects resulting from irradiation of the A15 compounds has been the subject of considerable discussion. Several models relating the effect of the defect state to the superconducting properties have been proposed. We discuss below some of these models, bearing in mind that at present there exists no single comprehensive model that satisfactorily describes all the observed phenomena.

One established feature of the radiation-induced defect state is the presence of site-exchange (or antisite) disorder [57, 76, 81, 82]. In this model, the principal, although not the only, defect that results from irradiation with energetic particles is taken to be atomic displacements and replacements resulting in a certain fraction of A atoms occupying B sites and vice versa. This interchange of atoms between the two different lattice sites in the A15 structure results in a decrease in the long-range-order parameter S, which is a measure of the defect state of the material.

The description above is based primarily on a series of diffraction experiments that have demonstrated the existence of such defects in neutron-irradiated A15 compounds. These experiments have been described in detail in Section V,B,5. Briefly stated, the results of neutron and x-ray diffraction on polycrystalline, irradiated A15 compounds show a significant decrease in the degree of long-range order. Alternative descriptions to explain the diffraction results could conceivably be based on a vacancy model, but the required number of vacancies is unrealistically large [84].

The general relationship between T_c and site-exchange disorder can be extended to compositionally induced antisite disorder as well as radiation-induced disorder. It is well known that the T_c of many Nb- and V-based A15 compounds is strongly dependent on the chemical composition, and that the maximum T_c generally occurs at the stoichiometric composition in systems where the single-phase A15 region includes this point [61–63, 93–96]. There exists some disagreement as to whether this relationship holds for certain thin-film samples [72, 97–99] but where well-characterized single-phase material has been prepared, the maximum T_c occurs at the 3:1 composition (see, for example, Fig. 42). There is, of course, a direct relationship between stoichiometry and order in the A15 phase; i.e., perfect order ($S = 1$) can only be obtained at the stoichiometric composition of 3:1, as can be seen from Eqs. (10) and (11). Thus, decreases in T_c due to changing composition may also be viewed as due to changes in order, and are of similar magnitude to the decreases produced by irradiation for comparable amounts of disorder [93].

The actual situation is more complicated in that a change of composition that affects the electron/atom ratio may alter the structure of the Fermi surface and influence the density of states. However, T_c and order measurements on irradiated samples of varying stoichiometry can help to separate out compositionally induced effects from the radiation-induced effects due to site exchange [57].

That there exists an intimate correlation between compositional and radiation-induced changes in T_c is illustrated by the following. Figure 18 shows the effect on T_c of neutron irradiation for the Nb-rich side of the Nb–Pt system for various compositions. The slopes $\Delta T_c/\Delta\Phi$ for these curves all have approximately the same value for a given T_c. In other

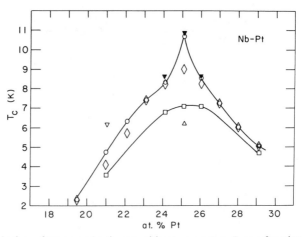

Fig. 42. Midpoints of superconducting transition temperature T_c as a function of composition across the A13 phase field of the Nb–Pt system. Solid lines are for visual aid. $\bigcirc$, 1800°C for 12 h then 900°C for 10 days; $\diamond$, 1800°C for 12 h; $\square$, 1800°C, rapidly quenched; $\triangledown$, 1800°C for 12 h then 900°C for 30 days; $\blacktriangledown$, 1800°C for 12 h then 900°C for 40 days; $\triangle$, splat cooled from melt.

words, the effect of irradiation on T_c of an off-stoichiometric sample (represented, for example, by point C in Fig. 18), which is described by the path CD, is equivalent to the effect on a stoichiometric sample, already irradiated to the point B, which follows the path BD. These two curves can be superimposed upon each other by a horizontal shift of origin. Thus as far as T_c is concerned the path AB, described by neutron irradiation of a stoichiometric sample to a fluence of 3.2×10^{18} n/cm^2 (point B in Fig. 18), is equivalent to the path AC obtained by a change in the composition of the stoichiometric sample to 24.1 at. % Pt (point C). Similar results have also been found for the Pt-rich side of stoichiometry in that the off-stoichiometric curves may be shifted to the right by a constant factor (Φ_0) until they match the stoichiometric curve as shown in Fig. 43. This curve, therefore, gives the dependence of T_c on fluence for the complete A15 range in the Nb–Pt system and contains implicitly the T_c variation with composition. It is thus possible to say that this curve suggests a general correlation between T_c and the defect state of the Nb–Pt system, irrespective of whether the defects are introduced by irradiation or by variations in composition. In the latter case, the initial defect state of a given sample may be characterized by the empirical factor Φ_0.

A similar analysis may be carried out for neutron-irradiated Nb–Al samples with varying compositions, and the "master curve" for this system, which contains the off-stoichiometric curves shifted to the right by appropriate amounts, is shown in Fig. 44.

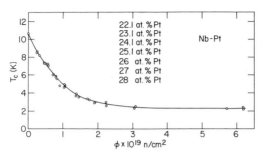

Fig. 43. Superconducting transition temperature T_c for the Nb–Pt system as a function of neutron fluence Φ. The curves for off-stoichiometric materials are matched to that for by factors Φ_0 as described in text. Solid line is for guiding the eye.

Furthermore, the initial slopes of the depression of T_c with fluence, $\Delta T_c/\Delta\Phi|_{\Phi=0}$, scale with T_c not only for different compositions of a given system but also for different systems (Fig. 20). Since the highest T_c and largest initial slope are currently found in the Nb_3Ge system, this is assumed to be the master curve for all the other Nb-based A15 systems. The results are shown in Fig. 45, where a striking overlap for these materials is observed. A good fit over most of the range of $T_c - T_{c\,sat}$ is provided by the relationship $T_c - T_{c\,sat} = (T_{c0} - T_{c\,sat})\exp(-c\Phi)$, where Φ contains the appropriate value of Φ_0 in each case, T_{c0} is the initial T_c of Nb_3Ge, $T_{c\,sat}$ is the saturation value of T_c for each system, and c is a constant equal to 8.6×10^{-20} cm$^2/n$.

Site-exchange disorder is not the only type of defect that has been observed following irradiation. Poate *et al.* [60, 72], Dynes *et al.* [99], and Burbank *et al.* [100] have observed a significant weakening of the x-ray intensities for the high-angle diffraction lines in alpha-irradiated Nb–Ge and

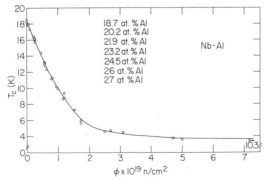

Fig. 44. Superconducting transition temperature T_c for the Nb–Al system as a function of neutron fluence Φ. The curves for Nb-rich materials are matched to that for $Nb_{75.5}Al_{24.5}$ by factors Φ_0 as described in text. Solid line is for guiding the eye.

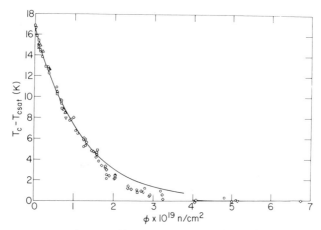

Fig. 45. Superconducting transition temperature T_c for various Nb-based A15 compounds as a function of neutron fluence Φ: □, Nb$_3$Ge; ▽, Nb$_3$Ga; ○, Nb$_3$Al; △, Nb$_3$Sn; ◇, Nb$_3$Pt. T_{csat} refers to the saturation value of T_c (~ 2–4 K) and has been subtracted from T_c to give $T_c - T_{csat} = 0$ at high fluences. The various curves are matched to that for Nb$_3$Ge by factors Φ_0 as described in text. The solid line depicts the relationship $T_c - T_{csat} = (T_{c0} - T_{csat}) e^{-c\Phi}$, where $T_{c0} = 20.3$ K and $c = 8.6 \times 10^{-20}$ cm^2/n. (After Sweedler *et al.* [57].)

Nb–Sn films, which cannot be accounted for by site exchange. Channeling experiments have been reported on single-crystal V$_3$Si irradiated with alpha particles by Meyer and Seeber [101], who have concluded that for [100] and [110] channeling directions, there are room-temperature rms displacements of the V atoms of 0.115 and 0.1 Å, respectively, in V$_3$Si irradiated to a T_c of 2.5 K, compared to displacements of 0.088 and 0.061 Å in the unirradiated sample. Similar experiments as a function of temperature have been reported by Testardi *et al.* [102], who observed additional rms displacements of approximately 0.1 Å. These channeling results have been interpreted in terms of static displacements of the V atoms from their lattice sites resulting from small bond distortions in addition to the normal thermal displacements.

Static displacements such as those noted above should show up as an increase in the Debye–Waller factors in diffraction measurments, which would account for the weakening of the high-angle x-ray intensities mentioned above. Neutron-diffraction studies on single-crystal V$_3$Si neutron irradiated to a T_c of 7.5 K [83] indicate a similar trend, but the increase is much smaller than 0.1 Å, the average (V and Si) rms displacements being 0.06 Å at 4.6 K and 0.04 Å for 30 K for the irradiated and unirradiated crystal, respectively. Diffraction measurements on neutron-irradiated polycrystalline Nb$_3$Al are also inconsistent with much larger values than this [81].

While the origin of the static displacements observed in the channeling experiments, and to a lesser degree in the diffraction experiments, is at present an open question, an effect of this type would certainly be expected to accompany site-exchange disorder. The exchanged atoms have different sizes and electronic structures, and there would accordingly be local relaxation of their neighbors from the ideal lattice positions due to size and bonding effects. It is thus very possible that both effects exist in the irradiated samples and the more difficult question remains as to which plays a more dominant role in affecting the superconducting properties.

The existence of another type of defect, consisting of highly disordered microregions about 20 Å in diameter, has been inferred by Karkin *et al.* [82] from small-angle neutron-scattering measurements on neutron-irradiated Nb_3Sn. From the simultaneous decrease in the long-range-order parameter they were led to conclude that the disordered microregions were coherent with the matrix crystal structure, and suggested that if the T_c of the disordered microregions was very small, an overall decrease in T_c could result from the proximity effect.

By the use of electron microscopy, Pande [103, 104] has recently confirmed the existence of small coherent regions of high disorder in the matrix of neutron-irradiated Nb_3Sn. He finds the average size of these regions to be about 40 Å, and has derived the following expression for T_c as a function of fluence within the framework of the proximity-effect model for the case of fast-neutron irradiation:

$$\ln\left(\frac{T_c}{T_c/T_{c0}}\right) = \left\{Q_0 + Q_1\left[\frac{1 - F_v}{(1 - F_v)/F_v}\right]\right\}^{-1}$$

where F_v is the volume fraction of disordered material and Q_0 and Q_1 are constants. A reasonable fit is obtained for neutron-irradiated Mo_3Os, Nb_3Al, Nb_3Ge, V_3Si, and oxygen-ion-irradiated Nb_3Sn.

Since F_v is taken to be approximately $1 - S$, the calculated volume fraction of highly disordered material is quite large (about 25% for Nb_3Al with a T_c of 9.6 K). In view of the wide range of incident neutron energies and the comparatively high irradiation temperature, the idea of such well-defined regions of low and high disorder represents a rather simplified picture. However, one would certainly expect significant short-range fluctuations in the amount of site-exchange disorder after fast-neutron irradiation.

No detectable line broadening has been observed in diffraction studies in the initial stage of irradiation, where T_c is most rapidly depressed. Line broadening has indeed been observed at sufficiently high fluences, but in these cases T_c is already close to the saturation value. This line broadening could arise from internal strains associated with the many atomic displacements that are produced at high fluences. For Nb_3Ge irradiated at

5×10^{19} n/cm^2, we noted earlier that loss of crystallinity occurs, whereas for Nb_3Sn, irradiation levels as high as $\sim 10^{22}$ n/cm^2 are required to produce appreciable line broadening [76]. If the regions of high disorder are considered to be analogous to the precipitation of particles of a new phase, then Krivoglaz [105] has shown that no line broadening is to be expected if the volume fraction of new particles is much less than unity. However, this is not the case in the presence of severe distortions or numerous dislocation loops.

It is not obvious that models based upon the proximity effect can account either for the similar T_c effects produced by electron irradiation, where the damaged regions are localized to a few atoms, or for the qualitatively similar depressions observed as a function of composition.

Other theoretical models have been proposed to account for the large T_c depressions observed in irradiated A15's. Appel [106] has calculated T_c as a function of long-range order and composition based on a uniform distribution of disorder through the lattice. His expression for T_c versus S follows from McMillan's equation for strong-coupled superconductors and contains implicitly the order dependence of the density of states. Reasonable agreement is obtained for oxygen-ion-irradiated Nb_3Sn. However, the positive curvature of the T_c versus S curves (Figs. 38 and 40) is not apparent in this model, and the order dependence of λ, the electron–phonon coupling parameter, must be independently determined from the order dependence of the low-temperature heat capacity, data that are not readily available.

A rather different approach from the models above has been taken by Farrell and Chandrasehkar [107]. They suggest that the large T_c depressions result from a reduction in a large gap anisotropy due to the irradiation-induced defects. An anistropy parameter $\langle a^2 \rangle$ of 0.6 gave good agreement with the reduction in T_c due to alpha-particle irradiation. However, Gurvitch et al. [108] have criticized this approach on the basis of their electron irradiations of Nb_3Sn. They have found that the T_c versus electron fluence curve does not follow the form suggested by the anisotropy model. Furthermore, if they attempt to fit their data to the anisotropy model, a value for $\langle a^2 \rangle$ of 0.03 is needed, much less than that required by the theory. The value of 0.03 is of the same order as has been determined for other type-II superconductors [109].

A still different approach has been proposed by Wiesmann et al. [70] and Ghosh et al. [74]. They point out that broadening of the density of states at the Fermi level, which results from the increase in the residual resistivity following irradiation, could account for the T_c decreases. From measurements of the critical field following irradiation, they conclude that the density of states is inversely proportional to the residual resistivity.

The model is qualitative in nature, and no quantitative calculations of T_c have been presented.

Voronova *et al.* [67] have suggested that T_c is more sensitive to the homogeneous distribution of small defects than to clusters. They consider T_c to be a measure of the fraction of superconducting chains that contain a minimum number of consecutive atoms of the transitional metal. The best fit to their data for Nb_3Sn is obtained with a value of 20 consecutive atoms, but the agreement is far from quantitative, and the model does not appear to throw any further light on the underlying physics.

It is clear that a great deal still remains to be learned about the nature of the irradiation-induced defects and the role they play in determining the normal and superconducting properties in these materials.

C.   Critical Fields and Currents

In a number of cold-worked Nb-based alloys Fietz and Webb [110] showed that a scaling law for the critical-current density (or volume pinning force) could be obtained in the form

$$J_c = [H_{c2}(T)]^m f(h, d) \qquad (13)$$

The exponent m was found to be >2. The function $f(h, d)$ is sensitive to changes in H_{c2} since $h = H - H_{c2}$, and to a parameter d that is representative of a spacing related to the pinning microstructure. The implication of this scaling law is that by measuring $J_c(h)$ at one temperature and determining $f(h, d)$ and m, one can calculate $J_c(H)$ at other temperatures by scaling by $[H_{c2}(T)]^m$. Conversely, it has widely been attempted to scale J_c according to Eq. (13) but by measuring at constant T and invoking information about either $[H_{c2}(T)]^m$ or $f(h, d)$ from the behavior of the critical current. From such an analysis one would hope to determine whether either microstructural (direct pinning) changes are taking place after, say, radiation damage, or whether the changes were magnetic in character and were vested in the $[H_{c2}(T)]^m$ term. Here again the great importance the ordered lattice of the A15 compounds plays in such analyses will be seen.

The phenomenology of Eq. (13) can be exploited further to elucidate the qualitative behavior of J_c upon changing properties induced by irradiation. Following Brown *et al.* [111] the variables in $f(h, d)$ may be separated as

$$f(h, d) = A(d)h^n(1 - h)^2 \qquad (14)$$

where n is a constant. Substituting in (13) gives

$$J_c = [H_{c2}(T)]^2 h^n(1 - h)^2 A(d) \qquad (15)$$

To find the behavior of J_c with Φ, where the fluence Φ is defined as the time integral of the flux φ, $\Phi = \int \varphi \, dt$, J_c is differentiated with respect to both H_{c2} and d:

$$\frac{dJ_c}{d\Phi} = \frac{\partial J_c}{\partial H_{c2}} \frac{\partial H_{c2}(T)}{d\Phi} + \frac{\partial J_c}{\partial d} \frac{\delta d}{\partial \Phi} \tag{16}$$

This yields

$$\frac{\partial J_c}{\partial \Phi} = C_1'(h) \frac{\partial H_{c2}(T)}{d\Phi} + C_3 \frac{\partial d}{\partial \Phi} \tag{17}$$

But in the "dirty limit" (in the introductory chapter by Dew-Hughes, this volume)

$$H_{c2}(0) = 3.06 \times 10^4 \gamma \rho_n T_c \tag{25}$$

where γ is the electronic specific heat coefficient. Substituting this for $H_{c2}(T)$ and differentiating gives

$$\frac{\partial J_c}{\partial \Phi} = C_1(h) \frac{\partial \rho_n}{\partial \Phi} + C_2(h) \frac{dT_c}{d\Phi} + C_3 \frac{\partial d}{\partial \Phi} \tag{18}$$

This approximation should be good so long as T_c is much greater than the measuring temperature of J_c, e.g., Nb_3Sn having a T_c of ~ 18 K and a J_c measured at 4.2 K.

The first two terms describe the "magnetic" effects of the irradiation on the critical current and the third describes the effects of microstructural changes. In this microstructural term, positive effects that would be expected to increase the pinning and therefore J_c would be caused by the introduction of large cascades or dislocation loops, smaller pinning defects, or growth of existing pinning centers such as precipitates. Negative effects would be expected for processes such as precipitation, dissolution, or grain growth. The magnetic terms are seen to be competitive, the first term containing $\partial \rho_n/\partial \Phi$ being an increasing function of Φ only (disregarding annealing effects) and the second term containing $\partial T_c/\partial \Phi$ a decreasing function only of Φ. A detailed discussion of the field dependence of these terms is given in Brown et al. [111]. The problem is then reduced to one of sorting out the relative importance of the third term with respect to the first two for a given type of irradiation and to determine the fluence ranges where the various terms are dominant.

For Nb–Ti the response of I_c to irradiation damage was seen to depend markedly on the preirradiation value of the critical-current density J_{c0}. For material with low J_{c0}, increases in I_c came about, but for "optimized" material with a high initial value of J_{c0}, only decreases come about. The situation with Nb_3Sn is analogous. Early ambient-temperature reactor irradiations by McEvoy *et al.* [112] and Cullen *et al.* [113, 114] to fluences of less than 10^{18} n/cm^2 produced increases in I_c that were for the most part independent of applied transverse field up to 20 kG. These increases were attributed to increased pinning at the defects formed by the irradiation. Cullen and Novak [115] later reported increases in I_c for low-J_{c0} material, but decreases in T_c for higher J_{c0} Nb$_3$Sn. The same behavior was observed following 20-K irradiation of vapor-deposited Nb$_3$Sn using 15-MeV deuterons [5]. Bode and Wohlleben [116] found increases in I_c for Nb$_3$Sn for room-temperature 3-MeV proton irradiations for applied fields up to 50 kG. Of special interest in this work was that the initial I_{c0} was recovered upon annealing between 700–800°C. They, too, attributed the changes to defect clusters produced by the irradiation. Continuing studies by Wohlleben [117] using 1-, 2-, and 3-MeV protons and 3-MeV deuterons on Nb$_3$Sn diffusion layers produced similar results, as did studies by Ischenko *et al.* using oxygen-ion irradiation [118]. Söll *et al.* [119] found increases in I_c in Nb$_3$Sn wires made by vapor-phase diffusion after reactor-neutron irradiation to 4×10^{18} n/cm^2 ($E > 0.1$ MeV) at 4.6 K. There was partial recovery of I_c upon annealing to room temperature, but a small further *decrease* in T_c was noted (a few tenths of a degree) during these anneals. They explained this phenomenon as resulting from clusters that are formed during annealing being more efficient at lowering T_c that isolated Frenkel pairs.

With the advent of Nb$_3$Sn with "maximized" J_{c0} ($\sim 2 \times 10^6$ A/cm^2 at 40 kG) made by the solid-state diffusion "bronze process" [120], radiation effects on what appears to be engineering-grade material could be determined. Although several of the investigators mentioned above observed small decreases in T_c and in I_c following the initial increases in I_c at relatively low fluences, large decreases in critical properties below the preirradiation values were not observed until the high-fluence work of Schweitzer and Parkin [121], Parkin and Schweitzer [122], and Bett [54]. As seen in Fig. 46 the critical current drops quite precipitously toward zero for neutron fluences greater than a few times 10^{18} n/cm^2 at ambient reactor temperature. The correlation with similar behavior of T_c over the same fluence range as seen in Fig. 16 led to the conclusion that reductions in I_c were caused by decreases in T_c. It has been seen earlier in this chapter that the T_c degradation has been attributed to the destruction of order in the A_3B compounds.

The behavior of the I_c degradations as a function of applied transverse magnetic field are shown in Fig. 47 for ambient reactor irradiations. Fields up to 40 kG and fluences to 6×10^{19} n/cm^2 were utilized. No consistent field dependence was obtained from this study except that at lower doses degradation is noted at lower fields sooner than at the higher fields inves-

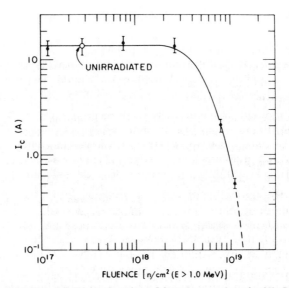

Fig. 46. I_c (40 kG) at 4.2 K as a function of fast-neutron dose for Nb$_3$Sn multifilament composites. (After Parkin and Schweitzer, [122].)

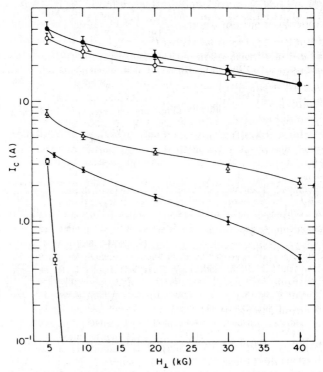

Fig. 47. I_c versus H_1 at 4.2 K for fast-neutron-irradiated and unirradiated Nb$_3$Sn multifilament composites: ●unirradiated; ○, 2.6 × 10^{18} n/cm^2, E] 1.0 MeV; △, 7.8 × 10^{18} Δ/cm^2; ●, , 1.1 × 10^{19} n/cm^2; □, 6 × 10^{19} n/cm^2. (After Parkin and Schweitzer, [122].)

tigated. This low-field (<40 kG) and low-fluence ($\lesssim 2 \times 10^{18}$ n/cm^2) regime was investigated in some detail by the group at Argonne National Laboratory using the cryogenic irradiation facility at the CP-5 reactor. This 4.2 K facility enables in situ measurements of neutron-irradiated samples in fields up to 35 kG. Brown et al. [111] observed decreases in I_c for fields above ~15 kG (see Fig. 48). Upon annealing to 295 K some recovery of the I_c enhancement was observed, but little recovery of the degradation below 15 kG (see Fig. 49). It is to be emphasized here that this enhancement above 15 kG is not the same type of enhancement observed at low fluences and fields in the earlier works since the earlier enhancements were essentially field independent, but the present ones are very field dependent. Figure 50 shows the field dependence of the enhancements of critical current over the fluence region investigated. It is apparent that the higher the applied field, the greater the enhancement.

Brown et al. interpreted the rise in I_c as due to increased flux pinning by radiation-induced cascades and the subsequent decrease of I_c at high fluence as due to a decrease in T_c (and correspondingly H_{c2}). The monotonic nature of the I_c decreases with dose below 15 kG was not discussed in detail. A similar interpretation was given the results of Colucci et al. [123–125] using similar Nb_3Sn wires at the same facility. The samples were warmed to nitrogen temperature after irradiation for transfer to a measuring Dewar with field capability of 100 kG. The same low-field crossover was observed, with enhancement of I_c an increasing function of

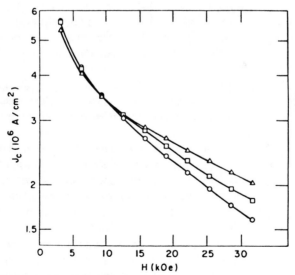

Fig. 48. Critical current of Nb_3Sn during irradiation with fast neutrons ($E > 0.1$ MeV) at 6 K (run 1): $T_{irrad} = 6$ K; $\bigcirc$, unirradiated; $\square$, after 2.1×10^{17} n/cm^2; $\triangle$, after 9.0×10^{17} n/cm^2.

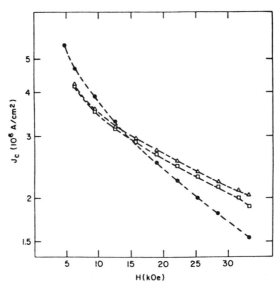

Fig. 49. Critical current of Nb$_3$Sn during irradiation and annealing (run 2): ●, unirradiated; △, after 1.82×10^{18} n/cm^2; □, after 295 k anneal. (After Brown *et al.*, [111].)

applied field. Annealing the specimens to room temperature produced the result that recovery of the I_c enhancement took place, the amount of recovery increasing in applied magnetic field up to the highest value of H (100 kG).

The above model to explain the increase and subsequent decrease of I_c with increasing neutron fluence invokes only changes in the second and third terms of Eq. (18). The effect that changes in the first term of Eq. (18) have on changes in I_c, especially through the relation coupling the normal-state resistivity to the upper critical field [Eq. (1-25)] will now be discussed. The first measurements above 100 kG of critical currents of Nb$_3$Sn wires with "modern" current densities and the extrapolation of the I_c versus H curves to get H_{c2} for these materials were made by Crow and Suenaga [126] and Suenaga *et al.* [127] The first radiation effects were reported on similar solid-state diffusion produced Nb$_3$Sn wires by Snead and Parkin [41, 128] following ~100°C fission reactor irradiations. For neutron fluences to 3.6×10^{19} n/cm^2 ($E > 1$ MeV) I_c was measured as a function of field up to 160 kG. The H_{c2} for the various fluences was obtained by extrapolating the high-field part of the curve to the H axis for a value of critical current of 0.01 A. These data are shown in Fig. 51. The striking feature of these data is the enhancement of I_c above I_{c0} for fluences below ~2×10^{18} n/cm^2. This enhancement is present above the "crossover" point at ~30 kG (in keeping with the observations of Brown *et al.*) and the relative increase in an increasing function of field, as seen in Fig. 52.

Since the value of H_{c2} for these low-fluence irradiations is seen to be higher than that of the preirradiation specimens (200 kG versus 170 kG), this increase in I_c is explained in part by the increase in H_{c2}. Since this increase in H_{c2} is caused by an increase in ρ_n owing to the irradiation, at low fluences we have the first term in Eq. (18) giving a positive change to $dJ_c/d\Phi$. Since T_c is changing very slightly over this low-fluence regime, the second term involving $\partial T_c/\partial \Phi$ can be set to zero. The increase in J_c is then due to an increase in ρ_n and to whatever effects the low-fluence irradiation have upon the third (microstructural) term involving $\partial d/\partial \Phi$. It must be remembered that increases in ρ_n in A15's with irradiation are caused by two different types of "defects," namely, the resistivity due to Frenkel-pair damage, which includes clusters and dislocation loops, and

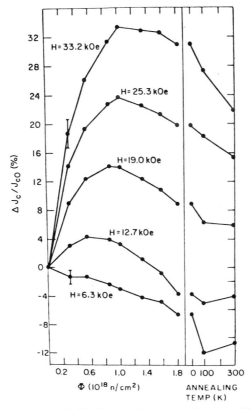

Fig. 50. Fractional change of critical current in run 1 as a function of field and dose. (After Brown *et al.*, [111].)

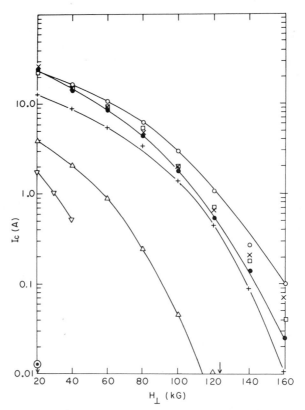

Fig. 51. Critical current versus applied magnetic field for 19-core Nb$_3$Sn multifilament conductor prior to and after irradiation: unirradiated (●); irradiated, $E > 1$ MeV, 0.6×10^{18} n/cm^2 (×), 1.0×10^{18} n/cm^2 (○), 1.5×10^{18} n/cm^2 (□), 9.0×10^{18} n/cm^2 (+), 15.6×10^{18} n/cm^2 (△), 22.0×10^{18} n/cm^2 (▽), 36.0×10^{18} n/cm^2 (⊙).

the resistivity due to disorder (the "antistite" defects). After neutron irradiation the disorder induced is not expected to be a function of irradiation temperature below self-diffusion temperatures (~700°C in Nb$_3$Sn). This is not the case, however, for Frenkel-pair damage. Differences in the contributions of the first term of Eq. (18) for neutron irradiations at, say, 6 and 400 K are expected. Annealing effects would also be expected to be present, especially at low fluences of helium-temperature irradiations when the ratio of Frenkel-pair resistivity to disorder resistivity is the highest.

The recovery behavior of I_c in Nb$_3$Sn following low-temperature neutron irradiations has been investigated in filamentary Nb$_3$Sn wires by Brown *et al.* [111] and Colucci *et al.* [125]. Generally, annealing to room temperature produced partial recovery of the increases in I_c for applied

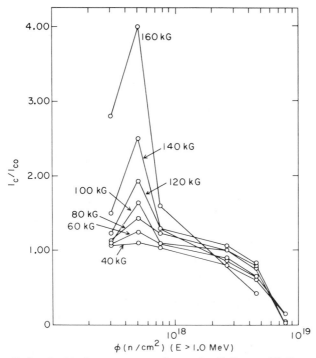

Fig. 52. Reduced critical current versus fluence for 19-filament Nb$_3$Sn conductor.

fields > 15 kG, and produced mixed recovery effects on the I_c decreases below 15 kG. Colucci *et al.* [123, 124] annealed to 415 K and measured I_c to 100 kG following neutron irradiation to 4.8×10^{17} and 6×10^{18} n/cm^2. Substantial recovery ($\sim 50\%$) in the I_c increases over most of the range of fields investigated in the higher-field regime was observed, with almost complete recovery seen at the highest field, 100 kG. Since little change in either the order of the irradiated samples or in the large defects (cascades) is expected for these anneals, the recovery in I_c changes must be attributed mainly to the resistivity recovery from Frenkel pairs.

At this point, however, with the limited annealing data available, and the phenomenon of higher recovery (almost 100%) seen at 100 kG relative to that between 15 and 100 kG, quantitative assignments of changes in I_c induced by H_{c2} changes are premature. Indeed, it seems that any such assignments must be field dependent. Later measurements in single-core Nb$_3$Sn wires by Colucci *et al.* [123–125] showed recovery upon annealing at 540 K of only $\sim 50\%$ of the increase in I_c measured at 100 kG.

Even though the low-field regime (below, say, 30 kG) is experimentally more accessible, the I_c change with irradiation and recovery with annealing is no less clear. That recovery upon annealing to room temperature of *decreases* in I_c is observed as before implies Frenkel-pair resistivity as the key mechanism of this recovery. This resistivity is the cause of I_c decreases at low fields.

This field dependence must be reflected in the sign of the $\partial \rho_n / \partial \Phi$ term in Eq. (18). Note that the "crossover" field is the field at which $\partial \rho_n / \partial \Phi = 0$ is roughly the same as the field where the pinning force is a maximum. This is consistent with the idea that the maximum in F_p occurs at the point where the flux-pinning and flux-shear contributions are equal. It would not be surprising if changing resistivity and therefore changing κ were evidenced in differing behavior for these two pinning mechanisms. One could speculate that the I_c decreases at low field with increasing κ might be due to the same mechanism responsible for I_c reductions in Nb–Ti; namely, a reduction of the pinning strength of the major pinning centers, the dislocation cell walls. In the case of Nb_3Sn, where the major pinners are grain boundaries, one would then say that increasing κ in the grains was responsible for the decreased pinning strength of the grain boundaries. It will take a great deal of careful work, however, before such a mechanism might be confirmed and before the irradiation-induced changes in I_c can be quantified throughout the entire range of applied fields. The contributions due to the term in Eq. (18) involving $\partial d / \partial \Phi$ will also have to be determined in this low-fluence regime where $\partial T_c / \partial \Phi$ is insignificant.

For higher fluences ($> 2 \times 10^{18}$ n/cm^2, $E > 1$ MeV) the reduction of T_c drives H_{c2} down and thus I_c with it. The high-fluence portion of Fig. 52 demonstrates that the I_c reduction is field independent in the high-fluence regime. With increasing fluence, I_c tends to zero (measured at 4.2 K) as T_c tends toward ~ 3 K for Nb_3Sn. One is probably safe in assigning all of the change in J_c to the second, $\partial T_c / \partial \Phi$, term in Eq. (18) in this high-fluence regime. There are very few data for irradiations at low temperature and for annealing in this high-fluence regime. Söll *et al.* [119] neutron-irradiated Nb_3Sn diffusion wires to 4×10^{18} n/cm^2 ($E > 0.1$ MeV) at helium temperature. The 150% enhancement of J_c observed at 50 kG was almost completely recovered upon annealing to 250 K. Since increases, rather than decreases, in J_c were observed, one concludes that the "high-fluence regime," where T_c changes dominate, has not been reached in this work.

In talking about fluence "regimes" and in comparing results for different fluences, especially between different reactor irradiation sources, one must constantly keep several things in mind. First, the neutron flux and the flux spectrum in most facilities are generally not known to very good precision. As dosimetry over the last few decades has improved, so has the precision of the flux determinations, and these have generally raised the flux values for a given reactor for the more recent determinations. Indeterminacy of flux as a function of time during a cycle, differences in fuel loading, and differing physical positioning of specimens in an irradiation hole all increase the fluence uncertainty. Smaller effects can arise from neutron spectrum changes at the specimen due to packaging, other sample packages in the vicinity, or the absorption of the specimen itself. These effects can give rise to fluence uncertainties of 20–100%. In comparing fluences one must also be cognizant of the neutron energy spectrum that is quoted. In general, fluences are quoted for total neutrons that have an energy greater than 0.1 or 1 MeV. In comparing an experiment where the fluence is quoted in terms of $E > 1.0$ MeV to one where the fluence is quoted in terms of $E > 0.1$ MeV, one has to know the neutron energy spectrum of one of the reactors in order to convert to the units quoted for the other. The fluences quoted for experiments where $E > 0.1$ MeV is used are always greater than those quoted when $E > 1.0$ MeV is the criterion. For instance, the flux of the Brookhaven High Flux Beam Reactor (HFBR) in the irradiation hole used for the work reported by the Brookhaven group has a flux of 1.3×10^{14} n/cm² sec ($E > 1$ MeV) and 5.3×10^{14} n/cm² sec ($E > 0.1$ MeV). To compare fluences reported by Brookhaven workers ($E > 1$ MeV) with, say, the work of Söll ($E > 0.1$ MeV), the Brookhaven fluences have to be multiplied by 4.1. The uncertainty in the flux determination and the differences in the treatment of the neutron spectrum in reporting fluences make for caution, then, in comparing reactor-irradiation results. To date, there are no high-fluence low-temperature neutron results for J_c measurements in Nb₃Sn.

Irradiations of high-J_{c0} Nb₃Sn at low temperatures into the T_c-limited high-fluence regime have been performed using 30-GeV protons at the Brookhaven Alternating Gradient Synchrotron. Single and multifilament wire specimens were irradiated at 4.2 K and I_c measurements made in situ in fields up to 40 kG. Figure 53 depicts the results of an irradiation of commercial Airco 361-core wire. The irradiation was interrupted at 7×10^{17} p/cm², the specimen annealed to room temperature, and the irradiation continued to a total of 1.6×10^{18} p/cm². The previously discussed behavior of larger decreases at low field in the high-fluence regime is to be seen for this case also. The damage energy for 30-GeV protons found empirically by comparison with reactor-neutron results is ∼400 b keV, roughly four times that of a spectrum-averaged HFBR neutron.

The large decreases of I_c at high fluences similar to the behavior of neutron-irradiated specimens are also seen here. At 40 kG at 99% reduction of I_c has taken place for the 1.6×10^{18} p/cm² exposure. The important thing to note here is the lack of recovery of the specimen's I_c with an-

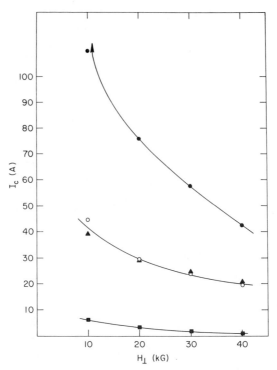

Fig. 53. Effect of irradiation on a commercial Nb_3Sn multifilament 361-core wire: ●, unirradiated; ○; 7×10^{17} p/cm^2; ▲, ~273 K anneal; ■, additional 9×10^{17} p/cm^2.

nealing. Subsequent to the 7×10^{17} p/cm^2 segment, the specimens were annealed to room temperature and I_c then remeasured. The solid triangles show that within experimental uncertainty no recovery is observed. Similar annealing (data not shown) following the next irradiation segment to 1.6×10^{18} p/cm^2 produced values of I_c identical to those prior to this latter anneal. In contrast to Nb–Ti, then, we see that no recovery of decreases in I_c takes place in Nb_3Sn with room-temperature annealing.

Contrast in the degradation rate and recovery between Nb_3Sn and Nb–Ti for low-temperature irradiations is seen in Fig. 54. Here the relative decrease in I_c measured at 40 kG for two specimens irradiated simultaneously with 30-GeV protons is plotted as a function of fluence. The much faster degradation rate of the Nb_3Sn versus that of the Nb–Ti is immediately evident. The recovery is represented by the vertical movement of the data points at 0.7 and 1.6×10^{18} p/cm^2. The three points at the higher fluence represent the samples irradiated, annealed to 125 K and then to 290 K. The lack of recovery in the Nb_3Sn specimen is the striking feature.

There has been little work on changes in I_c and H_{c2} in other high-J_{c0} A15 compounds. Only V_3Ga filamentary wires have received attention.

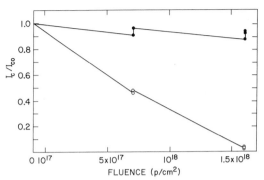

Fig. 54. Contrast in the degradation rate and recovery between Nb$_3$Sn (○) and Nb–Ti (●) for low-temperature irradiations.

Becker *et al.* [129] irradiated single-core V$_3$Ga specimens at ~15 K with 50-MeV deuterons. Increases in I_c at low fluence similar to Nb$_3$Sn behavior were observed with the usual "crossover" behavior in field; i.e., decreases in I_c below the crossover field with increases at all fields higher than the crossover. I_c decreases at high fluence were observed and were attributed to reduction of T_c.

Couach *et al.* [130], on the other hand, found only decreases in I_c with fluence for reactor neutron irradiations at 27 K for measurements up to 80 kG. For ambient-temperature reactor irradiations Snead [131] found increases in I_c in single-core V$_3$Ga wires at all fields from 20 to 180 kG for the low-fluence regime ($< \sim 2 \times 10^{18}$ n/cm^2, $E > 1$ MeV) and decreases in I_c in the high-fluence regime (see Fig. 55). One possible explanation for the lack of increases of I_c at low dose for the Couach *et al.* results could be that the fluences of their irradiations might be higher than stated and that the measurements are already in the high-fluence regime. The 2.3-K change in T_c for their reported fluence of 3×10^{18} n/cm^2 (neutron energy spectrum not specified) argues that the comparable fluence for $E > 1$ MeV was higher than that reported. The discussion earlier about the difficulty of comparing results from different reactors on the basis of total fluence is quite relevant here. Although some details seem different in comparing the radiation response of V$_3$Ga to that of Nb$_3$Sn, V$_3$Ga seems to behave very similarly with respect to I_c.

The response of H_{c2} to irradiation is, however, much different in V$_3$Ga than in Nb$_3$Sn. Measurements of H_{c2} as a function of reactor neutron irradiations (see Fig. 55) showed that in the low-fluence regime, H_{c2} of V$_3$Ga *decreases* slightly. The interpretation here is that the paramagnetically limited H_{c2} of V$_3$Ga is coupled to the normal-state resistivity in a much more complicated way than is H_{c2} of Nb$_3$Sn. Changes in H_{c2} in the low-fluence regime are small, but when T_c begins decreasing in the

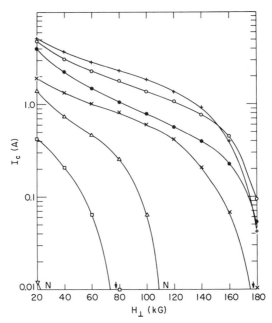

Fig. 55. Critical currents versus applied magnetic fields for a V_3Ga monofilament conductor: unirradiated (●); irradiated, $E > 1$ MeV; 0.4×10^{18} n/cm^2 (○), 1.0×10^{18} n/cm^2 (+), 2.0×10^{18} n/cm^2 (×), 10×10^{18} n/cm^2 (△), 20×10^{18} n/cm^2 (□), 34×10^{18} n/cm^2 (▽).

high-fluence regime, H_{c2} decreases in a manner similar to that of Nb_3Sn.

For the most up-to-date information on radiation effects in superconductors the reader should consult the *Proceedings of the International Discussion Meeting on Radiation Effects on Superconductivity* held at Argonne National Laboratory, June 13–16, 1977, published in *J. Nucl. Mater.* **72** (1978). In particular, those interested in flux pinning in A15s and the application of scaling to irradiated A15s should consult the papers by Colucci, Couach, and Kramer therein.

References

1. G. W. Cullen, *Proc. 1968 Summer Study Supercond. Devices and Accel.* p. 437. Brookhaven Nat. Lab. Rep. No. BNL 50155 (C-55) (1969).
2. L. Rinderer and E. Schmid, *Proc. Int. Conf. Low Temp. Phys., 7th* (G. H. Grahm and A. C. Hollis-Hallet, eds.), p. 395. Univ. of Toronto Press, Toronto, 1961.
3. D. P. Seraphin, C. Chiou, and D. J. Quin, *Acta Metall.* **9**, 861 (1961).
4. D. Markowitz and L. P. Kadanoff, *Phys. Rev.* **131**, 563 (1963).
5. H. T. Coffey, E. L. Keller, A. Patterson, and S. H. Autler, *Phys. Rev.* **155**, 355 (1967).
6. H. Berndt and F. Sernetz, *Phys. Lett.* **33A**, 427 (1970).
7. S. Klaumünzer, G. Ishenko, and P. Müller, *Z. Phys.* **268**, 189 (1974).

8. For a description of the GLAG theory see A. L. Fetter and P. C. Hohenberg, Theory of type II superconductors (R. Parks, ed.), Vol. IV, p. 817. New York. 1969.
9. G. Eilenberger, *Phys. Rev.* **153**, 584 (1967).
10. H. Ullmaier, "Festkörprobleme" (O. Madelung, ed.), Vol. 10, p. 367. Pergamon-Vieweg, Braunschweig, 1970.
11. H. Ullmaier, C. Papastaikoudis, and W. Schilling, *Phys. Status Solidi.* **38**, 189 (1970).
12. W. Dönitz, W. Hertz, W. Waidelich, H. Peisl, and K. Böning, *Phys. Status Solidi (a)* **22**, 501 (1974).
13. For the most recent review of flux-pinning effects in irradiated material see Edward J. Kramer, *J. Nucl. Mater.* **72**, 5 (1978).
14. H. Ullmaier, K. Papastaikudis, S. Takács, and W. Schilling, *Phys Status Solidi* **41**, 671 (1970).
15. H. Berndt, N. Kartascheff, and H. Wenzl, *Z. Angew. Phys.* **24**, 305 (1968).
16. H. C. Freyhardt, *Proc. Int. Disc. Meeting Flux Pinning Supercond. (IDMFDS), St. Andreasburg, Germany, September 1974*, p. 98. Akademic der Wissenschaften, Göttingen, 1975.
17. H. A. Ullmaier, *Proc. Int. Conf. Defects and Defect Clusters in BCC Met. and Their Alloys* (R. J. Arsenault, ed.), p. 363. Gaithersburg, Maryland, 1973.
18. See for instance, C. A. M. van der Klein, P. H. Kes, and D. de Klerk, *Philos. Mag.* **29**, 559 (1974).
19. See A. M. Campbell and T. E. Evetts, *Adv. Phys* **21**, 199 (1972) for a review.
20. For reviews and complete bibliographies, see, Cullen [1], Coffey *et al.* [5], B. S. Brown, in "Radiation Damage in Metals" (N. L. Petersen and S. D. Harkness, eds.), p. 330. American Society for Metals, Metals Park, Ohio, 1976, and S. T. Sekula, *J. Nucl. Mater* **72**, 91 (1978).
21. B. S. Brown, *Proc. Int. Disc. Meeting Flux Pinning Supercond.* (P. Haasen and H. C. Freyhardt, eds.), p. 200. Akademie der Wissenschaften, Göttingen, 1975.
22. B. S. Brown, H. C. Freyhardt, and T. H. Blewitt, *J. Appl. Phys.* **45**, 2724 (1974).
23. H. C. Freyhardt, A. Taylor, and B. A. Loomis, "Application of Ion Beams to Metals" (S. T. Picraux, E. P. Eer Nisse, and F. L. Vook, eds.), p. 47. Plenum, New York, 1974.
24. H. C. Freyhardt, B. A. Loomis, and A. Taylor, *Proc. Int. Conf. Low Temp. Phys. 14th,* (M. *LT14, Helsinki* Vol. 2, p. 481. North-Holland Publ., New York, 1975.
25. P. S. Swartz, H. R. Hartz, and R. L. Fleischer, *Appl. Phys. Lett.* **4**, 71 (1966).
26. T. Okada, T. Tsubakihara, S. Katoh, T. Horiuchi, Y. Monjha, and S. Tarutani, Radiation Effects and Tritium Technology, Conf. 750984, Vol. II, p. 436. Oak Ridge Nat. Lab., (1976).
27. K. Schmelz, G. Ischenko, B. Besslein, A. Greiner, S. Klaumünzer, P. Müller, and H. Neumüller, *Phys. Lett.* **55A**, 315 (1975).
28. K. Wohlleben, *J. Low Temp. Phys.* **13**, 269 (1973).
29. M. Söll, C. A. M. van der Klein, H. Bauer, and G. Vögel, *IEEE Trans. Magn.* **MAG-11**, 178 (1975).
30. M. Söll, PhD. thesis Technische Universität München (September 1974) (unpublished).
31. A. V. Narlikar and D. Dew-Hughes, *J. Mater. Sci.* **1**, 317 (1966).
32. D. F. Neal, A. C. Barber, A. Woolcock, and J. A. F. Gidley, *Acta Metall* **19**, 143 (1971).
33. R. G. Hampshire and M. T. Taylor, *J. Phys. F.* **2**, 89 (1972).
34. D. Dew-Hughes and M. J. Witcomb, *Philos. Mag.* **26**, 73 (1972).
35. M. Söll, S. L. Wipf and G. Vögl, *Proc. Appl. Supercond. Conf.* IEEE Publ. No. 72 Ch0682-5-TABSC, p. 434 (1972).
36. C. L. Snead, Jr., L. Nicolosi, and W. Tremel, *Appl. Phys. Lett.* **31**, 130 (1977).

37. C. L. Snead, Jr., *J. Nucl. Mater,* **72,** p. 192 (1978).
38. B. S. Brown, T. H. Blewitt, T. L. Scott, and A. C. Klank, *J. Nucl. Mater.* **52,** 215 (1974).
39. D. M. Parkin and A. R. Sweedler, *IEEE Trans. Magn.* **MAG-11,** 166 (1975).
40. See for instance, D. M. Parkin and C. L. Snead, Jr., *Proc. Int. Conf. Fundamental Aspects Radiat. Damage Met.* (M. T. Robinson and F. W. Young, Jr., eds.), p. 1162, CONF-751006 (1975).
41. K. Inoue and K. Tachikawa, *Appl. Phys. Lett.* **25,** 94 (1974).
42. A. R. Sweedler *et al., Int. Conf. Radiat. Effects Tritium Technol. Fusion Reactors, Gatlinberg, Tennessee* CONF-750989, Vol. II, p. 422 (1976).
43. B. S. Brown, J. W. Hafstrom, and T. E. Klippert, *J. Appl. Phys.* **48,** 1759 (1977).
44. R. Chevrel, M. Sergent, and J. Prigent, *J. Solid State Chem.* **3,** 515 (1971).
45. S. Foner, E. J. McNiff, Jr., and E. L. Alexander, *Phys. Lett.* **49A,** 269 (1974); R. Odermatt, O Fischer, H. Jones, and G. Bongi, *J. Phys. C* **7,** L13 (1974).
46. B. T. Matthias, M. Marezio, E. Corenzwit, A. S. Cooper, and H. E. Barz, *Science* **175,** 1465 (1972).
47. M. Marezio, P. D. Dernier, J. P. Remeika, E. Corenzwit, and B. T. Matthias, *Mater. Res. Bull.* **8,** 657 (1973).
48. T. Luhman and D. Dew-Hughes, *J. Appl. Phys.* **49,** 936 (1978).
49. Ye. A. Antonova, K. V. Kiseleva, and S. A. Medvedev, *Fiz. Met. Metall.* **27,** 441 (1969).
50. J. R. Gavaler, *Appl. Phys. Lett.* **23,** 480 (1973); L. R. Testardi, J. H. Wernick, and W. A. Royer, *Solid State Commun.* **15,** 1 (1974).
51. H. Üllmaier, Radiation Effects and Tritium Technology for Fusion Reactors, Oak Ridge Nat. Lab. Rep. No. ORNL-750989, Vol. II, p. 403 (1976).
52. C. P. Bean, R. L. Fleischer, P. S. Swartz, and H. R. Hart, Jr., *J. Appl. Phys.* **37,** 2218 (1966).
53. J. L. Cooper, *RCA Rev.* **25,** 405 (1964).
54. R. Bett, *Cryogenics* **14,** 361 (1974).
55. A. R. Sweedler, D. G. Schweitzer, and G. W. Webb, *Phys. Rev. Lett.* **33,** 168 (1974).
56. A. R. Sweedler, D. E. Cox, and L. Newkirk, *J. Electron, Mater.* **4,** 883 (1975).
57. A. R. Sweedler, D. E. Cox, and S. Moehlecke, *J. Nucl. Mater.* **72,** 50 (1978).
58. A. R. Sweedler, D. Cox, D. G. Schweitzer, and G. W. Webb, *IEEE Trans. Magn.* **MAG-11,** 163 (1975).
59. S. Moehlecke, Ph.D. thesis, Univ. of Campinas, Brazil (1977) (unpublished).
60. J. M. Poate, L. R. Testardi, A. R. Storm, and W. M. Augustyniak, *Phys. Rev. Lett.* **35,** 1290 (1975).
61. R. Flukiger, A. Paoli, and J. Muller, *Solid State Commun.* **14,** 443 (1974).
62. J. Muller, R. Flukiger, A. Junod, F. Heiniger, and C. Susz, *in* "Low Temperature Physics" (K. D. Timmerhaus, W. J. O'Sullivan, and E. F. Hammel, ed.), Vol. LT-13, p. 446. Plenum, New York, 1974.
63. S. Moehlecke, D. E. Cox, and A. R. Sweedler, *Solid State Commun.* **23,** 703 (1977).
64. R. M. Waterstrat and R. C. Manuszewski, Nobel Metal Constitution Diagrams, Part 2, No. 5, Natl. Bur. Stand. (U.S.) Report No. NBSIR 73-415 (1975).
65. T. L. Francavilla, B. N. Das, D. U. Gubser, R. S. Meussner, and S. T. Sekula, *J. Nucl. Mater.* **72,** 203 (1978).
66. H. Bauer, E. J. Saur, and D. G. Schweitzer, *J. Low Temp. Phys.* **19,** 171 (1975).
67. I. V. Voronova, N. N. Mihailov, G. V. Sotnikov, and V. Ju. Zaikin, *J. Nucl. Mater.* **72,** 129 (1978).
68. M. Söll, K. Böning, and H. Bauer, *J. Low Temp. Phys.* **24,** 631 (1976).

69. R. Viswanathan, R. Caton, and C. S. Pande, *J. Low Temp. Phys.* **30**, 503 (1978).
70. H. Wiesmann, M. Gurvitch, A. K. Ghosh, H. Lutz, K. W. Jones, A. N. Goland, and M. Strongin, *J. Low Temp. Phys.* **30**, 513 (1978).
71. M. Couach, J. Doulat, and E. Bonjour, *IEEE Trans. Magn.* **MAG-13**, 655 (1977).
72. J. M. Poate, R. C. Dynes, L. R. Testardi, and R. H. Hammond, "Superconductivity in d- and f-Metals" (D. H. Douglass, ed.), p. 489. Plenum, New York, 1976.
73. B. Besslein, G. Ischenko, S. Klaumüzer, P. Müller, H. Neumüller, K. Schmelz and H. Adrian, *Phys. Lett.* **53A**, 49 (1975).
74. A. K. Ghosh, H. Weismann, M. Gurvitch, H. Lutz, O. F. Kammerer, C. L. Snead, A. Goland and M. Strongin, *J. Nucl. Mater.* **72**, 70 (1978).
75. D. Dew-Hughes, S. Moehlecke, and D. O. Welch, *J. Nucl. Mater.* **72**, 225 (1978).
76. A. I. Skvortsov, Y. V. Shemeljov, V. E. Klepatski, and B. M. Levitski, *J. Nucl. Mater.* **72**, 198 (1978).
77. A. R. Sweedler, D. E. Cox, S. Moehlecke, R. H. Jones, L. R. Newkirk, and F. A. Valencia, *J. Low Temp. Phys.* **24**, 645 (1976).
78. T. L. Francavilla, R. A. Meussner, and S. T. Sekula, *Solid State Commun.* **23**, 207 (1977).
79. T. Luhman and A. R. Sweedler, *Phys. Lett.* **58A**, 355 (1976).
80. L. R. Newkirk, F. A. Valencia, A. L. Giorgi, E. G. Szklarz, and T. C. Wallace, *IEEE Trans. Magn.* **MAG-11**, 221 (1975).
81. A. R. Sweedler and D. E. Cox, *Phys. Rev.* **12**, 147 (1975).
82. A. E. Karkin, V. E. Arkhipov, B. N. Goshchitskii, E. P. Romanov, and S. K. Sidorov, *Phys. Status Solidi (a)* **38**, 433 (1976).
83. D. E. Cox and J. A. Tarvin, *Phys. Rev.* **18**, 22 (1978).
84. D. E. Cox, S. Moehlecke, A. R. Sweedler, L. R. Newkirk, and F. A. Valencia, "Superconductivity in d- and f-Band Metals" (D. H. Douglas, ed.), p. 461. Plenum, New York, 1976).
85. D. E. Cox, *Solid State Commun.* **23**, 709 (1977).
86. R. Flükiger, Private communication.
87. J. G. Kohr, T. W. Eagar, and R. M. Rose, *Metall. Trans.* **3**, 1177 (1972).
88. L. R. Aronin, *J. Appl. Phys.* **25**, 344 (1954).
89. G. R. Johnson and D. H. Douglass, *J. Low Temp. Phys.* **14**, 565 (1974).
90. L. J. Vieland, *RCA Rev.* **24**, 366 (1964).
91. A. Taylor, N. J. Doyle, and B. J. Kagle, *J. Less-Common Met.* **4**, 436 (1962).
92. A. R. Sweedler, S. Moehlecke, R. H. Jones, R. Viswanathan, and D. C. Johnson, *Solid State Commun.* **21**, 1007 (1977).
93. G. W. Webb, A. R. Sweedler, and S. Moehlecke, *Mater. Res. Bull.* **12**, 657 (1977).
94. A. H. Dayem, T. H. Geballe, R. B. Zubeck, A. B. Hallak, and G. W. Hull, *Appl. Phys. Lett.* **30**, 541 (1977).
95. S. M. Kuznetsova and G. S. Zhdanov, *Kristallografiya* **16**, 1230 (1971).
96. B. N. Das, J. E. Cox, R. W. Huber, and R. A. Meussner, *Metall Trans.* **8A**, 541 (1977).
97. L. R. Testardi, R. L. Meek, J. M. Poate, W. A. Royer, A. R. Storm, and J. H. Wernick, *Phys. Rev. B* **11**, 4304 (1975).
98. G. Ilonca, *Phys. Status Solidi (a)* **43**, 387 (1977).
99. R. C. Dynes, J. M. Poate, L. R. Testardi, A. R. Storm, and R. H. Hammond, *IEEE Trans. Magn.* **MAG-13**, 640 (1977).
100. R. D. Burbank, R. C. Dynes, and J. M. Poate, *Int. Disc. Meeting Radiat. Effects Supercond.* Argonne Nat. Lab., Argonne, Illinois (1977).
101. O. Meyer and B. Seeber, *Solid State Commun.* **22**, 603 (1977); O. Meyer, *J. Nucl. Mater.* **72**, 182 (1978).

102. L. R. Testardi, J. M. Poate, W. Weber, W. M. Augustyniak, and J. H. Barrett, *Phys. Rev. Lett.* **39**, 706 (1977).
103. C. Pande, *Solid State Commun.* **24**, 241 (1977).
104. C. Pande, *J. Nucl. Mater.* **72**, 83 (1978).
105. M. A. Krivoglaz, "Theory of X-ray and Thermal Neutron Scattering by Real Crystals," pp. 249–292. Plenum, New York, 1969.
106. J. Appel, *Phys. Rev.* **13B**, 3203 (1976).
107. P. F. Farrell and B. S. Chandrasehkar, *Phys. Rev. Lett.* **38**, 788 (1977).
108. M. Gurvitch, A. K. Ghosh, C. L. Snead, Jr., and M. Strongin, *Phys. Rev. Lett.* **39**, 1102 (1977).
109. A value for Nb of ~0.01 was obtained by S. Klaumünzer, A. Ischenko, and P. Müller, *Z. Phys.* **268**, 189 (1974).
110. W. A. Fietz and W. W. Webb, *Phys. Rev.* **178**, 657 (1969).
111. B. S. Brown, T. H. Blewitt, D. G. Wozniak, and M. Suenaga, *J. Appl. Phys.* **46**, 5163 (1975).
112. J. P. McEvoy, Jr., R. F. Decell, and R. L. Novak, *Appl. Phys. Lett.* **4**, 43 (1974).
113. G. W. Cullen and L. Novak, *Appl. Phys. Lett.* **4**, 147 (1964).
114. G. W. Cullen, R. L. Novak, and J. P. McEvoy, *RCA Rev.* **25**, 479 (1964).
115. G. W. Cullen and R. L. Novak, *J. Appl. Phys.* **37**, 3348 (1967).
116. H. J. Bode and K. Wohlleben, *Phys. Lett.* **24A**, 25 (1967).
117. K. Wohlleben, *Z. Angew. Phys.* **27**, 92 (1969).
118. G. Ischenko, H. Mayer, H. Voit, B. Besslein, and E. Haindl, *Z. Phys.* **256**, 176 (1972).
119. M. Söll, H. Bauer, K. Böning, and R. Bett, *Phys. Lett.* **51A**, 83 (1975).
120. M. Suenaga, T. S. Luhman, and W. B. Sampson, *J. Appl. Phys.* **45**, 4049 (1974).
121. D. G. Schweitzer and D. M. Parkin, *Appl. Phys. Lett.* **24**, 333 (1974).
122. D. M. Parkin and D. G. Schweitzer, *Nucl. Technol.* **22**, 108 (1974).
123. S. L. Colucci and H. Weinstock, *Proc. Int. Conf. Low Temp. Phys., 14th* (M. Krusius and M. Vuorio, eds.), Vol. 2, p. 9. North-Holland Publ., New York, 1975.
124. S. L. Colucci, H. Weinstock, and B. S. Brown, *Appl. Phys. Lett.* **28**, 667 (1976).
125. S. L. Colucci, H. Weinstock, and M. Suenaga, *J. Appl. Phys.* **48**, 837 (1977).
126. J. E. Crow and M. Suenaga, *Proc. Appl. Supercond. Conf., Annapolis, Maryland* p. 472 (1972) (unpublished).
127. M. Suenaga, C. Klamut, and W. B. Sampson, *IEEE Trans. Magn.* **MAG-11**, 231 (1975).
128. C. L. Snead, Jr. and D. M. Parkin, *Nucl. Technol.* **29**, 264 (1976).
129. H. Becker, P. Maier, J. Pytlik, H. Ruoss, and E. Seibt, Kernforschungzentrum Karlsruhe, Internal Rep. No. 75-92 (1975).
130. M. Couach, J. Doulat, and E. Bonjour, *IEEE Trans. Magn.* **MAG-13**, 665 (1977).
131. C. L. Snead, Jr., *Appl. Phys. Lett.* **30**, 662 (1977).

Future Materials Development

DAVID DEW-HUGHES and THOMAS LUHMAN

Department of Energy and Environment
Brookhaven National Laboratory
Upton, New York

I. Introduction 427
II. Development of Known Materials 428
   A. A15 Compounds 428
   B. Transition Metal Carbides and Nitrides 429
   C. Laves Phases 432
   D. Ternary Molybdenum Sulfides 433
   E. Other Materials 436
III. The Possibility of New Superconductors with Higher T_c's 437
IV. Conclusions 442
   References 445

I. Introduction

Materials with higher current densities, particularly at high fields, and higher operating temperatures than those currently available will always be advantageous. Because J_c is generally some function of H_{c2}, improvements in J_c are best sought in materials with increased H_{c2}. Since H_{c2} to some extent scales with T_c, and a higher T_c allows a higher operating temperature, and thus savings in refrigeration costs, it is not unnatural that in the search for improved superconductors attention is concentrated on raising the critical temperature T_c. This approach is largely, though not exclusively, justified by experience.

Better conductors can result from three approaches: the improvement of existing conductors; the fabrication of conductors from presently known but not yet utilized materials; and the search for entirely new materials. In this chapter the editors speculate briefly about the second approach, and very briefly about the third. The possibilities for improvement of existing conductors have been discussed in earlier chapters.

427

II. Development of Known Materials

A. A15 Compounds

This chapter would be incomplete without any mention of the possible development of some of the A15 compounds other than Nb_3Sn and V_3Ga into practical conductors. Since this topic has received attention in the chapter by Luhman, this volume, this section is purposely brief and merely summarizes some of the discussion in the earlier chapter.

Of the half dozen A15 compounds with interesting superconducting properties, only two, Nb_3Sn and V_3Ga, have so far been developed into practical conductors. The reasons for this, associated with problems of phase stability and thermodynamics of compound formation, have already been discussed in the chapter by Dew-Hughes on the physical metallurgy of A15 compounds and in the chapter by Luhman. Three compounds, Nb_3Ge, Nb_3Al, and $Nb_3(Al,Ge)$ offer promise of significant improvement over Nb_3Sn and V_3Ga, either with higher T_c or with higher H_{c2}, or both.

It has not yet been possible to produce any of these in the highly desirable form of a flexible multifilamentary conductor. Nb_3Ge can be prepared as a flexible tape by vapor deposition onto a suitable substrate. The chemical vapor deposition (CVD) process [1], described in the chapter by Luhman, this volume, is the process most likely, on currently available evidence, to result in a commercial material. Such an Nb_3Ge conductor, with $T_c \approx 22–23$ K, would be employed most readily in applications that call for an operating temperature in the range of 10–15 K, higher than can be achieved by V_3Ga or Nb_3Sn.

$Nb_3(Al,Ge)$ is of interest because of its much higher critical field (>40 T at 4.2 K). Again vapor deposition is at present the only satisfactory route for the fabrication of conductor. With this material, sputtering seems to offer the best promise [2], though other methods have met with some success (see the chapter by Luhman, this volume). To date, current densities, at one hundredth (for bulk systems) to one tenth (for thin-film samples) of those of Nb_3Sn, are disappointingly low [3]. However, considerable improvement is to be expected when as much effort is put into the development of this material as has been into Nb_3Sn.

The T_c of Nb_3Al, at 18.9 K, is only marginally higher than that of Nb_3Sn (18.3 K); but H_{c2} (4.2) at ~ 30 T represents a substantial improvement over those of Nb_3Sn and V_3Ga (~ 20–26 T). Several methods, mentioned in the chapter by Luhman, this volume, have been employed to fabricate Nb_3Al conductors. That due to Ceresara et al. [4] in which spiral rolls of Nb and Al foils are inserted in a copper matrix and reacted to form thin layers of Nb_3Al, is at present probably the closest to realizing an actual stable conductor.

It would seem that, should the need arise, conductors based on Nb_3Al, $Nb_3(Al,Ge)$, or Nb_3Ge can be developed. It is, however, unlikely (for the reasons explained in the chapter by Luhman, this volume) that these materials will readily be produced in multifilamentary form by some variation of the simple, versatile, and readily controllable bronze process.

B. Transition Metal Carbides and Nitrides

Transition metal carbides and nitrides that occur with the B1 (NaCl) crystal structure, and with an electron-to-atom ratio close to Matthias's favored value of 4.7 (see the introductory chapter by Dew-Hughes, Section II,D) may be superconductors with high critical temperatures. The critical temperatures of some carbides, nitrides, and carbonitrides of elements and alloys from groups IVA, VA, and VIA of the periodic table are given in Table I. The data are abstracted from Roberts's compilation [5]. T_c for the alloys and carbonitrides depend on the exact composition; in each case the maximum reported T_c is quoted. Note that T_c for the nitride

TABLE I

T_c FOR TRANSITION METAL CARBIDES AND NITRIDES WITH B1 (NaCl) STRUCTURE[a,b]

Metal or alloy[c]	Carbide	Nitride	Carbonitride[c]
Ti	3.4	5.8	—
Zr	—	9.8	—
Hf	—	6.6	—
V	—	7.9	—
Nb	11.1	17.3	17.8
Ta	10.1	14.3	—
Mo	14.3	14.8	—
W	10	—	—
Nb–Ti	—	18.0	18.0
Nb–Zr	—	13.8	—
Nb–Ti–Zr	—	—	17.7
Nb–Hf	—	—	17.6
Nb–Ta	13.9	16.5	—
Nb–W	13.5	—	—
Ta–W	10.2	—	—

[a] Data from Roberts [5].
[b] Compounds may not be stoichiometric.
[c] For composition that gives maximum T_c.

is always higher than that for the corresponding carbide. T_c can be drastically reduced by departures from stoichiometry, as shown in Fig. 1. T_c for the nitrides drops by ~ 2 K for every atomic percent deficiency of nitrogen; for the carbides the drop is ~ 5 K per atomic percent deficiency of carbon. As with the A15s, preparation of stoichiometric material is often difficult for these materials.

MoC has the highest T_c, 14.3 K, of the carbides, but it can only be prepared in the B1 structure by rapidly quenching from the melt. Whereas alloying enhances the T_c in the NbC–TaC, and NbC–WC systems, any additions to MoC lower T_c [7]. NbN has the highest T_c of the binary nitrides, 17.3 K, and this can be increased by alloying with NbC to a maximum of 17.8 K [8]. The highest T_c for a compound with the B1 structure occurs with the alloy nitride (Nb–Ti)N and the alloy carbonitride (Nb–Ti)(CN), both at 18 K [9]. All other attempts to produce a higher T_c in the B1 structure by alloying or mixing carbides and nitrides have failed, and 18 K currently represents a maximum in this system.

The compounds of practical interest are NbN ($T_c = 17.3$ K, $H_{c2} = 15$ T), and NbC$_{0.3}$N$_{0.7}$ ($T_c = 17.8$ K, $H_{c2} = 12$ T). Despite its lower critical field, the latter compound has greater possibilities than does pure NbN in that a technique is available for producing multifilamentary material [10]. A precursor carbon yarn is chemically converted to a mixture of NbC and Nb(CN) (actual compositions NbC$_{0.8}$ and NbC$_{0.1}$N$_{0.9}$). Though not specified, the carbon yarn is presumably the same as that used for carbon fiber reinforcement, prepared by the pyrolysis of a polymer, usually acrylonitrile, filament. The yarn is reacted for 2–5 min at a temperature of 1400–1600°C with a mixture of NbCl$_5$, H$_2$, and N$_2$. Reacted layers up to 1 μm thick have a $T_c \approx 16.5$ K, H_{c2} of 11 T, and a critical current density at 4 T of about 10^9 A/m^2 in the reacted layer. Stability can be conferred by copper plating.

This material is obviously capable of considerable development but it is not clear that it has any appreciable advantage over existing conductors. T_c is comparable with Nb$_3$Sn, H_{c2} with that of Nb–Ti, and J_c is at present only one tenth of that of Nb$_3$Sn, though here there is hope for improvement. Only limited studies of the effect of radiation damage on the superconducting properties of B1 compounds have been carried out, and it is not clear that there is reason to expect that Nb(CN) will be any more insensitive to radiation damage than Nb$_3$Sn, particularly since 1% of vacancies can lower T_c by as much as 3 K, as is shown in Fig. 2 [11]. It is also doubtful that the rather fragile yarn will be as easy to handle, and to wind into magnets, as multifilamentary Nb$_3$Sn, though a magnet made of this material has been demonstrated [12], nor is there any indication of a substantial saving in cost over any existing superconductors. It

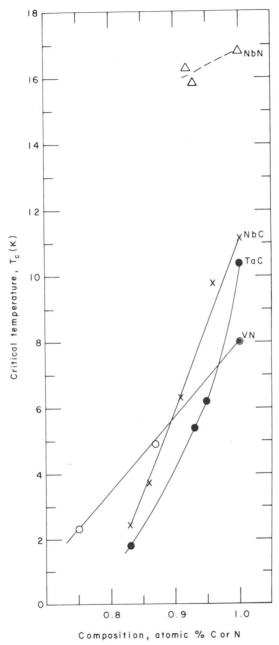

Fig. 1. Superconducting critical temperature versus composition for some transition metal carbides and nitrides. Data collected by Osipov [6].

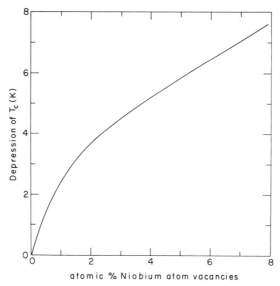

Fig. 2. The depression of superconducting critical temperature versus atom fraction of niobium vacancies for niobium oxynitrides, niobium carbonitrides, and niobium boronitrides. (After Storms *et al.* [11].)

must be concluded that the B1 structure is unlikely to provide a material that will supplant Nb_3Sn or other A15 compounds.

C. *Laves Phases*

Many of the compounds that crystallize in the three hexagonal phases C14 ($MgZn_2$ type), C15 ($MgCu_2$ type), and C36 ($MgNi_2$ type), known collectively as Laves phases, are superconductors. Of these, the binary compound with the best superconducting properties is V_2Hf, with the C15 structure. T_c and H_{c2} can be raised by substituting for some of the vanadium or some of the hafnium. T_c and H_{c2} (4.2 K) for V_2Hf and for the best ternary compound in each of four systems, from data of Inoue and Tachikawa [13], are given in Table II.

From this table it can be seen that compounds can be prepared that have critical temperatures superior to that of niobium–titanium and critical fields in excess of those for Nb_3Sn and V_3Ga. In addition, these compounds are less susceptible to radiation damage than are the A15s [14], and there is some indication that they are more ductile than the A15s and can be hot processed [15]. There is therefore a real incentive to develop these materials, and it is surprising that only one group of workers, Inoue and Tachikawa, seem to have given them any extensive attention.

TABLE II

T_c AND H_{c2} (4.2 K) FOR C15 COMPOUNDS BASED ON V_2Hf [13]

System	V_2Hf	V–Hf–Zr	V–Nb–Hf	V–Ta–Hf	V–Cr–Hf
Composition		$V_2(Hf_{0.5}Zr_{0.5})$	$(V_{1.83}Nb_{0.13})Hf_{0.75}$	$(V_{1.86}Ta_{0.14})Hf_{0.79}$	$(V_{1.8}Cr_{0.2})Hf$
Best T_c	9.2	10.1	10.4	10.0	9.9
Composition		$V_2(Hf_{0.6}Zr_{0.5})$	$(V_{1.89}Nb_{0.11})Hf_{0.83}$	$(V_{1.81}Ta_{0.19})Hf_{0.71}$	$(V_{1.83}Cr_{0.17})Hf_{1.33}$
Best H_{c2} (T)	20	24	25.7	26.1	23.4

Tapes have been fabricated in both the $V_2(Hf,Zr)$ and $V_2(NbHf)$ systems. In the former case, a core of Hf–Zr alloy is placed in a vanadium tube and then rolled into tape [15]. The Laves phase was produced by heat treatment at temperatures between 750 and 1300°C. Optimum properties were found with a $Hf_{0.35}Zr_{0.65}$ alloy core after heat treating in the temperature range 900–1000°C. Current densities in the reacted layer were 10^9 A/m² at 14 T. This is comparable with data for Nb_3Sn.

$V_2(NbHf)$ alloy tapes were cold rolled directly from arc-melted ingots, with intermediate anneals at 1500°C [16]. After reducing to finished thickness (200 μm), the tapes were given a final heat treatment at a temperature between 700 and 1500°C. Optimum properties are reported for the composition $V_2(Hf_{0.51}Nb_{0.49})$ after a final heat treatment at 1150°C. The microstructure consists of a mixture of Laves phase and bcc vanadium solid solution. For hafnium contents below 0.15 (5%) no C15 phase is observed. When the hafnium content is greater than 0.60 (20 at. %), the material is too brittle to be cold rolled. The current density as a function of composition and heat treatment is shown in Fig. 3. The best current density is 10^8 A/m² T, over the entire tape cross section.

These materials would appear to have considerable development potential, and could represent a serious challenge to existing commercial conductors. Their resistance to radiation damage and ability to be fabricated by simple deformation processing make them very attractive when compared to the A15s.

D.   Ternary Molybdenum Sulfides

In 1971, Chevrel *et al.* discovered a new class of ternary compounds based on molybdenum sulfide [17]. The following year Matthias *et al.* showed that many of these Chevrel phases were superconducting [18]. The superconducting Chevrel phases are derived from the ideal stoichiometry compound MMo_6S_8, where M is Ag, Sn, Ca, Sr, Pb, Ba, Cd, Zn, Mg, Cu, Mn, Cr, Fe, Co, Ni, Li, Na, Sc, or Y. An isomorphous series of selenides has also been found [19]. The remarkable feature of these com-

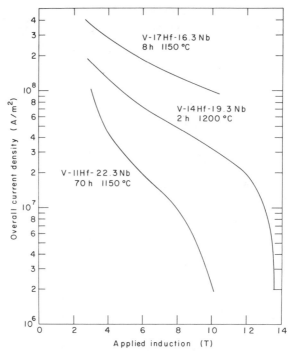

Fig. 3.   Critical current density J_c versus applied magnetic field H for several $(VNb)_2Hf$ tapes of indicated composition and heat treatment. (After Inoue and Tachikawa [16].)

pounds is that they have extremely high upper critical fields, in excess of 50 T [20].

Foner *et al.* [21, 22] systematically studied the relationship between composition, synthesis conditions, and the resultant superconducting properties T_c and $H_{c2}(T)$. Some of their results are shown in Fig. 4. The compounds $PbMo_5S_6$, $SnMo_5S_6$, and $SnAl_{0.5}Mo_5S_6$ were found to be the best superconductors. The compounds were synthesized from mixed elemental powders by reaction in evacuated quartz ampoules heated to various temperatures for 24 h and then slowly cooled. $Pb_{1.0}Mo_{5.1}S_6$, heated to 1100°C, gave the maximum T_c of 14.4 K, dH_{c2}/dT of 6 T/K, $H_{c2}(4.2\ K)$ of 51 T, and estimated $H_{c2}(0)$ of 60 T. This is of course the highest critical field of any bulk superconductor.

The ternary sulfides are extremely brittle, and their T_c is severely degraded by mechanical damage [23]. They are even more sensitive to radiation damage than the A15 compounds [14, 24]. However, their extremely high critical fields do not allow them to be ignored as potential conductor materials.

The authors [25] have successfully fabricated wires of lead molybdenum sulfide by a modified Kunzler technique. The sulfide was prepared as
described above, ground to a fine powder, placed in a metal tube, and reduced to a wire by drawing. Subsequent heat treatment sintered the sulfide powder, and restored the superconducting properties to those of the
original material. The most serious problem lay in the choice of a suitable
metal for the tube. Of those tried, silver was the only one that did not
react with and leach out the sulfur from the sulfide. The use of a silver
tube, however, limited the sintering temperature to ~925°C (below the

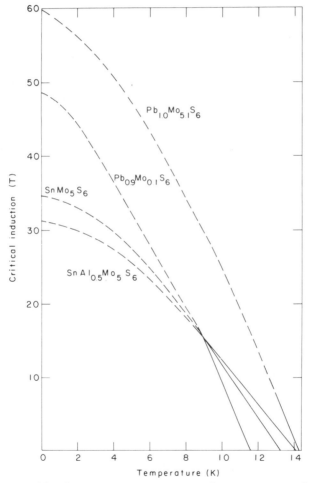

Fig. 4.   Upper critical field $H_{c2}(T)$ versus temperature for some ternary molybdenum sulfides. (After Foner *et al.* [21].)

melting point of silver at 960°C) and this proved insufficient to properly sinter the sulfide grains into a continuous wire. As a result, measured current densities were $1-5 \times 10^7$ A/m² at 4.2 K and 4 T; this is several hundred times less than for Nb_3Sn.

Alekseevskii *et al.* [26] prepared superconducting tapes of Pb–Mo sulfides and Sn–Mo sulfides by reacting molybdenum ribbon in the vapor of the two other components for several hours at 800–1000°C. J_c (4.2 K, 4 T) was $\sim 2 \times 10^7$ A/m² for a Pb–Mo–S sample. A similar method was used by Decroux *et al.* [27] to produce $PbMo_6S_8$ wires. A molybdenum wire was first reacted with sulfur at 68°C for 12 h. This formed a layer of molybdenum sulfide approximately 0.02 mm thick. This wire then underwent a second reaction in lead vapor for 24–96 h at temperatures between 950 and 1100°C. The final product was a Mo core wire, 1 mm in diameter, coated with a 0.03-mm thick layer of $PbMo_6S_8$. J_c for this material was almost identical to that of the tapes described above; J_c(4.2 K, 4.2 T) = 2×10^7 A/m².

Films of Chevrel phase copper molybdenum sulfide ($Cu_xMo_6S_8$) have been prepared by sputtering onto hot sapphire substrates [28]. This material has T_c varying from 8 to nearly 11 K, and H_{c2} (4.2 K) ≈ 9 T. The critical current density J_c was found to be in the range $3-7 \times 10^7$ A/m² in an applied field of 4 T [29]. This pinning is associated with excess molybdenum precipitates as a second phase.

A similar technique has been used to produce sputtered films of lead molybdenum sulfide [30]. The results on this material are rather disappointing. H_{c2} is only ~ 39 T, and J_c at 15 T and 1.8 K is only $\sim 5 \times 10^7$ A/m².

It is clear that Chevrel phase material is capable of further improvement and developments are eagerly awaited.

It is interesting to note that another, unrelated ternary sulfide, $Li_xTi_{1.1}S_2$ ($0.1 \leq 0.3$) with the hexagonal Ti_2S_4 structure, is superconducting with T_c in the range 10–13 K and an onset as high as 15 K [31].

E. Other Materials

High-T_c materials are here interpreted as those with a critical temperature in excess of 10 K and, apart from the bcc transition metal alloys, A15, B1, C15, and Chevrel compounds, which have already been discussed, either in this or in earlier chapters, there are only a few other superconductors that may be called high-T_c materials. Of these, three classes deserve mention. Sesquicarbides with the body-centered cubic $D5_c$ structure, isomorphous with plutonium sesquicarbide, Pu_2C_3, may be stabilized by high-pressure, high-temperature treatments. Both yttrium [32] and lanthanum [33] sesquicarbides produced in this way are superconducting, with T_c's of 11.5 K and 11 K, respectively. Higher T_c's are achieved in pseudobinaries of yttrium thorium sesquicarbide, 17 K [34], lutetium thorium sesquicarbide, 11.7 K [35], and lanthanum thorium sesquicarbide, 14.3 K [36]. The latter is noteworthy in that it is formed by arc melting alone, and does not require a high-pressure (~20 kbar) treatment for its synthesis. Critical fields and critical currents do not yet appear to have been measured for these materials, and their potential as practical conductors cannot be assessed.

The first high-T_c oxide material is lithium titanate, $LiTi_2O_4$, with the face-centered-cubic spinel structure. This has been reported to have a T_c onset of 13.7 K [37]. H_{c2} for this material is severely paramagnetically limited at ~20 T [38]. Attempts have been made to fabricate wire from this material. Best critical current densities in the superconductor are low, 5×10^7 A/m² at 5 T [39]. Again it is far too early in its development to assess the potential of this material.

The hydrides and deuterides of palladium and some of its alloys are superconductors. The highest reported T_c is 16.6 K for $Pd_{0.55}Cu_{0.45}H_{0.7}$ [40]. This high hydrogen ratio has been achieved by ion implantation. H_{c2} and J_c values have not been reported.

III. The Possibility of New Superconductors with Higher T_c's

It is reasonable to begin any search for new high-T_c superconductors by speculation as to the possibility of finding new A15 materials, since the highest critical temperatures to date are to be found in compounds with this structure. In fact all proven superconductors with a T_c in excess of 18 K are niobium-based A15 compounds. T_c is enhanced by the closest possible approach to the stoichiometric A_3B composition and by maximizing the degree of long-range crystallographic order (see the chapter by Dew-Hughes on the physical metallurgy of A15 compounds, this volume). The thermodynamically stable range of homogeneity of the A15 phase in those binary compounds with T_c greater than 20 K, Nb_3Ga (20.3 K) and Nb_3Ge (23 K), does not include the stoichiometric composition. The high-T_c material is metastable and can only be formed by nonequilibrium techniques such as splat-cooling, sputtering, chemical vapor deposition, or high pressures. It is possible that metastability is an inherent characteristic of very high temperature superconductors [41].

After Nb_3Ge, the next compound that is expected to have a high T_c

is Nb_3Si. Its critical temperature has variously been predicted to be 22.6–30.9 [42], 38 [43], 25.4 [44], and 31–35 K [45]. However, as can be seen from the portion of the phase diagram reproduced in Fig. 5, there is no equilibrium A15 structure in the Nb–Si system; the phase labeled Nb_3Si has the tetragonal Ti_3P structure [46]. Various nonequilibrium methods have been tried in order to produce metastable samples with the A15 structure. Electron beam vapor deposition first resulted in Nb_3Si with the cubic Cu_3Au structure [47]. Later attempts produced the A15 phase, but with only 21 at. % Si; T_c was 9 K and the lattice parameter a_0 was 5.17 Å [48]. Films of A15 Nb_3Si with up to 22.5 at. % Si and $a_0 = 5.16$ Å have been formed by chemical vapor deposition, but T_c was never higher than 8 K [49]. Using a dc getter sputtering technique, Somekh and Evetts have produced A15 Nb_3Si with an onset T_c of 14 K and a lattice parameter of 5.18 Å [50]. This T_c has since been raised to 17.6 K [51]. The highest onset T_c for Nb_3Si, 18.5–19 K, was obtained by explosive compression, at a peak pressure estimated to exceed 1 Mbar, of the high-temperature tetragonal Ti_3P phase [52]. The lattice parameter, deduced from rather inconclusive x-ray data, was only 5.03 Å, compared to that predicted on the Geller scheme (see the chapter by Dew-Hughes on the physical metallurgy of A15 compounds, this volume) of 5.08 Å. The explosive compression method was subsequently confirmed by Dew-Hughes and Linse, who found that the onset T_c was 18 K and the lattice parameter 5.12 Å [53].

The discrepancies between predicted and experimental values of T_c for A15 Nb_3Si are not surprising, since the nonequilibrium fabrication

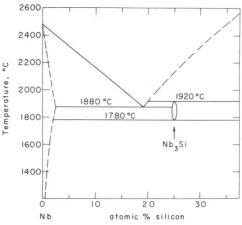

Fig. 5. A niobium-rich high-temperature portion of the niobium–silicon phase diagram showing the extremely limited range of the tetragonal Nb_3Si phase. (After Pan *et al.*, [46].)

methods are likely to result in material that is off-stoichiometry and disordered, both of which will reduce T_c from its maximum value (see the chapter by Dew-Hughes on the physical metallurgy of A15 compounds, this volume). There is every possibility that, with continued effort, either by high-pressure or improved thin-film deposition techniques, the experimental T_c of Nb_3Si will be raised above that of Nb_3Ge. It is worthwhile speculating as to the possible maximum T_c to be expected for fully ordered stoichiometric A15 Nb_3Si. This can be done by extrapolating trends in existing A15 compounds. Patterns have emerged from empirical correlations established between T_c values and the chemical, crystallographic, and electronic characteristics of A15 compounds. These correlations are only approximate since they depend for their validity upon accurately characterized samples.

Hatt et al. noted a correlation between the T_c of stoichiometric A_3B (A15) compounds and the inverse of the atomic mass M_B of the B element [54]. This led to a prediction that T_c for Nb_3Ge should be greater than 20 K, compared with the then experimental value of 17 K for quenched material. This prediction was subsequently confirmed when sputtering produced thin-film Nb_3Ge with $T_c \approx 23$ K. In the light of this result, Dew-Hughes and Rivlin reexamined this correlation and concluded that for A15 compounds based on V, Nb, or Ta, in which the B element is from group IVB of the periodic table, T_c varies as $(M_B)^{-1/2}$ [43]. The T_c predicted for Nb_3Si on this scheme was 38 K. In an attempt to find a correlation with more universal validity for A15 compounds, Dew-Hughes [44] suggested that

$$T_c = 19.6T_A V_0(A)[V_0(A15)\tilde{M}^{1/2}]^{-1}$$

where T_A is the superconducting transition temperature and $V_0(A)$ is the atomic volume of the A element, $V_0(A15)$ the atomic volume of the A15 compound, and $\tilde{M}$ the average atomic mass of the compound. This equation, as can be seen from Fig. 6, seems adequately to describe T_c for compounds based on V, Nb, and Ta in which the B element comes from either group IIIB or group IVB of the periodic table. If the B element is from any other group, and in particular if it is a transition metal, the T_c of the compound falls considerably below that predicted by the foregoing equation. When the A element is from other than group VA, for example if it is Mo (the only other A element for which a series of superconducting A15 compounds exists), the correlation fails completely.

This particular correlation would be of no very great interest but for two subsequent experimental facts. By a repetition of their earlier method with Nb_3Si, Pan et al. [55] have subjected Ta_3Si, which also has the tetragonal Ti_3P structure, to explosive compression to produce super-

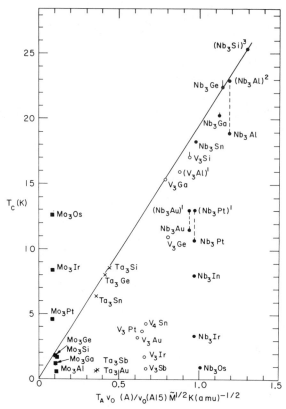

Fig. 6. The critical temperature of superconducting A15 compounds, based on V, Nb, Ta, and Mo, plotted as a function of the critical temperature of the A element (T_A), of the fractional change in atomic volume between the A element and the compound $[V_0(A)/V_0(A15)]$, and the average atomic mass of the compound $\tilde{M}$. (1) Extrapolation due to Flükiger. (2) Sweedler's extrapolation for Nb₃Al. (3) Estimate for stoichiometric Nb₃Si. (After Dew-Hughes [44].)

conducting A15 Ta₃Si. This has a T_c of 8.6 K, as compared to that predicted by the Dew-Hughes equation, 8.7 K. Also, their A15 Nb₃Si, with $T_c = 19$ K, contains only 23 at. % Si, as confirmed by microprobe [56]. The largest change in T_c with composition observed in A15 compounds is ~3 K/ at. % (see the chapter by Dew-Hughes on the physical metallurgy of A15 compounds, this volume), suggesting that stoichiometric A15 Nb₃Si, with the same (undetermined) degree of order as the Pan et al. sample, would have a $T_c \approx 25$ K. Again, this is close to that predicted by the foregoing equation, 25.4 K. If the validity of this correlation is accepted, one is led to the pessimistic conclusion that no binary A15 compound is likely to have a T_c in excess of 25–30 K. The addition of a third element to form a ternary A15 compound is unlikely to produce any substantial increase in T_c. No ternary has yet been made in which T_c is greater than the appropriately weighted average of the T_c's of the pure stoichiometric binaries. The only reason for the success of $Nb_3(Al,Ge)$ was that, as described in the chapter by Dew-Hughes on the physical metallurgy of A15 compounds, this volume, it allowed a closer approach to stoichiometry than had at that time been attained in Nb_3Ge. It seems reasonable to conclude that, in the light of what has up to now been achieved, T_c in the A15 structure is probably limited to a maximum of $\sim$25–30 K.

Though ternaries of A15 structures do not seem to hold out any hope for significant increases in T_c, they can show substantial improvements in H_{c2}. The addition of Ga to Nb_3Sn produced by the bronze process raises H_{c2} at 4.2 K from 18–19 to $\sim$25 T [57]. Small quantities of Al, Ga, In, Tl, or Pb in arc-melted samples of Nb_3Sn raised H_{c2} at 4.2 K from 23 to $\sim$30 T [58]. Similar results are obtained for doped Nb tapes dipped into tin. These increases are striking when it is realized that only a few percent of the tin is replaced by the third element. Such additions give a valuable enhancement in current density at high fields.

In a rather curious patent, Winter and Sethna [59] claim that by incorporating fine powders of an A15 compound coated with copper (100-Å particles of either Nb_3Sn or V_3Si with a 40-Å layer of copper) into a copper matrix, they can increase both T_c and H_{c2} to figures well above those values usually accepted for these materials. For example, one specimen of Nb_3Sn has $T_c = 24.7$ K and $H_{c2} = 71$ T, and a specimen of V_3Si has $T_c \approx 29$ K and an H_{c2} of 100 T. Critical current densities in these conductors (no field values are quoted, so they are presumably self-field values) are $\sim$9 × 10^9 A/m^2. This is less than self-field values quoted for Nb_3Sn (see Table III of the chapter by Bussière, this volume). The high values of T_c and H_{c2} may actually be observations of superconducting fluctuations above the true T_c and H_{c2}, which are more likely to occur in small samples.

From this brief look at high-T_c materials, it is clear that no other class of materials appears to offer a serious challenge to the A15 compounds insofar as T_c and J_c are concerned. Only the ternary molybdenum sulfides have higher critical fields. The reason for the supremacy of the A15 compounds is not fully understood, but as mentioned in the introductory chapter by Dew-Hughes, Section II,D, this structure may represent the optimum compromise between a high density of states and strongly covalent bonds. If this is so, it looks as if the highest critical temperature to be found in an electron–phonon superconductor will be about 25 K. Higher critical temperatures will require a different mechanism for superconductivity.

Two other types of superconductors have been proposed. Ginzburg [60] suggested an exciton mechanism, which in theory could lead to very high T_c's. His mechanism requires thin (~ 10 Å) layers of metal sandwiched between layers of a highly polarizable dielectric. When certain transition metal dichalcogenides, e.g., TaS_2 and $NbSe_2$, are intercalated with electron-donating organic compounds (Lewis bases) such as pyridene, they have superconducting critical temperatures of up to 7 K. These were at first believed to be examples of excitonic superconductors, but are now known to be phononic. The existence of excitonic superconductivity has yet to be demonstrated.

In 1964, Little proposed that organic molecules with highly polarizable side groups might be superconductors and could, in principle, have critical temperatures in excess of room temperature [61]. Despite an intensive research effort, no superconducting polymer was found until 1974, when $(SN)_x$ was found to be superconducting below 0.3 K [62]. This, the first and only superconducting polymer, is believed to be a conventional BCS electron–phonon superconductor, though considerable interest was rekindled in the possibility of polymer superconductors when it was announced that the organic charge-transfer complex tetrathiafulvalene–tetracyanoquinodimethane (TTF–TCNQ) showed imminent superconductivity at 58 K [63]. A conductivity maximum just before a transition to an insulator was erroneously ascribed to superconducting fluctuations. A more recent report of superconductivity at 140 K in CuCl under pressure [64] probably has as an explanation that it undergoes an insulator-to-metal transition.

IV. Conclusions

Experimentally no mechanism other than the electron–phonon mechanism of superconductivity has been confirmed. The highest experimental critical temperature is 23.5 K in the A15 compound Nb_3Ge [65]. It would be extremely foolish to state either that no mechanism other than the electron–phonon mechanism exists, or that no better superconducting compounds than the A15s will ever be found. It is, however, correct to conclude in the light of our current experimental knowledge that new mechanisms or better compounds are extremely unlikely. This, rightly, will not deter others in the search for new superconducting materials.

Since the discovery of superconductivity in 1911, the average rate at which the known T_c has been raised is a disappointingly low 0.3 K per year [66] (see Fig. 7). If progress continues at this average rate, 24 K should be reached in about 1980–84 and 30 K not until after the year 2000. Short-term efforts should be directed toward making usable conductors from the high-T_c A15s Nb_3Al, Nb_3Ga, Nb_3Ge, and $Nb_3(Al,Ge)$. For service in higher fields, 20–40 T, the ternary molybdenum sulfides must be developed. At lower fields, and where resistance to radiation damage is important, the Laves phase materials could offer an economic alternative. The properties of these candidate materials are listed in Table III, and the temperature variation of their critical fields compared in Fig. 8.

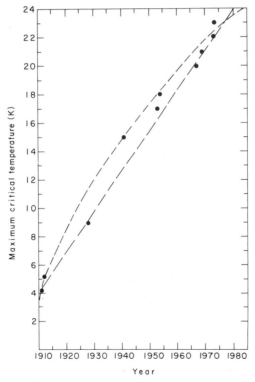

Fig. 7. Maximum critical temperature of superconductors versus year of discovery. (After Mathias [66].)

TABLE III

<small>SUPERCONDUCTING MATERIALS FOR FUTURE DEVELOPMENT</small>

Materials	T_c (K)	H_{c2} at 4.2 K (T)	Fabrication
A15			
Nb_3Al	18.9	30	Diffusion (poor), sputtering
Nb_3Ga	20.3	33	Evaporation
Nb_3Ge	22.5	37	CVD, sputtering
$Nb_3(Al,Ge)$	21.0	41	Sputtering
Ternary sulfide			
$PbMo_{5.1}S_6$	14.6	50	Powder metallurgy, sputtering, CVD
C15 (Laves phase)			
$V_2(Hf,Zr)$	10	24	Diffusion
$(V,Nb)_2Hf$	10.4	26	Diffusion

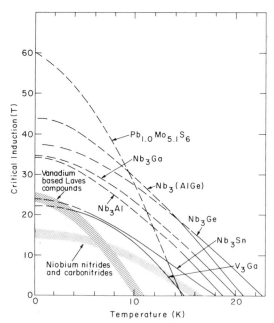

Fig. 8.   Upper critical field H_{c2} versus temperature for the various materials that may be regarded as prime candidates for development. (After Dew-Hughes [67].)

References

1. A. I. Braginski, J. R. Gavaler, G. W. Roland, M. R. Daniel, M. A. Janocko, and A. T. Santhanam, *IEEE Trans. Magn.* **MAG-13**, 300 (1977).
2. S. D. Dahlgren, M. Suenaga, and T. S. Luhman, *J. Appl. Phys.* **45**, 5462 (1974).
3. J. Ruzicka, *Cryogenics* **14**, 434 (1974).
4. S. Ceresara, M. V. Ricci, N. Sacchetti, and G. Sacerdoti, *IEEE Trans Magn.* **MAG-11**, 263 (1975).
5. B. W. Roberts, *J. Phys. Chem. Ref. Data* **51**, 581 (1976).
6. K. A. Osipov, *Dokl. Akad. Nauk SSSR Phys. Chem.* **230**, 645 (original), 900 (translation) (1976).
7. R. H. Willens, É. Buehler, and B. T. Matthias, *Phys. Rev.* **159**, 327 (1967).
8. M. W. Williams, K. M. Ralls, and M. R. Pickus, *J. Phys. Chem. Solids* **28**, 333 (1967).
9. N. Pessall and J. K. Hulm, *Physics* **2**, 311 (1966).
10. W. D. Smith, R. Y. Liu, J. A. Coppola, and J. Economy, *IEEE Trans. Magn.* **MAG-11**, 182 (1975).
11. E. K. Storms, A. L. Giorgi, and E. G. Szklarz, *J. Phys. Chem. Solids* **36**, 689 (1975).
12. M. C. Ohmer, J. J. Wollan, and J. C. Ho, *IEEE Trans. Magn.* **MAG-11**, 159 (1975).
13. K. Inoue and K. Tachikawa, *Proc. Appl. Supercond. Conf., Annapolis, Maryland.* IEEE Cat. No. 72-CHO 682-5 TABSL, p. 415 (1972).
14. A. R. Sweedler *et al., Int. Conf. Radiat. Effects Tritium Technol. for Fusion Reactors, Gatlinburg, Tennessee.* Conf. 750989, Vol. III, p. 422 (1976).
15. K. Inoue and K. Tachikawa, *Appl. Phys. Lett.* **18**, 235 (1971).
16. K. Inoue and K. Tachikawa, *IEEE Trans. Magn.* **MAG-13**, 840 (1977).
17. R. Chevrel, M. Sergent, and J. Prigent, *J. Solid State Chem.* **3**, 515 (1971).
18. B. T. Matthias, M. Marezio, E. Corenzwit, A. S. Cooper, and E. Barz, *Science* **175**, 1465 (1972).
19. M. Sergent and R. Chevrel, *C. R. Acad. Sci. Paris* **24**, 1965 (1972).
20. R. Odermatt, Ø. Fischer, H. Jones, and G. Bongi, *J. Phys. C.* **7**, 213 (1974).
21. S. Foner, E. J. McNiff, Jr., and E. J. Alexander, *Phys. Lett.* **49A**, 269 (1974).
22. S. Foner, E. J. McNiff, Jr., and E. J. Alexander, *IEEE Trans. Magn.* **MAG-11**, 155 (1975).
23. P. Müller and M. Rohr, *Phys. Status Solidi.* (*a*) **43**, K19 (1977).
24. B. S. Brown, J. W. Hafstrom, and T. E. Klipper, *J. Appl. Phys.* **48**, 1759 (1977).
25. T. S. Luhman and D. Dew-Hughes, *J. Appl. Phys.* **49**, 936 (1978).
26. N. E. Alekseevskii, M. Glinski, N. M. Dobrovolskii, and V. I. Tsebro, *JETP Lett.* **23**, 412 (1976).
27. M. Decroux, Ø. Fischer, and R. Chevrel, *Cryogenics* **17**, 291 (1977).
28. C. K. Banks, L. Kammerdiner, and H. L. Luo, *J. Solid State Chem.* **15**, 271 (1975).
29. S. A. Alterovitz, J. A. Woollam, L. Kammerdiner, and H. L. Luo, *Appl. Phys. Lett.* **31**, 233 (1977).
30. S. A. Alterovitz, J. A. Woollam, L. Kammerdiner, and H. L. Luo, *Appl. Phys. Lett.* **33**, 264 (1978).
31. H. E. Barz, A. J. Cooper, E. Corenzwit, M. Marezio, B. T. Matthias, and P. H. Schmidt, *Science* **175**, 884 (1972).
32. M. C. Krupka, A. L. Giorgi, N. H. Krikorian, and E. G. Sklarz, *J. Less-Common Met.* **17**, 91 (1969).
33. A. L. Giorgi, E. G. Sklarz, M. C. Krupka, and N. H. Krikorian, *J. Less-Common Met.* **17**, 121 (1969).

34. M. C. Krupka, A. L. Giorgi, N. H. Krikorian, and E. G. Sklarz, *J. Less-Common Met.* **19,** 113 (1969).
35. M. C. Krupka, A. L. Giorgi, and E. G. Sklarz, *J. Less-Common Met.* **30,** 217 (1973).
36. A. L. Giogi, E. G. Sklarz, N. H. Krikorian, and M. C. Krupka, *J. Less-Common Met.* **22,** 131 (1970).
37. D. C. Johnston, H. Prakash, W. H. Zachariasen, and R. Viswanathan, *Mater. Res. Bull.* **8,** 777 (1973).
38. S. Foner and E. J. McNiff, Jr., *Solid State Commun.* **20,** 995 (1976).
39. U. Roy, A. Das Gupta, and C. C. Koch, *IEEE Trans.* **MAG-13,** 836 (1977).
40. B. Stritzker, *Z. Phys.* **268,** 261 (1974).
41. B. T. Matthias, E. Corenzwit, A. S. Cooper, and L. D. Longinotti, *Proc. Nat. Acad. Sci. U.S.* **68,** 56 (1971).
42. L. Gold, *Phys. Status Solidi* **4,** 261 (1964).
43. D. Dew-Hughes and V. G. Rivlin, *Nature (London)* **250,** 723 (1974).
44. D. Dew-Hughes, *Cryogenics* **15,** 435 (1975).
45. S. Geller, *Appl. Phys.* **7,** 321 (1975).
46. V. M. Pan, V. V. Pet'kov, and O. G. Kulik, "Physics and Metallurgy of Superconductors, Moscow (1965–1966)" (E. M. Savitskii and V. V. Baron, eds.). Consultants Bureau, New York, 1970.
47. F. Galasso and J. Pyle, *Acta Crystallogr.* **16,** 228 (1963).
48. R. H. Hammond, *IEEE Trans. Magn.* **MAG-11,** 201 (1975).
49. H. Kawamura and K. Tachikawa, *Phys. Lett.* **55A,** 65 (1975).
50. R. E. Somekh and J. E. Evetts, *Solid State Commun.* **24,** 733 (1977).
51. R. E. Somekh and J. E. Evetts, Private communication (1978).
52. V. M. Pan *et al., JETP Lett.* **21,** 228 (1975).
53. D. Dew-Hughes and V. D. Linse, *J. Appl. Phys.* **50** (1979).
54. B. A. Hatt, J. K. R. Page, and V. G. Rivlin, *J. Low Temp. Phys.* **10,** 285 (1973).
55. V. M. Pan, A. G. Popov, V. P. Alekseevskii, O. G. Kulik, and V. V. Yarosh, *Fiz. Nizk. Temp.* **3,** 801 (1977).
56. V. M. Pan, Private communication (1978).
57. D. Dew-Hughes and M. Suenaga, *J. Appl. Phys.* **49,** 357 (1978).
58. R. Akihama, K. Yasukochi, and T. Ogasawara, *IEEE Trans. Magn.* **MAG-13,** 803 (1977).
59. H. Winter and D. Sethna, German Patent Disclosure No. 2516, 747, October 28, 1976; US Patent No. 4,050,147, September 27, 1977.
60. V. L. Ginzburg, *Contemp. Phys.* **9,** 355 (1968).
61. W. A. Little, *Phys. Rev. A* **134,** 1416 (1964).
62. R. L. Greene, G. B. Street, and L. J. Suter, *Phys. Rev. Lett.* **34,** 577 (1975).
63. A. B. Coleman, M. J. Cohen, D. J. Sandman, F. G. Yamagishi, A. F. Garito, and A. J. Heeger, *Solid State Commun.* **12,** 1125 (1973).
64. H. B. Brandt, S. V. Kubschinnikov, A. P. Rusakov, and M. V. Semerov, *Pisma JETP* **27,** 37 (1978).
65. R. E. Somekh and J. E. Evetts, Applied Superconductivity Conference, Pittsburgh (1978). *IEEE Trans. Magn.* **MAG 15,** 494 (1979).
66. B. T. Marthias, *in* "Critical Materials Problems in Energy Production" (C. Stein, ed.), p. 663. Academic Press, New York, 1976.
67. D. Dew-Hughes, *Adv. Cryogen. Eng.* **22,** 316 (1977).

Index

A

A15 (Cr$_3$Si) crystal structure, 138
A15 compound formation, 226
A15 compounds
   see also A15 conductors; A15 superconductors
   addition of third element to, 155
   atomic radii for, 140
   calculated vs. measured lattice parameters for, 141
   critical current density for, 162–167
   critical temperatures for, 142–143
   defect state in, 402–409
   dislocations in, 201–202
   electron diffraction from, 200
   fabrication of, 222
   fast-particle irradiation of, 142–143
   flux pinning theory in, 164–167
   future development of, 428–429
   impurity effects and pseudobinaries in, 153–155
   ion beam milling for, 179
   irradiation effects in, 372–422
   long-range crystallographic order for, 142–143
   martensitic type shear transformation in, 140
   microstructure and critical currents in, 162–164
   Nb$_3$Al, 149
   Nb$_3$Ga, 149–150
   Nb$_3$Ge, 151–153
   Nb$_3$Sn, 145–146
   phase diagrams and T_c for, 144–153
   physical metallurgy of, 137–167
   solid state technique in, 229–231
   stability of, 224–231
   superconducting critical temperature for, 142–148
   T_c theory in, 156–158
   thermodynamics and phase stability of, 144–145
   thinning techniques for, 179–181
   upper critical field H_{c2} for, 158–162
   V$_3$Ga, 147
   V$_3$Si, 146–147
A15 conductor processing
   chemical vapor deposition in, 233–234
   electron beam vapor deposition in, 234–235
   liquid state diffusion in, 232–233
   physical vapor deposition in, 234–235
   solid state diffusion process in, 235–238
   sputtering deposition in, 234–235
   variations on solid state and liquid solute diffusion techniques in, 238–239
A15 conductors
   see also Bronze-processed conductors; A15 superconductors
   A15 phase in, 227
   bronze process for, 223–224, 227, 235–238
   compressive strains on superconducting compound, 262
   conductor processing methods for, 231–254
   critical current density and microstructure in, 244–250
   critical magnetic field for, 222
   cubic Cr$_3$Si(A$_3$B) structure of, 221
   metallurgy of, 221–263

nomenclature in, 221
phenomenological flux-pinning theories
    250–254
stable phase in, 227
stress–strain relationship in, 254–256, 262
and superconducting properties vs. micro-
    structure in bronze-processed conductors,
    240–254
technological importance of, 222
under tensile strain, 254–263
ternary phase in, 228
"wind and react" method for, 222
wire vs. tape types, 222–223
A15 phases, stability of, 139, 227
A15 structure
    see also A15 compounds; A15 conductors
    Bragg peaks in, 395
    chains of A atoms in, 138
    fundamental and superlattice spots in, 201
    radius ratio in, 139
    structure cell of, 139
A15 superconductors
    see also A15 conductors; Transmission
        electron microscopy
    diffraction contrast studies of, 188
    flux pinning by grain boundaries in, 194–
        195
    grain growth in, 198–200
    with higher critical temperatures, 437–442
    ion milling techniques for, 202
    radiation damage to, 196–198
    stacking faults in, 195–196
Accelerator magnets, configurations of, 77
Ac losses
    in composite conductors, 304–307
    for conductors in cable configuration,
        302–303
    edge effects in, 302
    surface currents and, 298–300
    in Meissner state, 300–302
    metallurgical effects in, 303–304
    of pure niobium, 303–309
    temperature dependence, 308–309
    trapped magnetic flux, 307–308
    in type II superconductors, 296–303.
Annealing studies, irradiation effects and,
    381–388
Antistite defects, 416
Argonne National Laboratory, 73
Astron sheath, 94

Auger spectroscopy, for niobium surfaces,
    332–333

B

Bardeen–Cooper–Schrieffer theory
    for strong-coupling superconductors, 26
    of superconductivity, 23–24
Beam transport systems, 77
BEBC, see European bubble chamber mag-
    net
Berkeley Bevatron, 79
Body-centered-cubic alloys, irradiation ef-
    fects in, 359–370
Bose–Einstein statistics, Cooper pairs and,
    23
Bragg peaks, in A15 structures, 395
Brittleness problem, in multifilamentary A15
    conductors, 64
Bronze process
    for A15 conductors, 223–224, 235–238
    irradiation effects and, 411
    single-phase A15 compounds and, 227
    variation in, 238
Bronze-processed A15 conductors, magni-
    tude and field dependence of, 253
Bronze processed conductors
    bronze-to-niobium ratio in, 259
    compound growth and its effect on T_c of,
        240–244
    compressive strain in, 256
    critical temperature and, 242–244
    superconducting properties vs. micro-
        structure in, 240–254, 257–258
Bronze-processed niobium–tin multifila-
    ment conductors, 224–225
Bronze-processed tapes, electrical stability
    of, 247
Brookhaven Alternating Gradient Synchro-
    tron, 80, 419
Brookhaven High Flux Beam Reactor, 419
Brookhaven National Laboratory, 237, 277,
    337
Bubble chamber magnet (CERN), 75
Bulk pinning, 288
    see also Flux pinning; Pinning

C

Cable design
    ac losses and, 302–303
    coaxial cables and, 271–272

Cell wall, superconducting condensation energy of, 103
Central Electricity Research Laboratories, 272, 276
CERN, *see* European Organization for Nuclear Research
Charged-particle irradiation, 379–381
Chemical thinning process
   for A15 compounds, 179–181
   vs. electropolishing, 177
Chemical vapor deposition process
   in A15 conductor processing, 233–234
   future development of, 428
   in power transmission superconductor fabrication, 281
Chevrel phases, 370–371
Closed toroidal systems, 94
Coaxial cables, for power transmission, 271
Composite conductors, ac losses in, 304–307
   see also A15 conductors; A15 superconductors
Conductors
   A15, *see* A15 conductors; A15 superconductors; Superconductors
   bronze-processed, *see* Bronze-processed conductors
   intrinsically stable, 54–59
Controlled thermonuclear reactor, superconductors in, 92
Cooper pairs
   Bardeen–Cooper–Schrieffer theory and, 23–24
   Bose–Einstein statistics and, 23
   condensation of electrons into, 23–24
   energy levels and, 24
Cos θ configuration, of transverse-field magnets, 77, 82
Critical current
   defined, 28
   external field and, 27–28
   flux pinning and, 28–31
   irradiation effects and, 355–359
   magnetization and, 29
   for type I superconductors, 27
Critical current density
   microstructure and, 244–250
   of niobium–titanium conductors, 104–116
   pinning force and, 28
   scaling law for, 409–422

Critical field, irradiation effects and, 353–355
Critical state
   concept of, 29
   for type II superconductors, 27–30
Critical temperature
   heat treatment and, 242–244
   superconductivity and, 22–26
Cryostat, superconducting lenses and, 213–215
Cryostrand cable, 223
Crystal structures, superconductivity in, 25–26
CTR, *see* Controlled thermonuclear reactor
Culham superconducting levitron, 93, 96
CVD, *see* Chemical vapor deposition
Cyclotrons and synchrotrons, beam transport systems for, 80

D

Debye–Scherrer x-ray patterns, irradiation effects and, 390
Dynamic stabilization, defined, 53

E

EBD, *see* Electron beam vapor deposition
Electrolytic thinning techniques, 174–176
Electron beam evaporation method, 316
Electron beam vapor deposition, in A15 conductor processing, 234–235
Electron lenses, design and performance of, 208–209
Electron microscopy
   see also Transmission electron microscopy
   high resolution, 172
   high-resolution Bitter technique in, 204–205
   high-voltage, 207
   lattice resolution in, 183–184
   liquid helium stages in, 181–183
   magnetic flux line observation in, 203–207
   martensitic phase transformation in, 192–193
   nature of micrographs in, 187
   of niobium and alloys, 188–191
   preparation of thin foils for, 173
   scanning transmission instruments and, 172

superconducting lenses and, 207–215
superconducting magnet technology and, 172
superconductivity and, 171–216
unconventional techniques in, 205–206
Electron–phonon interaction
origin of, 24
in superconductors, 23
Electrons, condensation of into Cooper pairs, 23–24
Electron-to-atom ratio, in superconductivity, 25
Electropolishing techniques, 174–176
hydrogen-free, 176–177
Energy Doubler/Saver Project, 81
EPR, *see* Experimental power reactor
European bubble chamber magnet
conductor lengths required for, 73–74
cryostat for, 74
European Organization for Nuclear Research, 73, 75
Experimental power reactor, 94, 96
"External diffusion" process, in multifilamentary composites, 63

F

Fermi National Accelerator Laboratory, 81
Fernández–Moran superconductivity lens, 209
Ferromagnetism, superconductivity and, 5, 25
Filamentary stabilization, in superconducting strips, 42–44
Filling factors, in high current capacity macrocomposites, 59
Flexible cables
of niobium, 276–278
in power transmission, 275–278
FLL, *see* Flux-line lattice
Fluence regimes, irradiation effects and, 418
Flux, across superconducting strip, 40–41
see also Flux lines; Flux pinning
Flux avalanche, 39
Flux disturbances, origin and propagation of, 42
Flux jumping
in high-field superconductors, 47
instabilities of, 49–51
stabilization against, 51–59
Flux lattice elasticity, in flux pinning, 36

Flux line investigation, unconventional techniques in, 205–206
Flux-line lattice, 204
plastic shear of, 251
replica technique in, 205
spacing of, 252
Flux lines
direct resolution of, 206–207
grain boundaries and, 164
pinning of, 28
pinning centers and, 32
Fluxoids, pinning of, 358
Flux pinning
see also Pinning; Pinning force; Surface pinning
in A15 compounds, 164–167, 250–254
competing mechanisms in, 102–103
critical current and, 28–31
effective defect density for, 364–365
flux-lattice elasticity and, 36
flux-line density in, 33–34
microstructure and, 37–39
in niobium-tantalum alloys, 117–120
strength of, 38
theories of, 30–37
three types of, 33
for type II superconductors, 27–30
Frenkel pairs
density distribution of, 357
resistivity due to damage in, 415–416
Fuji Electric Company, 89
Future materials, development of, 427–444

G

Garrett Corporation, 86
General Electric Company, 86, 89
Gibbs free energy, in type II superconductors, 286–287
Gibbs function
of magnetic system, 13
Meissner effect and, 14
Ginzburg–Landau constant, 19
Ginzburg–Landau equations, 14–15
Ginzburg–Landau parameter, 353
Ginzburg–Landau theory, 13–16
Grain boundaries
elastic strain field of, 166
flux lines and, 164–165
Grain growth, carbon removal and, 334

H

HACS, *see* High-amperage conductors
Helium refrigerator, cost of, 65
Helmholtz coil, for neutron scattering experiments, 70
HFBR, *see* Brookhaven High Flux Beam Reactor
High-amperage conductors
  categories of, 125
  fabrication of, 125–134
  of niobium-tantalum, 125–134
  vs. rapidly changing magnetic fields, 128
High-energy accelerators, superconducting main ring magnets and, 79
High-energy particles, irradiation with, 375
High-field magnets, rf devices and, 328
High-field superconductors, 21
  discovery of, 47
  flux pumping in, 66
High-resolution Bitter technique, in flux-line lattice investigation, 204–205
High-voltage electron microscopy, 207
High-voltage overhead lines, long-distance transmission over, 267–268
Homopolar superconducting machines, 85–86
HVEM, *see* High-voltage electron microscopy
HYBUC chamber magnet, 75
Hydrofluoric acid, in chemical thinning, 180
Hysteretic losses, in type II superconductors, 296

I

Image contrast, from lattice defects, 183
Institute for Electrical Machines (Leningrad), 89
International Research and Development Company, 86
Interstitial-cluster formation, 355
Interstitial trapping at impurities, 355
Interstitial-vacancy annihilation, 355
Ion beam milling, 178–179
Ion milling chamber, 178
Irradiation effects
  for A15 compound, 372–422
  annealing studies and, 381–388
  body-centered-cubic alloys and, 359–370
  "bronze process" in, 411

  charged-particle type, 379–381
  critical field and currents in, 355–359, 409–422
  Debye–Scherrer x-ray patterns in, 390
  fluence regimes and, 418
  lattice parameter in, 388–394
  "magnetic" effects in, 410
  in niobium–titanium conductors, 367–368
  in non-A15 compounds, 370–372
  in superconducting materials, 349–422
  transition temperatures and, 373–409
Isabelle dipole magnets, 83–84
Isochronal anneals, 382–383
  temperature of, 384

K

Kirkendall porosity, tin diffusion and, 243
Kunzler technique, for lead-molybdenum sulfide wires, 435
Kurchatov Institute of Atomic Energy (Moscow), 89

L

Laboratoire de Génie Electrique de Paris, 86
Lattice defects
  see also Flux lattice elasticity; Flux-line lattice
  absorption contrast and, 185
  diffraction contrast and, 186–188
  electron diffraction patterns for, 187
  image contrast from, 183–188
  lattice resolution in, 183–184
  Moiré fringes and, 184
  out-of-focus contrast in, 185
Lattice parameter, transition temperatures and, 388–394
Laves phases, future development of, 432–433
Lawrence Livermore Laboratory, 96
Lead molybdenum sulfide
  fabrication of by Kunzler technique, 435
  ternary, 7
LEED, *see* Low-energy electron diffraction
Leningrad Polytechnic Institute, 86
Lenz's law
  diamagnetic current loops and, 50
  in power transmission flow, 285–286
Levitron, *see* Culham superconducting levitron

Liquid helium stages
    coolant circulation in, 181
    in electron microscopy, 181–183
Liquid hydrogen bubble chamber magnet, 73
Liquid solute diffusion, in A15 conductor
        processing, 232–233
London limit, 12
London theory, of superconductors, 12
Long-range crystallographic order, for A15
        compounds, 142, 150, 153, 156–158
Lorentz force, vs. applied magnetic field for
        niobium–titanium conductors, 115
Low-energy electron diffraction, 330
LRO, see Long-range crystallographic order

                              M

McMillan limit, 27
Magnet(s)
    beam transport, 77
    conventional power for, 73
    superconducting vs. conventional, 65–66
Magnet–Dewar–refrigerator systems, 79
Magnetic design, 49–65
    filling factor in, 59
    flux jumping instabilities and, 49–51
    high current capacity macrocomposites in,
        59–60
    Gibbs function per unit volume for, 20
    homogeneity of, 67–68
Magnetic field intensity, surface currents
        and, 290–293
Magnetic field orientation, 290–293
Magnetic flux lines, electron microscopy
        and, 203–207
Magnetic ore separation, superconductors
        and, 8
Magnetic system, Gibbs function and, 13
Magnetohydrodynamics, 8–9, 92
Magnet performance
    braiding of strands in, 61
    conductor motion in, 61
    degradation and training as obstacles to,
        49–50, 60–61
Martensitic phase transformation, 192–193
Mechanical strain, degradation of supercon-
        ducting properties under, 256–263
    see also Stress–strain relationships
Meissner currents, in type II superconduc-
        tors, 285

Meissner effect, 5, 287
    absence of, 6
    and externally applied field, 17
    Gibbs function and, 14
    incomplete, 14
    Maxwell's equations for, 12
Meissner state, ac losses in, 300–302
MHD, see Magnetohydrodynamics
Mirror electron microscopy, 205
Molybdenum–osmium compound, as A15
        structure, 399–401
Motors and generators, superconductor ap-
        plications in, 84–90
Multifilamentary A15 composites, 62–65
Multifilamentary conductors, manufacture
        of, 63
Multifilamentary superconductors, grain
        growth in, 198–200
Multifilament niobium wire, transmission
        electron micrographs of, 192
Mylar–copper–Mylar cooling layers, in
        superconducting magnets, 68

                              N

National Laboratory for High Energy Phys-
        ics (Japan), 79, 81
Niobium
    ac losses of, 303–309
    chemical polishing techniques for, 177
    ductile alloy superalloys based on, 99
    electron microscopy of, 188–191
    high-temperature annealing of, 328
    irradiated, 557
    neutral-irradiation studies on, 357–358
    for rf devices, 328
    surface preparation of, 329
    in type II superconductors, 279–280
Niobium–aluminum A15 alloys, 376, 386
Niobium–aluminum phase diagram, 149
Niobium conductors
    see also A15 conductors
    temperature dependence in, 308–309
    trapped magnetic flux and, 307–308
Niobium dioxide, flux of, 330
Niobium flexible cables, for power transmis-
        sion, 276–278
Niobium–gallium phase diagram, 149–150
Niobium–germanium compounds, x-ray dif-
        fractometer scan for, 392

Niobium–germanium conductors
  ac losses of, 309–322
  loss behavior in, 318–321
  in type II superconductors, 280–282
Niobium–germanium–copper ternary phase
    diagram, 229
Niobium–germanium phase diagram, 151
Niobium–germanium superconductor, high-
    est critical temperature in, 442–443
Niobium monoxide, flux of, 330
Niobium–platinum compound, as interme-
    diate A15 compound, 385
Niobium–platinum system, 377, 384
  A15 range in, 404
Niobium–selenium conductors, as magnet
    material, 371–372
Niobium surfaces
  see also Niobium
  Auger spectroscopy for, 332–333
  exposed and anodized, 336–339
  high-vacuum results at, 330–336
  impurities at, 330–339
  metallurgy of, 327–346
  penetration depth and surface critical field
    for, 340–345
  rf superconductivity and magnetic field
    breakdown for, 345–346
  superconducting parameters and, 327,
    339–346
  superconductivity measurements for,
    339–340
Niobium–tin compounds, in type II super-
    conductors, 280–282
Niobium-tin conductors
  see also A15 conductors
  ac losses of, 309–322
  bronze processing vs. superconducting
    properties in, 257–258
  bronze-to-niobium ratio in, 259
  critical magnetic field for, 222
  electron beam evaporation and, 316
  fabrication of, 311–313
  increased reliability of, 262
  loss behavior in, 311–313
  metallurgical effects in, 310–321
  processing methods for, 231–254
  stress–strain relationships in, 254–256
  as superconducting A15 compound, 224–
    227
  superconducting critical currents involv-
    ing, 254–263

  temperatures and magnetic fields vs. ten-
    sile strain in, 254–263
Niobium–tin–copper ternary phase dia-
    gram, 228
Niobium–tin crystal foil, transverse electron
    micrographs of, 199
Niobium–tin flexible cables, 277–278
Niobium–tin layer thickness, heat treat-
    ment and, 242–243
Niobium–tin materials, polycrystalline,
    201–202
Niobium–tin multifilament conductor,
    bronze-processed, 225, 235–238
  powder metallurgy and, 238–239
Niobium–tin phase diagram, 145–146
Niobium–tin seven-strand cable, 223
Niobium–tin structures
  see also Niobium-tin conductors
  crack propagation in, 202
  grain size in, 245
  neutron-irradiated, 196–198
  superdislocations in, 203
Niobium–tin tapes, 223
  see also A15 conductors
  electron microscopy of, 194
Niobium–tin–zircon structures, 194
Niobium–titanium alloys
  in multifilamentary superconductors, 62–
    65
  transmission electron microscopy of, 191
  in type II superconductors, 279–280
Niobium–titanium conductors
  available commercial types (1977), 134–
    135
  BNL type, 132
  cable form of, 129
  circular braid, 131–132
  cold area reduction schedule for, 115
  cold working of, 110
  composite multifilamentary conductor
    fabrication in, 124–125
  composite of photographs of, 130
  conductor fabrication and, 123–124
  copper and, 127
  critical current of in transverse magnetic
    field, 365–366
  cryogenic stabilization of, 121
  diagram of, 120–123
  dynamic stabilization of, 121
  enthalpy stabilization of, 121
  flat braid, 132

flattened single-layer cable, 129–131
flux pinning theory for, 117–120
Formvar strand insulation in, 131
interstitial addition of various elements to, 116
Lorentz force vs. applied magnetic field, 115
metallurgy of, 99–135
microstructure and critical current density of, 102–117
monolithic, 132–134
optimum heat treatment temperature for, 109
oxygen-free high-conductivity copper billet for, 124
physical properties of, 109
production schedule for, 123–124
Rutherford type, 129–131
stranded composites and, 126–132
upper critical temperature and critical current densities for, 104–116, 361
zirconium addition to, 114
Niobium–titanium filaments, 57
Niobium–titanium lenses, 209
Niobium–titanium samples, critical current densities for, 104–116, 361
see also Niobium–titanium conductors
Niobium–titanium system, characteristics of, 100–102
see also Niobium–titanium conductors
Niobium–zirconium alloys, in type II superconductors, 279–280
Niobium–zirconium lenses, 209
Normal metal layers, surface currents of, 293–294

O

Oak Ridge experimental power reactor, 95–96
Omega spectrometer magnet, 75
Optique à grande acceptance (OGA), 79

P

Pauli paramagnetism, 20
Phillips 300 microscope, specimen chamber of, 182
PHT (precipitation heat treatments), 100, 108–109

Physical vapor deposition, in A15 conductor processing, 234–235
Pinning
see also Flux pinning; Surface pinning
bulk, 288
for power transmission superconductors, 282
neutron-induced, 307
surface, see Surface pinning
Pinning center, in flux pinning, 35
Pinning effect
irradiation and, 357
Pinning force
see also Flux pinning
critical current density and, 28
reduced, 253
total, 35
Pinning force density, 28, 252
Pippard limit, 13
Pippard theory, of superconductors, 12
Powder metallurgy, for niobium-tin conductors, 238–239
Power transmission
cable designs for, 268–271
conductor configurations and fabrication in, 271–278
flexible cables for, 275–278
niobium flexible cable in, 276–278
rigid cables (niobium composites) for, 272–275
superconductors for, 267–322
temperature dependence in, 282–283
Precipitates, role played by in A15 compounds, 162
Precipitation heat treatments, in niobium–titanium system, 100
Prototype synchrotron magnets, 80–82
Pseudobinary systems, A15 compounds and, 154–155

R

"Race track" magnets, 77
Radiation damage, electron microscopic techniques used in, 198
Radiation-induced resistivity, critical-current decreases and, 368
Rotating electrical machinery, superconductor applications in, 84–90
Rutherford type induction, 129–131

S

Scanning transmission electron microscopy, 172, 206
SD, *see* Sputtering deposition
Secondary beam line magnets, 79–80
Shadow electron microscopy, in flux line imaging, 205
Silsbee hypothesis, for type I superconductors, 27
SIN cyclotron (Switzerland), 80
Solid state diffusion (bronze) process, in A15 conductor processing, 235–238
Sommerfeld constant, 19
Spin–flip scattering frequency parameter, for niobium-based compounds, 161
Spin–orbit coupling, spin–flip scattering induced by, 161
Sputtering deposition, in A15 conductor processing, 234–235
Stabilized conductors, flux jumping prevention and, 51–59
Stacking faults, in A15 structures, 195–196
Stanford University, 282
Stellarator closed toroidal system, 94
STEM, *see* Scanning transmission electron microscopy
Stress–strain relationships, in A15 conductors, 254–256
Superconducting alternators, 89
Superconducting beam line system, 79–80
Superconducting condensation energy, at pinning center, 33
Superconducting critical currents, of niobium–tin wire conductors under tensile strain, 253–263
Superconducting devices
  see also Superconductors
  for alternating circuits, 9–10
  materials requirements for, 9–10
Superconducting lenses, 207–209
  advantages and disadvantages of, 209–213
  design of, 209
  high-resolution electron microscope with, 213–215
  cryostat and, 213–215
  types and descriptions of, 210–212
Superconducting machinery, homopolar machines and, 85–86
Superconducting magnets, 8–9
  see also A15 superconductors; Magnet(s); Magnet performance
  ac losses from pulsing in, 77
  applications of, 65–96
  early experiments with, 48
  electron microscopy and, 172
  field homogeneity in, 79
  fusion reactions and, 75
  highest-field type, 70–72
  inductive energy storage in, 75
  largest, 73–76
  nuclear irradiation effect on, 62
  Omega spectrometer magnet and, 75
  "race track" type, 77
  stabilizing techniques for, 68–69
  strong field of, 75
  stresses in, 74
  transverse-field type, 76–84
Superconducting materials
  annealing studies in, 383–388
  for future development, 444
  introduction to, 1–44
  irradiation effects in, 349–422
  selection of, 270–271
  upper critical field in, 16–22
Superconducting motors and generators, 8
Superconducting quadrupole doublet, 79
Superconductor rotor, shielding of, 87
Superconducting solenoid magnets, 66–67
Superconducting state
  Gibbs function and, 13
  Meissner effect and, 5
Superconducting transmission-line design, 268–270
Superconductivity
  see also Electron microscopy
  antiferromagnetism and, 25
  applications of, 7–9
  Bardeen–Cooper–Schrieffer theory of, 23–24
  chemical thinning techniques in, 177
  critical current and, 3
  critical temperature in, 22–27
  crystal structures and, 25
  defined, 1
  discovery of, 1–2
  electron microscopy and, 171–216
  electron-to-atom ratio in, 25
  electropolishing techniques in, 174–176
  elementary theories of, 10–13

ferromagnetism and, 5, 25
highest critical temperature in, 442
phenomena and applications in, 1–7
superconducting lenses and, 207–215
Superconductor design, ductility in, 100
Superconductor filament diameter, formula
for, 53
Superconductors
see also A15 superconductors; Conductors; Superconducting magnets; Superconducting materials
ac losses in, 54, 56
chemical vapor deposition for, 281–282
commercially available, 7
critical temperature of, 25, 437–442
dynamic stabilization of, 53
electrodynamics of, 11
in electronic circuitry, 8
electron–photon interaction in, 23
and externally applied field, 17
fabrication of, 10
filamentary stabilization in, 42
finely divided, 56
flux avalanche and, 39
flux jump and, 39
flux pinning and, 28–32
flux redistribution in, 39
fully stabilized, 51–59
high current densities of, 9
with higher critical temperatures, 437–442
high-field, 21, 47, 66
irradiation and pinning effects in, 350–353,
357
for large magnetic fields, 7–8
London theory of, 12
magnetic properties of, 182–183
most widely used, 62–63
multifilamentary form of, 62–63
Pippard theory of, 12
in power generation, 8
for power transmission, 7, 267–322
properties of, 9–10
pure metal, 4
for rotary electrical machinery, 84–90
stability in, 40–41
triangular flux array for, 18
type II, 17–18, 20, 278–284
upper critical fields for, 21
zero resistance of, 269
Surface barrier, in type II superconductors,
286–287

Surface currents
asymmetries in, 290–293
magnetic field intensity and, 290–293
and normal metal at surface, 293–294
surface roughness and, 294–295
temperature dependence and, 295
total current and, 298
in type II superconductors, 290–295
Surface pinning, in type II superconductors,
288–290
see also Flux pinning; Pinning
Surface roughness, surface currents and,
294–295
Surface sheath, in type II superconductors,
286
Synchrotron magnets, prototype, 80–82
Synchrotrons and cyclotrons, beam transport systems for, 80

T

Tape conductors
early problems with, 222–223
quenching and burnout of, 222–223
TEM, *see* Transmission electron microscopy
Temperature
critical, *see* Critical temperature
irradiation effects and, 350–353
transition, *see* Transition temperature
Temperature dependence
of loss behavior in type II superconductors, 321–322
in superconductor power transmission,
282–283
Ternary molybdenum sulfides
future development of, 433–436
higher critical fields of, 441
Ternary systems, A15 compounds and, 155
Thermonuclear fusion, power generation by,
8
Thin foils, for electron microscopy, 173
Tin compounds, LEED studies on, 346
see also Niobium–tin conductors
Tokamak closed toroidal system, 94–96
Toshiba Electric Company, 86
Transition metal carbides and nitrides, future
development of, 429–432
Transition temperature
and defect state in A15 compounds, 402–
409

lattice parameter in, 388–394
order parameter and, 394–402
Transmission electron microscopy, 173–188
see also Electron microscopy
of A15 superconductors, 191–203
absorption contrast in, 185
diffraction contrast in, 186–188
electrolytic thinning technique in, 174–176
image contrast from lattice defects in, 183–188
Moiré fringes in, 184
out-of-focus contrast in, 185
radiation damage in, 196–198
scanning, *see* Scanning transmission electron microscopy
specimen preparation in, 173–181
Transmission line, operating temperature of, 269
Transverse field magnets, 76–84
configurations of, 77
field homogeneity in, 79
Isabelle dipole magnet and, 83–84
T_c theory, in A15 compounds, 156–158
see also Critical temperature
β-Tungsten structure, 137
Type I superconductor(s)
critical current for, 27
Silsbee hypothesis for, 27
Type II superconductor(s)
see also Superconductivity; Superconductors
ac losses in, 278, 296–303
bulk critical currents of, 278–284
bulk losses in, 296–297
composite conductors and, 304–307
currents circulating around, 302–303
defined, 17
flux array in, 18
flux distribution across, 30
flux jumping process and, 50
flux pinning and critical state for, 27–39
hysteretic losses in, 296
instability in, 39–44
irreversible properties of, 27–44
low ac losses in, 278
Meissner currents in, 285–287

niobium–tin and niobium germanium in, 309–322
normal metal layers at surface of, 293
surface barrier in, 286–287
surface currents in, 284–295, 298–300
surface pinning in, 288–290
surface sheath in, 286
temperature dependence of losses in, 321–322
trapped magnetic flux and, 307–308
upper critical fields for, 21

U

Unknown universal defect, 197
U.S. Naval Ships Research and Development Center, 86

V

Vanadium-based compounds, $H_p(0)$ field for, 159
Vanadium–copper bronze diffusion couple, 230
Vanadium–gallium conductors, 224–227
see also A15 conductors
processing methods for, 231–254
Vanadium–gallium multifilamentary conductors, bronze process in, 235–238
Vanadium–gallium phase diagram, 147–148
Vanadium–gallium samples, neutron-irradiated, 387
Vanadium–gallium structures
grain boundary in, 246
grain size in, 245
Vanadium–silicon compounds, isochromal annealing curves for, 386
Vanadium–silicon phase diagram, 147

W

"Wind and react" method, for A15 conductors, 222

Z

Zero resistivity, 4